Pembroke Thom

This book is dedicated to Kate, Malcolm, Beth, Cindy and Connie
It could not have been completed without their love and support.

Published by Pembroke Thom © 2024

January 1st, 2024

I started writing this book in 1974 at the tender age of 27. I thought that I knew everything there was to know about car repair and that writing it down would take about three years. Now 76, I have finally finished its third iteration. Forty-nine years. For me, aging has offered profound lessons in humility . . .

Don't think for a moment that this was a continuous process. I was only able to work on it for two to five years out of each of the ensuing decades. Life got in the way.

I was asked recently why I would bother with such a complex task at my age, particularly with Chilton, Haynes and You-Tube already in that space. First of all, I promised my children that I would. Second, Haynes owns Chilton. No source is perfect, and Chilton manuals are very good, but in my opinion, where to go for repairs, what needs to be done and when, are either incomplete or non-existent in every one of the above sources. Manuals typically cover just one car. And as for You-Tube, I have found the material there is often incomplete or flat-out wrong.

The manual you are holding covers all makes and models, all years, and therefore, is, by its very nature, incomplete. But because it provides a generalized overview from the perspective of a mechanic with over fifty years experience, a shop owner who got into the business of fixing and owning cars at the age of fourteen, it offers insights that most if not all available manuscripts cannot provide.

This book was designed to address both the needs of the newcomer and those who want to get their hands dirty. Chapters 1-5 address where to go for service, what to ask for, and how often that might be necessary. Chapter 6 is meant for *all* readers to provide instruction in owner-specific, regular maintenance requirements. Should you be so inclined, Chapter 7 covers buying a starter kit of tools. The remainder of the book delves into the many aspects of car design – oil changes, coolant level and servicing, performance issues, brakes, tires and alignment, etc. – to be read as necessary to increase your knowledge of whatever problem you might be confronting. Each of these chapters has a do-it-yourself (DIY) section at the end.

The book's first two iterations were devoted to learning how to craft clarity and economy into technically complex topics, and to creating consistency across multiple chapters. Those first two versions were written while I was still working, both before dawn and after the kids went to bed. The version you are holding was only undertaken after years of study and research.

I believe I've made great strides since writing became my profession, enough so that producing copies for the general public seems worthwhile to me. There is just far too much useful knowledge contained in these pages that to not print them would, in my opinion, be wasting a lifetime of effort. I hope you'll agree.

Acknowledgments

I owe a tremendous debt of gratitude to those who've helped me on my journey. George Armstrong gave me my first summer job in his Renault / Peugeot dealership, based solely on the strength of my relationship with his daughter Susie. George opened one of the first Toyota dealerships on the East Coast when both Renault and Peugeot opted out of the American market. I was there. I continued to work on Toyotas during summer breaks, after college, and for the next thirty years.

I opened my own independent shop after George retired, and gradually developed into a master mechanic. My shop grew rapidly over the next sixteen years, thanks in large part to three great mechanics: Rick, Italo, and Stevie. I was lucky enough to find them and still feel a deep sense of gratitude for their help in creating Pembroke Service Center. Thank you!

Dr. Alain Rook saved my life. By the time I met him, he had spent years studying Sezary syndrome. Unlike every other dermatologist I had ever seen, he knew exactly what I was suffering from. The treatments I have received at the Perelman Center for Advanced Medicine, part of the University of Pennsylvania Medical School, have since then steadily improved my health to the point that I am actually capable of writing again. Thank you to every one of the nurses and doctors responsible for that care. Words cannot express my gratitude!

The following page contains a summary of the Chapters. Every chapter opens with a much more complete Table of Contents. Each chapter is identified by its number, placed in a green box located along every page's outer margin. I believe that this approach makes a great deal more sense than a mind-numbing list of pages devoted to indexing. I would appreciate your comments; it's quite conceivable that your support will lead to my hiring a devoted InDesign specialist to integrate their skills with my work.

If any of you find a misdirected reference, a misspelled word, or an inaccurate statement, please don't hesitate to let me know. Whatever the problem might be, I'll try to correct it by the next printing.

Chapters

C

If working on a car is not something you do every day, please read Section 2, p. 113 (For the Novice Interested in Tinkering), and Section Three, p. 114 (Safety Considerations) before you go anywhere near your car. Thank You.

Notes

Chapter 1. Choosing a Shop

1

Notes

Choosing a Shop

1. Introduction

When your car needs work, the automotive service business offers a considerable number of options from which to choose. Even if you are inclined to leave the care of your car in the hands of one particular mechanic or shop – the most sensible approach for most purposes – there will be plenty of times when the need for tires, a quick oil change, or a catastrophe could lead you into one of the following:

- One of the national or regional chain stores that specialize in oil changes, mufflers, brakes, tires, or minor servicing
- New car dealership
- Gas station
- Independent garage
- Shade-tree mechanic

We'll consider each of these and explore practical means to gauge the shop that you've chosen.

Because of my bias toward smaller, independent shops, I felt an obligation to research what others had said before I launched into my own discourse on these options. I was surprised to find how closely my thoughts meshed with the survey takers. Because of these similarities, I have peppered the following pages with a few statistics and quotations from *The Center for the Study of Services, The Center for Auto Safety, and Consumer Reports.* Many thanks for their kind permission to reprint their findings here.

The Consumers' Checkbook series is produced by *The Center for the Study of Services*, an independent non-profit that studies automotive repair facilities (among many other things) in eight metropolitan regions across the United States. Their most recent surveys[1] found that 85% of customers were satisfied with the way independent shops had fixed their cars, compared to only 64% at the average dealership. As for cost, the dealers were charging (on average) $124 per labor hour versus $100 at the independent garages. Note, however, that a weighted score comparing a group of specific *labor* costs (water pump replacement, timing belt, and front brakes, for example) showed a smaller price differential between the two groups: $108 per hour for the dealers versus $97 for the independents. That being said, dealerships use only factory parts and are notorious for markups of 25% and more; most independent shops use OEM[2] parts supplied by aftermarket auto parts stores.

---

1   checkbook.org Auto Repair Shop Articles; *Is an Independent Shop or a Dealer Better for Car Repairs?* © Center for the Study of Services.

  Obviously, there were variations in results across these regions. Therefore, the numbers cited above are nationwide averaged values; but in every region, the dealerships' satisfaction scores were in the 60th percentile (median score 63%), the independents all clustered in the 80th percentile (median score 85%).

2   Original Equipment Manufacturer: A large percentage of the parts that comprise your car were manufactured by global suppliers, built to design specifications provided by the factory. These companies produce comparable parts for the aftermarket as well, often with varying degrees of quality and price. Parts built to the factories' original specs are referred to as OE or OEM parts. More on that later (§4, p. 12).

1

These numbers have not varied much over the years. *The Center for Auto Safety* had this to say in 1990:

> "Many experts on consumer auto repair abuses agree that small, independent garages that pay their mechanics a straight salary, as opposed to following the "flat-rate" manual, often offer the best repair work. Since, on the whole, independent shops are usually the most reliable, consider one even if the mechanics are paid by "flat-rate."[3] The reliability of independent garages is increased because they usually survive on reputation only, as opposed to a "big-name" shop like Sears or Goodyear or a car dealership which attracts many customers on the basis of its name alone.

> "Most surveys show greater consumer satisfaction with the performance of independent shops when compared to dealerships . . . Large retail outlets such as Sears Roebuck and Montgomery Ward received a 74 percent satisfactory rating."[4]

2. New Car Dealerships

> "NEW CAR DEALERS – Most car dealers sell cars, not service, to make money. The service department primarily supplements the car sales business. But where a relatively new car requires complicated repairs to the engine, transmission, or electrical system, often the dealer's service department has the most expertise. A dealer's service department generally has the right equipment, as well as familiarity with manufacturer service bulletins and circuit diagrams not always available to non-dealer shops. If you suspect that your car's problem is the subject of a "secret warranty," you will find it difficult to take advantage of that policy unless you go to a dealer."[5]

On the other hand, the latest edition of the *Consumer's Checkbook*[6] says the following:

> "Except for getting free in-warranty repairs, there's little reason to use a dealer's repair shop. Many consumers believe dealers have access to proprietary knowledge, sophisticated diagnostic software, and better tools than independent garages. That's not true. Both kinds of shops subscribe to the same databases—for example, Identifix and ShopKey—that detail repair instructions, diagrams, and news from manufacturers. Although many car dealerships feature spacious, nifty-looking workstations, independents have access to the same tools and equipment. Despite what dealerships would have you believe, local garages can access the same information, software, and equipment."

---

3 Where the mechanic is paid a set amount for every procedure he performs, regardless of how long it takes for him to do it. For example, front brake pad replacement might pay him 1.5 hours even if the mechanic completed the work in 25 minutes.

4 The Lemon Book, Third Edition, pp. 230–231. Authored by Clarence Ditlow and Ralph Nader. Published by Moyer Bell Limited, Wakefield, R.I. 1990 © The Center for Auto Safety.

5 The Lemon Book, p.232

6 Delaware Valley Edition, *"How to Find a Good Auto Repair Shop"* 2023

2A. Pros and Cons

New car dealerships offer a number of advantages found in short supply elsewhere:

- They use factory parts exclusively.

- Their factories provide in-house training for their mechanics, as well as the occasional day- or week-long seminar.

- The dealerships stock factory manuals and wiring diagrams, typically for each model they service. The factories also send out bulletins detailing the problems endemic to each model, with suggestions on how to deal with them.

- The dealer also has a direct line to the factory that includes routine visits by the factory rep (one of the factory's traveling technical staff) and a technical hotline that can be used as needed.

- The good dealerships even have a few technicians that actually use these resources: their team leaders, the shop foreman, and the wanta-bees. Unfortunately, most dealerships don't require that their mechanics gain familiarity with these materials; their study is typically viewed as extra-curricular.

- Finally, these shops offer state-of-the-art diagnostic tools, decent pay scales, clean bays, really big flags, and lots of balloons . . .

A good dealership is one of your best choices should you be cursed with driving something that just won't fix, even if its warranty has long since crumbled to dust. It's the *only* place to go for warranty repairs: They're required by law to fix these problems free of charge. No one else in the cosmos will do that for you . . .

What might be viewed as an enigma at the gas station might turn out to have an absurdly simple solution at the dealership: You can find a great deal of knowledge there regarding the common problems (and quirks) of your particular make and model. Look for those mechanics who have worked in the dealership for five years or more; they have a feel for your make of car that's unattainable anywhere else except in the small specialty shops.

Unfortunately, all the goodies that the dealerships possess do not necessarily translate into an environment that's conducive to customer satisfaction.

One reason is their size: I used to have my hands full maintaining quality control with a staff of six; imagine trying that with twenty or thirty! The dealerships might respond that they accomplish this by using a system of teams, essentially subdividing their staff into manageable clusters of five and six: one service writer, one team leader, and a group of mechanics with varying degrees of skill. The dealerships might also point to their shop foreman, service manager, and their ties to the factory as further means of ensuring quality. But the fact remains that many individuals within the team setting operate with little or no supervision. As one consequence, their work will not meet uniform standards.

Another problem for the dealerships is that most pay their employees on commission: The number of labor hours that a team can bill for in a given week determines their pay, divvied up according to the mechanics' seniority and/or skill level within the team. Consequently,

1

each mechanic has strong incentives built into his job to move any and all "gravy tickets" through his bay just as quickly as he can. Particularly on the lower-mileage services, the less that's touched, the less chance there is that the car will come back with a problem of the mechanic's making. It should be evident that these services actually pay better for those mechanics who do less! Is it realistic to expect a mechanic to work for his customer's well-being when the two of them never see each other, kept separate by concrete walls and several layers of support staff?

> "One important policy is letting you speak directly with the technician who will be working on your car. Service write-up personnel at dealerships and large shops often know little about repairs, and may not be able to describe your car's symptoms to a mechanic as well as you can. If you can't easily explain a weird noise or problem, take the technician for a ride to point it out. A nationwide auto repair shop study found that vehicle return rates (to fix improper repairs) were about one-third lower if a customer had dealt with the repair technician rather than a service writer."[7]

So "productivity" can easily become an impediment to quality service. So too can the ill will that breeds within certain shops as a direct result of competition for a decent paycheck. Consider this: Car repair is a seasonal business, which means that large shops can run out of work quite early in the day. Those mechanics that can "produce" problems that *must be addressed today* create income not only for themselves, but for the shop as well. Increasing your bosses' income plays well with most folk. The favoritism that can stem from that runs rampant in some dealerships, fomenting conflict and frustration. This pattern was pervasive in three of the five dealerships in which I worked.

Those that produce often become the routing manager's pet, fed a staple diet of 30K services and brake jobs, while those who take the time to check the customers' battery water levels and spare tire inflation end up fixing the problem cars. And that's completely understandable: Problem cars that cycle repeatedly through the shop ultimately create customer meltdown. (The letters NPF – no problem found – can only grace a repair order so many times before a customer hits the roof!) That's when the service manager or team leader is forced into assigning the car to someone else. The "flat-rater" wins twofold: First, he's freed from the problem, and second, he gets another brake job. The man who ends up having to fix the squeak or overhaul the transmission gets the pep talk, the hearty "Thank you!" and the shaft: His paycheck could easily be hundreds of dollars less for his efforts. (Warranty work seldom allots money for time spent diagnosing the problem, nor does it pay at the same rate as straight retail work).

In my opinion, this type of inequity is endemic to the majority of large shops that operate on commission, be it a dealership, national chain, or independent. When you couple the issue of size – a lack of adequate supervision, the physical barriers separating customer from mechanic – to the very real problems that result from paying these men on commission, you are left with conditions ripe for those motivated by greed. One – or even two – shop foremen simply cannot handle all the problem cars when paychecks are determined by output rather than by an employee's skill or commitment to quality. These dynamics had everything to do with my decision to start my own business, and also explained my commitment to paying employees a fixed hourly wage.

---

7 *Consumers' Checkbook*, Delaware Valley Edition, 2023

In all fairness, there are good dealerships out there. Each was built by management's commitment to placing the quality of their workmanship and a real interest in their employees' well-being above the almighty buck. Typically, these shops have several strong, experienced team leaders to oversee the efforts of a small group of others. These men in turn have close bonds to the shop foreman and service manager. You can identify a good dealership – or any good business, for that matter – both by what other customers in their waiting room say about them, and by how their employees interact with each other – and with you.

2B. Warranty Issues

This topic is both complex and variable; consequently, a comprehensive treatment is well beyond the scope of this book. But this much is certain: **Every new car owner should take the time to read their owner's manual to become familiar with both the benefits and requirements described by that warranty.**

Any time you find yourself in the dealership for servicing, ask if there are any warranty campaigns currently under way that are relevant to your car. This very same question should be asked even after the warranty period has expired. If any problems have been developing in your model that have led to either a public recall campaign or to "under the table" warranty repairs, the dealer will know about them. The issue then becomes whether or not the factory reimburses its mechanics well enough for them to mention the repair to you if you haven't initiated the discussion by describing the problem to them.[8]

A warrantable item could involve any of the following:

- ❑ A part that has failed while the car is still within its warranty period. The labor to diagnose the problem, the cost of the part, and the labor to put it in should all be covered. Naturally, the warranty will not apply if the owner caused the failure. For example, if your muffler is leaking exhaust because you bent it while backing over a curb, then all bets are off.

- ❑ A part or system that stands a good chance of failing.

  - When these problems are safety-related, they are usually handled by means of a recall campaign: The factory sends out letters to individual owners as well as to the media. These problems do not necessarily have to arise within the warranty period. Therefore, it is always a good idea to keep your manufacturer informed of any change of address. Your owner's manual or its packet should contain one of these forms.

---

8 The dealerships I worked for typically paid their mechanics about 65% of retail for warranty work. These sums were calculated using a factory-supplied manual and were ultimately paid by the auto maker. If a job took one hour to complete, the mechanic would receive about 40 minutes pay. Naturally, we were never thrilled by the prospect of doing warranty work. In fact, the only time most of the mechanics would "discover" warranty problems was on very slow days after the retail work had dried up.

1

- Where a part or system failure is not related to safety, the factory need not advertise its (your) misfortune. So-called secret warranties are used to cover these areas.[9] In certain cases, dealers have authorization to rectify the problem on every car that comes through the door; in others, they have permission to address the problem only if the customer has described the problem. Again, these problems do not necessarily have to arise within the warranty period.

❑ Seatbelts and Airbags

- The Passive-Restraint Warranty covers seatbelts and airbags for five years or 50,000 miles, whichever comes first. However, in cases where a manufacturers' defect puts an entire car line at risk, the government can extend that warranty to all affected vehicles. The Takata airbag failures are a case in point.

❑ Emission Control Components

- Federal law requires automobile manufacturers to warrant the following components for two years or 24,000 miles, whichever occurs first: Virtually any part that is responsible for producing abnormal engine emissions, exhaust leakage, or emissions from the fuel tank. These include, but are not limited to, all fuel-injection components, electronic control boxes, thermoswitches, vacuum switching valves, exhaust pipes and gaskets, etc. Standard tuning gear (air or fuel filters, spark plugs, etc.) are not covered by this warranty.

- The following three components – catalytic converter, electronic emissions control unit (ECU), and onboard emissions diagnostic device (OBD-II) – must be warranted for eight years or 80,000 miles. (Again, Federal law.)

❑ **Many new car warranties limit certain portions of their coverage to a brief period of time – squeaks and rattles are often covered for just 90 days.** Again, it's wise to read the fine print, particularly if you own a brand-new car.

Typically, the cost of standard service work is not covered under warranty. Routine oil changes and maintenance work set forth in your owner's manual are usually the owner's responsibility. **Your warranty does not require that you return to the dealer for maintenance work. The Federal Clean Air Act guarantees consumers their freedom to select where this service work is performed, as well as the brand of parts to be used.** This aspect of the law was included to promote competition in the marketplace, in order to hamper any effort by factory or dealer to arbitrarily set prices. Naturally, **the manufacturer has the right to reject a warranty claim if the owner cannot produce adequate documentation that the factory's schedule of maintenance (as defined in the owner's manual) has been performed.**[10] Furthermore, the factory can refuse to honor its warranty if it feels certain that an aftermarket part was flawed and so caused the problem. But in a court of law,

---

9 The Center for Auto Safety (www.autosafety.org) is a great resource for information on both secret and standard warranty campaigns specific to all affected makes and models. The National Highway Traffic Safety Administration (NHTSA) is another valuable resource, but I'm not sure how complete their secret warranty information actually is. (www.nhtsa.gov)

10 **Keep your receipts for oil, oil filters, and all other maintenance items! This is especially true if you do any of your own maintenance work. Keep all service records!**

the burden of proof is on the factory, not the consumer. Naturally, very few of us want to tangle with the courts, even if they are on our side. As a practical matter, you just want your car to run right; and if it does develop a problem, you don't want your dealer telling you to take a hike.

> "Dealers' quality scores may suffer to some extent because they take on more difficult jobs. Dealers argue that they are blamed for manufacturing defects, tend to work on cars when they are new and owners are especially critical, and get jobs too difficult for independents to handle. But in an analysis of actual success rates on emissions-related repairs (as evaluated by state inspectors), independents perform substantially better than dealerships. Our advice: If the work you need is not covered by a new-car warranty, use an independent shop."[11]

3. The Independents

3A. Overview

> "Independent garages are usually the best place for consumers to get their cars serviced because they do not sell anything but service."[12]

> "Unless the work you need is covered by a new-car warranty or manufacturer recall, use an independent shop, not a dealership. Unless the work you need is covered by a new-car warranty or manufacturer recall, use an independent shop, not a dealership."[13]

> "We consistently find that prices charged by dealers are significantly higher than those at independent shops, and that dealers don't offer better quality service."[14]

> "Our annual survey of Consumer Reports subscribers found that independents outscored dealership service once again for overall satisfaction, price, quality, courteousness of the staff, and work being completed when promised. With few exceptions, the entire list of independent shops got high marks on those factors. The same couldn't be said for franchised new-car dealers."[15]

There are a number of different types of independents, including some of the smaller tire stores, alignment shops, transmission specialists, general purpose auto repair garages, and specialists in a particular brand. There is also a whole industry of shops that occupy supporting roles: machinists, radiator shops, upholstery shops, and detailers.

---

11 checkbook.org Auto Repair Shop Articles; *Is an Independent Shop or a Dealer Better for Car Repairs?* © Center for the Study of Services.
12 The Lemon Book, pp. 230–231.
13 *Consumers' Checkbook*, Delaware Valley Edition, 2023
14 *Consumers' Checkbook*, Delaware Valley Edition, 2023
15 Consumer Reports: *Independent vs. Dealer Shops for Car Repair.* © Consumer Reports. January, 2015.

Virtually all of these are fairly small, no-frills operations; virtually none are backed by the deep pockets of some parent organization. Most began through the efforts and talents of a single owner, and in most cases that owner will be quite visible in the day-to-day operation of the place. Look for that! A business will not grow past infancy without leadership, which includes technical expertise, some savvy, consistent quality control, and good relations with both customer and employee. Although an effective leader can hire the right people, train them, and initiate the procedures necessary to maintain quality, there is a qualitative difference between a staff composed solely of employees and one that includes the owner. Simply put, the employee has less invested in the operation; an owner's whole life is entwined there.

The relatively small size of most independents is a characteristic that is of considerable benefit to the customer. There are three principal reasons for this:

❑   The mechanics will be fairly accessible. But to benefit from this, you need to be there. **Whenever you drop off your car for servicing, wait around until they present their list of recommendations.** This strategy is beneficial to both sides. You'll get a personal tour of your car's problems – if it's not offered, ask for it! – and they won't have to wait for you to return their call before knowing how to proceed. This conversation, adjacent to your car, will permit you to communicate any particular problems or quirks that your car might have on a much more detailed level than you could with the service writer when you dropped off the car. **The principal benefit is that the communication is taking place while the mechanic still has your car on his lift.** If there are any other avenues of investigation that he needs to pursue, he'll have the information when he needs it. His checkover, on average, should take about a half hour to forty-five minutes from the moment your car rolls into his bay.

❑   Consistent quality control is the trademark of any successful business. Auto repair is about fixing one car at a time, one problem at a time. Even though each model possesses its own collection of idiosyncratic problems, no two cars come through the door presenting the same list of issues. While it would have been nice if each car that went through my shop came out the other side with all its problems resolved, that's just not real life. Someone – my shop foreman, another mechanic responsible for quality control, or I – had to road test and recheck that work or my company's reputation would have gradually gone straight down the tube. How many cars can one individual check out in a day? Realistically, no more than two an hour, especially with all the other demands placed on our time. Active quality control remains a manageable goal in shops of four to six mechanics. Past that, you might as well hang it up unless management commits one person exclusively to that task. And that's a very strong commitment, because in the world of dollars and cents, that salary generates absolutely no income (although it certainly creates a tremendous amount of goodwill). I've seen this job description only once, in a dealership with nine mechanics. He was a very busy man. . . In a perfect world, a smaller shop should translate to better quality control.

❑   Your business is more valuable to a small shop than to a large one. An independent's customer base builds one car at a time, mostly by word of mouth. Misunderstandings over the cost of a particular job or problems with a given repair can and do lead to disagreements. In smaller shops, the owner or his manager are far more likely to be available to discuss the problem with an unhappy customer. That was definitely not my experience in four of the five dealerships in which I worked. In the impersonal world of most large dealerships, the owner is upstairs managing his money while the foot-soldier standing before you has been listening to customers kvetch all day. And don't forget: That employee's commission check depends upon you paying your bill – all of it!

While small size and an owner on board certainly make the independents an attractive alternative when compared with other options, there is another, more salient reason for choosing an independent: the level of knowledge and skill that you are likely to find there. To avoid generalization, the following discussion is limited to those independents that specialize in one or two car lines and those that specialize in one particular aspect of repair, such as alignment work or transmission repair. The independent shops engaged in general purpose auto repair are discussed in *§6, pp. 16 - 17.*

3B. Specialists in a Particular Brand of Car

Most of these shops were started by a dealership mechanic who, for whatever reason, grew tired of the company line. A mechanic who has worked within the dealership system for five to ten years [16] can leave with a tremendous amount of knowledge regarding the idiosyncrasies of that factory's various model lines: the characteristic problems of a particular model that require a "look-see" with every service interval, the part failures endemic to a certain model year. This sort of knowledge saves lots of time both in diagnostics and repair. It also makes for safer cars. And although your bill might not seem to reflect any real savings, consider the millions of people who have paid exorbitant sums to the head-scratchers for nothing more than time spent in the ignorant pursuit of a solution, still unresolved . . .

Another benefit of these shops is that most use either factory parts or their OEM counterparts. But because there can be a significant price differential between factory parts and the aftermarket, you should always ask or state your preference.

Three or four times a year, my shop came across a problem that we had never seen before, usually on some new model. The friendships that I maintained over the years with men who remained with the dealers were invaluable in these situations.

The independent specialists have one final quality to offer – their independence. Once I felt that I knew enough about Toyotas to break free from the dealers, I had the opportunity and the privilege to run my business my own way. It's been my experience that the interpersonal contact between customer and mechanic within the small shop environment made all the difference in my own – and my employees – outlook on our profession. I think that's true as a general statement within most smaller shops.

3C. Specialists in a Particular Kind of Work

I would never entrust a transmission overhaul to anyone except a transmission specialist (unless you are still under warranty). Nor would I have my wheels aligned in a department store garage or a typical gas station. Valve work, engine rebuilding – these should be the exclusive domain of the machinist. However, the removal and reinstallation (R&R) of the

---

16 It takes every bit of five years to make a good mechanic. This is not to say that a person, properly trained, can't be putting out good service work within a year; but the nuances of the job – troubleshooting the problem car, the various alternatives to a particular repair – take years of seasoning.

1

cylinder head for valve work, as well as engine R&R for a total rebuild, is typically performed by a mechanic. The component parts are then delivered to the machinist. I would not recommend this same procedure for transmission work, because the transmission shop needs to perform certain tests (fluid pressures, shift points) both before and after its repair.

3D. An Aside on Costs

The typical independent, like most small businessmen, runs a hand-to-mouth operation. Labor figures seem outrageous until you realize just how expensive the shop's overhead can be: Garage space, properly zoned, is becoming an increasingly rare commodity. The insurance industry takes a hefty chunk: I had no choice but to carry garagekeeper's insurance, liability insurance, health insurance, and workers' compensation. Naturally, the Feds, the state and the town took their fair share. Then there's the cost of parts, oil, uniforms, and power. These bills roll in like waves on a beach. And yet this business is seasonal: Those months of good weather in spring and fall act like an anchor to drag down the checkbook balance – usually far into the red – but salaried employees must be paid regardless of the season. Believe me, a highly trained mechanic is worth every bit of his paycheck – so long as he's honest – regardless of how much (or little) business is being generated. In light of the above, when next you see a $500 bill for repairs, reconsider the knee-jerk reaction ("These guys must be making a fortune!") because in most cases it just ain't so.

4. Parts Supply – Dealerships and Parts Jobbers

All automobile manufacturers produce many of their own parts – fenders and frame components, for example – and contract with outside manufacturers for the rest. These subsidiary companies might specialize in electronics, seatbelts, or driveshafts. Together, the factory and its satellites, constitute the original equipment manufacturers (OEM) for that particular brand. All of these components are built to design specifications provided by the factory, whose engineers spend a great deal of time and money designing quality into their parts. The muffler on a Ford Escort was engineered specifically to fit that car; it has a specified amount of back pressure, a specific amount of resonance. Ford designs them to very tight tolerances, and they fit well. Likewise, the brake pads on a Corolla have a precise size and composition; spring clips and shims specific to that model are included in every box; they all fit together perfectly, and they're silent. These are the parts that comprise your car as it rolls off the assembly line, and these are the parts that you will be handed should you purchase them from your local dealership.

The original equipment manufacturers produce parts for the aftermarket as well; most produce several product lines of varying quality for different sectors of the market. Their best, built to meet or exceed the factory's original specs, are referred to as OE or OEM parts. Their less expensive product lines might be perfectly satisfactory, but then again . . . Maybe those spring clips and shims aren't included in the box, perhaps their friction material is a bit too hard or soft. Consequently, they squeal, they screech, they chatter – they do everything but wear out soon enough. And usually when those cars are returned for some resolution to the problem, the owners are sold new calipers, new rotors, everything but what they needed in the first place – a set of factory pads or OEM equivalent with the appropriate brake pad shim kit. Spring clips and shims work hand-in-glove with brake pads to eliminate noise. Spring clips lose their springiness and shims tend to bubble up with rust over the lifespan of most brake pad sets. Ideally, they should be replaced every time the pads are replaced, but more often than not they are reused or thrown away, because (a) they aren't included in most brake pad kits, and (b) because it's faster and far cheaper to use an adhesive spray or other goo to dampen vibrations – a short-term solution at best.

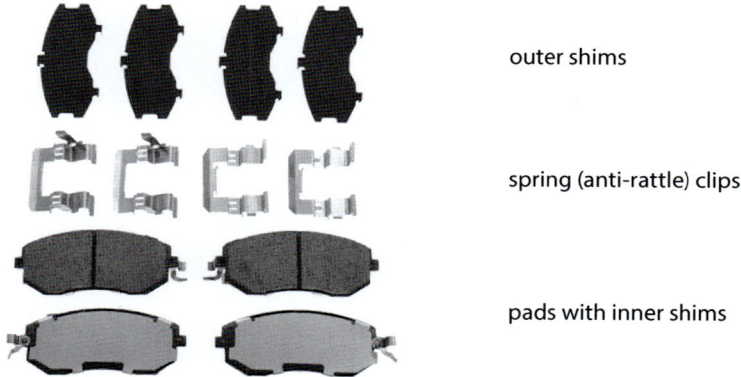

outer shims

spring (anti-rattle) clips

pads with inner shims

Fig. 1–1. OE & OEM brake pad kits include all hardware
Courtesy of Amazon

For every manufacturer that spends the money needed to ensure a high quality product, there's at least one or two others willing to knock off a copy of that particular part. By not "wasting" time on research and development, they have a part that's not only cheap, but very profitable. Consider the following: If the muffler for a Ford Escape has roughly the same dimensions as that for a certain Mazda, then simply using a different inlet or outlet pipe enables it to fit both. That very same inlet pipe might be used on an old Chevy, providing a veritable smorgasbord of mix and match combinations! But what about the hangers that suspend the muffler beneath the car – are they the same on both models? No . . . What about the positioning of those bends in the pipe? If things don't line up exactly, you'll get a vibration whenever adults ride in the back seat. And what about those mufflers that start out their lives suspended by only one hanger? (More the rule than the exception.) When that inlet pipe rusts through, the front of the muffler drops to the pavement and for a few brief moments beats itself against body and ground, until it breaks free to take up residence in the middle of the freeway . . .

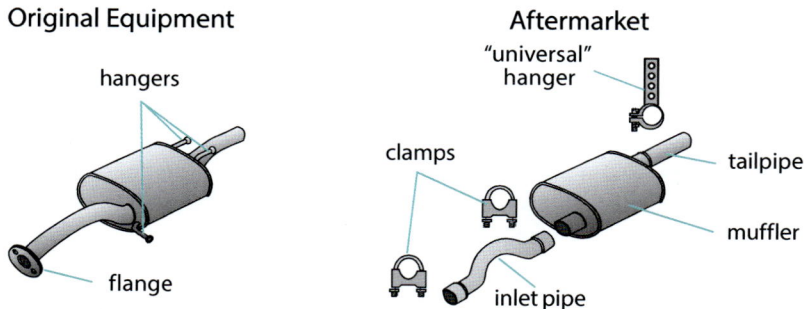

Original Equipment

hangers

flange

Aftermarket

"universal" hanger

clamps

tailpipe

muffler

inlet pipe

Fig. 1–2. Differences between one OE muffler and many aftermarket substitute. © PT

1

Shops that work on most anything usually stock very few parts. There are just too many kinds of cars with too many possible problems to anticipate what might be needed next. Instead, the general purpose shop customarily turns to the parts jobbers – companies that evolved out of the general mechanics' need for one source of supply. Imagine the alternative: Calling all over town, first to the Chevy dealer, then to Dodge City, keeping track of who is sending the clutch assembly and when the VW parts truck might arrive. Talk about a logistical nightmare!

Parts jobbers stock all kinds of parts and have access to millions more. They typically offer a range of price points, from name-brand OEM products to generic lines of indeterminate quality. Just look on the NAPA website, for example (napaonline.com). It is currently offering 14 different possibilities for front brake pads on a 2014 Subaru Outback, ranging in price from $20 to $98. If I were to order the original-equipment pad kit online from the closest Subaru dealer, I'd pay $85. At SubaruPartsDepot.com, I'd pay just under $65 for the exact same thing. Naturally, taxes and shipping costs will bump up those numbers.

I mentioned on page one that dealerships are notorious for marking up their parts prices. That's true both on their shops' repair orders and across their parts counters. I'm not saying that every dealer does that, nor am I saying that other types of repair shops don't. What I am saying is that it pays to do your homework. A quick search of the internet might just save you 25%.

Engine and cabin air filters, oil and oil filters can all be purchased from your local parts jobber for considerably less than you'd pay at the dealership. Just make sure you are buying name brand products from companies that supply OE parts. Fram comes to mind, as do AC-Delco, Bosch, and a dozen others. Here again, you get what you pay for. If you are presented with two air filters, one for $8, the other for $14, I'd suggest you go with the more expensive choice if you can afford it, because in all probability it will do a better job filtering dirty air.

As for wiper blades . . . That's where I draw the line (well, one of them). I have never met an aftermarket wiper blade that can compete with the factory's brand. As with the muffler discussion above, a 15-inch blade from any aftermarket supplier is designed to fit virtually every car that employs a 15-inch blade. But all those variations in windshield contours means that the generic blade won't fit any one of them particularly well. Often, the result is that the rubber insert won't flip direction when the blade does – particularly in a light rain or mist – producing chatter and lousy visibility exactly half the time.[17]

Other parts that I would insist on buying from the factory include any and all internal engine or transmission components – valves, camshafts, pistons; gaskets and seals, gears, clutch packs, etc. (And may all of you be blessed with never having to encounter the need!) While aftermarket parts may fit properly and perform satisfactorily to begin with, it's quite possible that they won't hold up as well. Premature failure might occur while you're still holding some kind of warranty, but typically, only the part will be warranted – not the labor to put it in. You'll be right back where you started from.

---

17 Some manufacturers sell wiper inserts – the rubber squeegie portion of the blade assembly – as a stand-alone part. Unless last winter's ice bent the blade, a bit more labor will save you a bundle.

Until you have a good working relationship with your chosen shop, be sure to ask what kind of parts they intend to use. Insist that they install either original factory parts or OEM equivalents from a reputable jobber. While you're at it, ask what brand of lubricants they use. Or bring your own!

How can you tell what they've used? Until you get to know them, you might ask them to save your used parts along with the boxes or bags that the new parts came in. (By the way, you definitely don't want your oil filter back!)

5. While-U-Wait Oil Change Franchises

These can be great for those on the go; they're well-equipped and they're set up to do one thing well. That being said, it's the personnel that operate the equipment who ultimately make the difference between a worthwhile experience and possible disaster:

❑ A lightning-fast oil change will be of benefit only if:

- They manage to put in the right amount of oil (we're talking close to minimum wage here); and

- It doesn't leak back out. Many of these outfits use what I consider to be inferior oil filters and questionable drain plug gaskets made of hard plastic rather than brass or fiber. Every one of you should read Sections 1 through 3 (pp. 125-130) regardless of where you get your oil changed.

- I grant you that what follows is an old quote, but in my opinion it is still worth reading: The July 1996 issue of Consumer Reports stated that "in the cases where we could compare the oil grade we received with the grade the car's owner's manual listed as preferred, the quick-lube shop used a different grade half the time."[18] *Half the time?* Now it's conceivable–perhaps probable–that things have improved since then, but if you consider just how many grades of oil are specified by the manufacturers–synthetic, non-synthetic, thinner/thicker combinations, seasonal variations, then it seems clear to me that there's just not enough room in any one of those shops for all of those 55 gallon drums. To be fair, the same should be said of any general purpose shop. All the more reason to bring your own oil and filter with you to a shop that you trust.

❑ All those nifty checklists that you get with your oil change, painstakingly filled out, would lead one to believe that absolutely every level on your car had been checked . . . Don't believe it. I can assure you that no one has opened your radiator cap to investigate whether your coolant level is low. Hot engines simply do not lend themselves to this; you'll find a warning printed atop every radiator cap on this planet cautioning against opening the system when hot – *it's under pressure.* These shops can tell you whether the coolant level in your *overflow* bottle is low and advise you of the protection level of the antifreeze contained within that bottle, but that's the extent of it on a while-u-wait basis. For more on this, see §3, p. 151.

❑ A growing number of these stores will try to sell you air filters, wiper blades, transmission services, a cup of coffee . . . Before you authorize any of this, take a look at what they're selling. None of these shops carry factory replacement parts, and some sell stuff that's not even OEM.

18 Consumer Reports, Volume 61, No. 7, p. 17

1

□ An oil change is one thing; inspections or servicing of transmissions, differentials, and cooling systems is another matter altogether. Please read 3G, p. 107 and §5, p. 153 before proceeding down that path.

6. Gas Stations

6A. Advantages and Limitations

Virtually all gas stations meet one of my central tenets: They are small enough to maintain control over the quality of their workmanship. But most fall down sharply on another criteria that I hold dear, namely: They'll work on most anything.

As discussed in §3A, p. 09, having the owner on premises is a real plus. He can provide management skills as yet undeveloped in your typical post-adolescent service writer (I was one of them), and he has a very real commitment to making his business succeed. Again, a hands-on type of guy is easy to spot by the dirt under his nails; in my book these men are leagues ahead of those who brought money but have no real repair experience.

"Jack of all trades, master of some." The general purpose mechanic might have certain limitations to his field of knowledge that cannot be buttressed by factory manuals or other materials. (To be fair, the very same thing can be said about many independents and the less resourceful dealership mechanics.) As a general rule, your garden-variety oil change, brake job, minor service, water pump, clutch replacement, and the like are all within the domain of a good gas station mechanic. But diagnostics on a late-model stalling problem, an emissions problem, or an electrical "gremlin" are best left for a more specialized shop. There is no reason why you should pay for the education of a general purpose mechanic when a specialist might have already seen your particular problem a dozen times.

That being said, the internet has altered this landscape in profound ways. Companies like Chilton (chilton.cengage) and AllData (alldatadiy.com) offer on-line diagnostic support specific to individual makes and models. These services don't come cheap, but they have had a profound impact on the quality of auto repair within those smaller shops that can afford to pay for these resources.

How can you differentiate between a gas station that has turned repairing and service into a successful aspect of its business from one that relies primarily on its fuel sales? Look for a gaggle of cars ringing their lot throughout much of the year. This implies that there is at least one good mechanic on the premises, a man who has built a following for himself. Stop in and snoop around. These shops usually have a service manager to oversee the operation. If you like what you see, check them out with your local consumer group.

> "GASOLINE STATIONS – If you can establish yourself as a regular customer at such a service station, your chances of being satisfied increase – if there are competent mechanics employed there." [19]

---

19 The Lemon Book, p. 232.

6B. The Highwaymen – (Certain) Gas Stations and Repair Shops Along the Interstates

There are, no doubt, thousands of perfectly respectable shops sitting just off the Interstate. Unfortunately, there are others that are positioned like predators waiting to strike, and most if not all of these just happen to operate a towing service. How could you tell, stuck there on the freeway waiting for the tow truck to arrive? You really can't without a smartphone and sufficient composure to use it well.

Under these circumstances, I'd rather pay to be towed to the nearest dealer, where I can rent a car and get on with my life (or back home if that's a reasonable expense). Naturally, if you just ran out of gas, all of this is a moot point. But if you need a replacement tire, remember that gas station rubber will generally be generic and expensive. It will get you where you need to go, but it will probably have a short life span and will occasionally not even match the size of the other three. A tire specialty shop is a much better bet.

So what about AAA? Back to *Consumers' Checkbook*:

> "The American Automobile Association (AAA) has a program to "approve" entire shops. The AAA inspects for a broad range of equipment and customer conveniences, examines staffing and quality control procedures, surveys a sample of customers, and checks complaint records at local consumer agencies. Approved shops must guarantee their work for a minimum of 12 months or 12,000 miles (whichever comes first), and agree to let AAA arbitrate members' complaints (at no charge to the member). Approved shops display the official AAA approval sign."

> "Although the AAA program appears to be a well-conceived effort, we found no correlation between AAA approval and scores on our surveys of area consumers. In fact, both AAA-approved dealer and non-dealer repair shops evaluated on our Ratings Tables rated slightly lower than non-approved shops on our customer survey. And AAA-approved shops may charge more than non-approved ones: Approved shops had slightly higher price comparison scores, on average, than non-approved shops."[20]

6C. State Inspection Stations

Many (most?) states keep an eye on the condition of their citizens' wheels through automotive safety inspection programs administered through gas stations and other independents, rather than through a network of state-managed facilities. In most of these states, the only requirements for a station's certification are the application fee, an employee who can pass a brief course and who has sufficient respectability to be licensed. I don't like these places – kind of like going to the dentist.

Make it clear when you drop off the car that you intend to take your car elsewhere for any repairs they might deem necessary. Be nice about it; you might say that your brother is a dealership mechanic. Knowing in advance that someone else will be following in their footsteps

---

20 checkbook.org Auto Repair Shop Articles; *How to Find a Good Auto Repair Shop.* © Center for the Study of Services.

will keep all but the most brazen of them on the objective side of honesty. Furthermore, the knowledge that they can't make a buck on your car will hasten their moving on to the next.

> "Shops that are authorized to do government-authorized inspections or emission repair work had customer satisfaction scores that were on average seven percentage points lower for 'doing work properly' than the scores of shops that are not government authorized. Prices at authorized shops were eight percent higher than prices at nonauthorized shops."[21]

> "Apparently many good shops choose to stay out of the inspection business and some second-rate shops make an effort to qualify as an easy way to get more customers."[22]

7. National or Regional Specialty Chains

Virtually every aspect of the automobile that can be quickly swapped has at least one national chain dedicated to that specialty. I've split these into three categories: those that sell tires, those engaged in other specialties such as exhaust work or brakes, and the department store garages.

7A. Tire Stores

These are of two types: those that offer just one brand of tire, and those that carry a wide variety of different manufacturers. At the risk of over-generalizing, the one-brand stores usually have to supplement their tire sales with other related work to survive against the competition they're getting from the low-overhead, high-volume tire stores to be discussed next. To make ends meet, most rely upon alignment work, strut and steering component replacement, drive-belt installation, and oil changes. This is not to say that you can't buy a good tire from them, nor do I mean to imply that the convenience of an alignment or an oil change at the same time is not a great idea. But be watchful of advertised specials that lead to the hard sell on a new set of struts or selected front-end components, particularly on a car with lower mileage. Should that occur, ask them to button the car up, because you'll be seeking a second opinion.

The second group of tire stores – those selling a number of different brands – occupies a somewhat different niche within the marketplace. They are usually more competitive in their tire pricing and are therefore less apt to sell you everything but the kitchen sink. Many of these stores do alignment work as well – a true convenience considering its importance to your tires' longevity; but again, be wary of a laundry list of problems, particularly if your car has less than 80,000 miles on it.

---

21 Washington Consumers' Checkbook, Volume 11, Number 4, p.27. This study was conducted in the Washington, D.C. metropolitan area and reflects local test results only. However, these findings seem to have universal validity.

22 Washington Consumers' Checkbook: Volume 11, Number 4, p.29.

7B. Parts Swappers

Meineke and Midas focus on exhaust work and brakes; AAMCO does transmissions; Precision Tune handles maintenance work. These and at least a 100 other chains share one thing in common – all make their living swapping high-profit parts that are easily changed. This approach can offer several very real benefits:

- If you know exactly what you need, you can get it fast and cheap.

- Their warranties are valid across large sections of the country – a distinct advantage to those who travel extensively.

- These companies typically offer generous warranties – some even represent them as being good for the "lifetime" of your car. But read the fine print. For example, at least one of the major players in exhaust system work warrants its mufflers but not the piping that connects it to the rest of the car (Fig. 1–2, p.13). Since this piping is usually fabricated from thinner gauge steel than their factory counterparts, these pipes rust through rather quickly, often producing a new exhaust leak within two to three years. That "lifetime" warranty on the muffler ensures that most of the herd heads back to the pen for at least one more roundup. And just like sheep, a lack of knowledge about what happens next leads many to getting sheared once again!

Unfortunately, many of the nationals use aftermarket parts that do not meet OEM specifications. For them, durability, reliability, and safety are secondary considerations to cost. By not "wasting" money on research and development, their suppliers can produce parts that are not only cheap, but very profitable. Sad to say, these manufacturers fill a very real need: A large portion of our driving public is either eager to embrace the bargain-basement approach, or they have no other choice. These chains would not be in business if that weren't true.

Whenever junk is installed instead of a factory original or comparable OEM component, much more often than not, at least one additional trip through the repair mill becomes essential to correct the problem.

I don't want to leave you with the impression that all these chains should be avoided. Some use better parts than others, some are willing to substitute the better grades of OEM equivalents when asked, some are largely owner-operated. A franchise business that can boast an owner on the premises implies to me that management has a very real stake in the quality of work being produced. That owner has his home or considerable cash tied up in the business. But be advised, some owners have brought plenty of money to the table but little or no experience. This can produce its own set of problems, starting from a lack of common ground between management and employees. The ideal franchise would be owned by a mechanic who earned his way in. What's that refrain? "He'll be wearing a uniform, and he'll have some dirt under his nails."

One final note regarding the national chains. People tend to choose a particular store because of its advertising and location. But be advised: Convenience should not be the determining factor. If you live in a metropolitan area, most of these franchises will have

1

several stores. Local consumer groups will not necessarily be able to certify that a particular shop is any good, but they will certainly be able to identify the rotten fruit – each will have an "odor" at least ten complaints deep.

> "FRANCHISE OPERATIONS – . . . such as AAMCO Transmissions, Midas Muffler and General Tire. When dealing with one of these franchise shops, make sure to check complaints lodged at local consumer affairs offices; another of the franchise's shops in your area may have a better record. Be careful with advertised specials. They often attempt to sell you more expensive repairs . . . If you travel alot, these offer the advantage of warranty coverage at each of their locations."[23]

7C. Department Store Garages

If you know that you need a battery and you're determined to use their credit card, fine. But don't go in expecting them to diagnose a bad starter motor or alternator. As far as having them service your car or fix your brakes, remember the old adage that you get what you pay for.

> "MASS MERCHANDISERS – . . . such as Sears, Penney's, Montgomery Ward, and the major tire manufacturers . . . push fast-moving and profitable items such as tires, batteries and accessories. They are usually interested in selling parts – not service. Although the price may be right . . . watch out for unneeded repairs or more repairs than you anticipated."[24]

8. Shade-Tree Mechanics

A shade-tree mechanic is an individual who works without benefit of a "real" shop. He might work out of his own garage, the boss's shop on an after-hours basis, or on the street. In all probability, he's working without a license to do business. Having started my own business from these humble roots, I would suggest that there are some mechanics out there doing pretty good work. But you'd better find a good one, because having that guy make good on a comeback can be tricky at best.

Why would you want to find such a mechanic? Because compared to the labor charges you would pay in a "real" shop, their prices can seem like a gift from the gods! How can you find a good one? Strictly by word of mouth. Is there any way to evaluate one of them? Not really until you can see his bill and road test his work. But there are a number of qualities to look for:

- He should be employed in a well-respected shop or dealership by day. His daytime specialty and your brand of car should be one and the same.

- His "raison d'être" is to provide extra income for his family or to build a base clientele from which to launch his own business. Never get entangled with a mechanic who's unemployed. Good mechanics are never out of a job. I mean that literally.

- He works out of a garage. (There are distinct disadvantages to working on the street. Believe me, I've done it both ways . . .)

---

23 The Lemon Book, p.232.
24 The Lemon Book, p.232

If he works out of the same bay that he uses by day, he'll have all his tools on hand. But this can be tricky: Some owners aren't particularly enamored of having their space used in this fashion, so if the guy asks you to show up at 7:00 PM or later, be sure to inquire about this point. The last thing you'd want to see is an angry exchange between owner and wrench, with the result that your car is pushed out in the street . . . in the dark . . . undriveable. I've seen it happen.

A better alternative would be a mechanic who works out of a garage that's attached to his home. You'll have much more control over the situation, and you'll know where to find him if a problem should arise. A man who has set up shop in his home will probably be quite serious about his work. And you can expect that he'll be relatively well-equipped.

Working with a shade-tree mechanic has its problems:

- If the car doesn't fix the night that he's got it – either because of time constraints or due to the absence of a critical part – then you'll be walking until the following evening at the earliest. It's hard for a mechanic to predict what he might need for a particular car, so even if the car is not disabled, you may well be returning for round number two, round number three, etc.

- If you have problems with his work, the standard pathways to resolution will be of no avail. The money you've given up will probably be lost.

9. Looking for a Good Shop

How do you find a good repair shop? Well, if you're lucky enough to live in any of the metropolitan areas listed below, then most of your homework has already been done for you. Yes, you'll need to buy a subscription, but for that price, you can also research local plumbers, physicians, and a wealth of other services. (checkbook.org)

Table 1-1:   *Consumers' Checkbook* Automotive Reviews are Available in the Following Metropolitan Areas (2017)

Washington DC	Delaware Valley
Boston	Chicago
Twin Cities	San Francisco
Puget Sound	National

For the rest of you, word of mouth, local news outlets and consumer groups are your best option. Forget about convenient location; an extra half-hour drive could save you considerable money and stress. Ask which shops had the fewest complaints over the past year. Then ask about the past three months: Has one of them improved their record? Has one picked up a string of complaints? Finally, ask for their own opinion: Which of these shops seems to resolve their customers' problems with the least amount of rancor? Where do *they* go for service?

1

Next, call each service department and describe what you need to have done. You are looking for intelligence in their responses, a reasonable attitude, and answers that you can understand. If you're taking the car in for routine maintenance work, then price is certainly a factor, but don't get hung up on costs over the phone. After all, no price that you hear is valid until a mechanic has actually checked out your car. If you like who you're talking to, *get his/her name.*

If you're taking the car in for diagnostics and repair of a specific problem, then access to the mechanic who will be working on your car is essential to assuring that there will be no breakdown in communication. If that appears to present a problem, either keep looking or ask for clarification from the service manager.

Make it clear that you will be waiting on your car. Equally as important, arrive on time. There are a number of reasons for this approach, all of which are detailed in *§*1, p. 51, but for now, let it suffice that an adequate evaluation of personnel requires that you interact with them. While you're waiting, the conversations that you overhear will speak volumes.

Obviously, if your car needs warranty work, you'll be headed to one of your local dealers. **It need not be where you bought your car.** For the sake of convenience, you might choose to have other work performed while you're there, but **you are under no obligation to do so.**

> "If the work you need is not covered by a new-car warranty, use an independent shop. We consistently find that prices charged by dealers are significantly higher than those at independent shops, and that dealers don't offer better quality service."[25]

> "When we compare dealers to non-dealers on results from our surveys of consumers and on price, non-dealers scored better on all measures."[26]

Independents that specialize in one or two car lines usually tailor their services quite closely to the manufacturer's maintenance requirements. Moreover, they are more likely to have closer ties to the dealerships and their parts departments than those shops that work on anything. Most of these independent specialists worked their way up through the dealership ranks but grew tired of the company line. (My career in a nutshell.) But just as with any other segment of the workforce, you need to question, observe, and follow up the work performed with your own inspections.

Independents that work on anything – this includes the gas stations – are much more likely to use a parts jobber for their parts supply. A busy manager takes tremendous comfort in the knowledge that he can count on one or two suppliers to provide all his needs. If you intend to use one of these shops for service work, insist on OEM parts.

Another issue to keep in mind with shops that work on anything: Their definition of standard servicing may not fit your car's factory definition of routine maintenance. You will need to ask what the shop includes in its standard service, and how they document

---

25 checkbook.org Auto Repair Shop Articles; *How to Save Money on Car Repairs.* © Center for the Study of Services.

26 *Consumers' Checkbook*, Delaware Valley Edition, 2023

that work. If your maintenance requirements call for a procedure outside the scope of their standard servicing, would they be willing to perform that procedure and include it in writing on their repair order?

Because most of the national chains limit themselves to a particular aspect of servicing – oil change and levels, brake inspection, or tuning – the responsibility for keeping up with your warranty requirements falls much more squarely onto your shoulders than in the other two types of shop.

10. Evaluating a Shop

You can't get a handle on any shop in just one visit. There are just too many variables – including your own preconceptions – as well as your mood! It might take as many as three visits through the mill before you have an accurate line on a particular shop. Take them your oil changes as well as one annual service, and refer to the following list each time you visit.

❑ How busy are they? If the shop is on your way to work, an occasional glance in their direction will give you some idea. A parking lot that's full to capacity is always a good sign. But since most of the action takes place indoors, an accurate appraisal is hard to make without actually dropping in. Don't expect a shop to be busy every day; this is a seasonal business in which the idyllic days of spring and fall leave many shops wondering and worrying about how to fill their time. But good shops are busy most of the time; a good mechanic is like a magnet for cars.

❑ How many different types of cars do they work on? A shop that works on all foreign cars or the three major American car lines should be much busier – and proportionately larger – than a shop that limits itself to one manufacturer.

❑ What kinds of parts do they use? OEM parts exclusively? Are they willing to supply factory originals?

❑ Do they keep records on your car? Do they bother to refer back to them when you come in? The process of reviewing previous repair orders can be incredibly informative. When written with sufficient detail, they speak to what a car might require this time in, and more importantly, what will not need to be checked.

❑ Do they seem to have an interest in the whole car, or do they limit themselves solely to the issues that you've mentioned? Clearly, there's a fine line here: Many a customer would feel that a mechanic was being intrusive if he uncovered a problem with the brake system when all the customer came in for was a simple oil change. Others view this type of thoroughness with gratitude – as long as they don't have to pay for an inspection that they had not requested. There's no simple answer to this issue from either the customer's perspective or the shop's. In my shop, we preferred to err on the side of being thorough; it protected us as well as our customers. If the customer didn't want the repair, then we didn't charge for the time we had spent to perform the inspection.

❑ Do they invariably try to sell you the whole ball of wax every time you bring in your car, or are they inclined to present you with options and prioritize them for you? We always found a steady stream of work at our door, in large part because our philosophy was to limit our sales to those issues that the car actually needed that day, and to take the time necessary to educate our customers to their cars' future needs – at the next oil change interval, for example, or six months later. The trust that develops from not "jamming" a customer every time she or he walks through the door generates far more business than could possibly be accomplished any other way. This idea leads back to the matter of record keeping discussed above. I never worked for a dealership that bothered to consult its previous records – unless the car was a comeback and the customer was raising a stink. If Mary Jones brought her car in every six months, she'd get a six-month service each time, regardless of the number of miles she might have driven. Every other visit would end up as a one-year service, even if she had only driven 6,000 miles that year. Now clearly, part of this is Mary's fault: Her ignorance and timidity are costing her $300 to $600 per year. But it also indicates the malaise within this industry; the lack of concern for a customer's economic well-being, and the attitude that whatever flows from your pocket into mine can somehow be justified.

Now that computers have made record retrieval so easy, one might be lulled into believing that these records are routinely examined by your mechanic when in fact they might not be. Part of your job as a consumer is to evaluate whether these records are actually being put to good use. Ask a question or two about information that you know was contained on your last repair order or your most recent annual service: your brake percentages, antifreeze reading, or a recommendation listed for your next trip in. Your service writer/mechanic will either have that information directly in front of him – in which case it would seem likely that your records have at least been scanned – or he might ask you to wait a moment while he pulls them up. Your inquisitiveness will ensure that, at least on your car, he'll be looking them over. Unfortunately, many of you will be able to discern a certain amount of resistance at this point, perhaps an artful dodge. Those of you in this predicament will have learned that the responsibility for good record keeping is about to fall directly on your shoulders, which under most circumstances will be in your best interest anyway!

❑ When you drop off the car, does your service writer actually listen to what you have to say? Does he ask a pertinent question from time to time to elicit more information or to clarify your statements?

Let's take the opposite perspective on this for a moment. Countless customers have come up to me after spending ten minutes with their owner's manuals, to recite the list of scheduled maintenance back to me as if they had some in-depth knowledge of what they were talking about. They give themselves away by their word-for-word, item-by-item rote recitation. There is a great deal of difference between telling somebody how to do their job, and defining a very real problem with the driveability of your car. In the former case, I would try to direct the discussion away from wasting precious time toward the real questions: When was the car last serviced? When did the problem first start? Does the problem occur upon first starting the car or after it's been driven for ten miles? Does weather have anything to do with it? For more on this, see Tables 3–1 through 3–3, pp. 56-63.

If you come in with a well-thought-out list of issues concerning your car and the work you would like to have performed, and are met with interest and intelligence, be sure to get that person's name, carve it into your memory, maybe even bake him some cookies! On the other hand, should you be met by indifference, find another solution: either another service writer, the service manager or shop foreman, or a different shop.

"Service write-up personnel at large shops often know very little about car repair, and those who do know car repair may not be able to describe your car's symptoms to a repair technician as well as you can . . . A nationwide study of auto repair found that vehicle return rates (to fix improper repairs) were about one-third lower if a customer had dealt with the repair technician rather than a service writer.[27]

❑ Once your car has been checked over, are they willing to take the time to explain to your satisfaction what your car needs, and why? Again, look at this issue from the other side. It is far easier to explain a physical problem clearly if you can point at the evidence. If you have left the garage, getting a decent verbal description of the car's problems is not a realistic expectation. After all, an individual skilled in the art of painting word pictures would be more likely found on Madison Avenue than the auto park. To my mind, it is always worth waiting until the diagnostics have been performed and the estimate written.

❑ Can they provide an accurate quote? Unfortunately, you'll have to wait on the answer to that question until you pick up the car. Several thoughts:

  • It's very easy to provide an estimate for work that's performed with great regularity. Oil changes, brake work, annual services and the like all have numbers attached that should roll right off the tongue of an experienced service writer. If they blow this kind of a quote, you have either been dealing with an inexperienced service writer, a con artist, or a mechanic who was just trying to "help out" on a busy day. And it's always possible that your experience is just one more demonstration of the fact that everybody makes mistakes. Mistakes can happen to the best of us – you included – and are most prevalent on the busiest of days, those that give true meaning to the term "pressure cooker."

  • It's very hard to estimate the diagnostics involved in emission control problems, certain kinds of performance problems, and virtually all electrical work. Likewise, repairs involving deep engine or transmission work can be very hard to call with any precision.

  • It is also true that a certain percentage of routine jobs with predictable bills can turn into a nightmare for both customer and shop. Several examples:

    – The broken bolt, particularly common in front exhaust pipe work and wheel removal.

    – The hidden problem that can't rear its ugly head until after the engine (transmission) has been disassembled to that point.

    – The problem that develops through no fault of the mechanic as a direct result of performing a related task – akin to the patient whose lung collapses as a result of open-heart surgery.

---

27 checkbook.org Auto Repair Shop Articles; *How to Find a Good Auto Repair Shop.* . © Center for the Study of Services.

1

–And the problem that develops as the direct result of an inexperienced mechanic who has gotten in way over his head! (Just one more good reason for choosing a smaller shop in which you can tell exactly who's working on your car).

❑ Is the owner on premises, or at least actively involved in the operation?

❑ Is the shop neat and clean in a dirty sort of way, or is it a pigsty? With the exception of the glitzier dealerships, no garage is likely to glisten. Grease, oil, and road dirt make for brown to black color schemes; exhaust fumes add a touch of grey. (I've always suspected that the dealerships feel some bizarre compulsion to support the illusion that car repair and surgery have more than just the size of their bills in common.) Most of the independent shops don't have the resources to pay even one additional salary, much less an entire crew of wipers and sweepers. In the typical shop, the mechanics are the ones who add whatever luster you might encounter, and they do this after their real work is done. Look for neatness and organization, order within chaos, paint that's fairly fresh. Shops that offer old motors and other useless parts for you to trip over, shops that are covered in layers of grime, shops with poor lighting and little or no equipment – these are places that should send you right back through the door.

❑ Awards and local recognition: It took my shop ten years of growth before we were formally recognized for our achievements. Then they came all in a flurry. The only real change during that period was our size. So although local awards do provide the customer with some assurance, in some cases the company's longevity might be the most significant criterion being measured. Every dealership in the galaxy has twenty or more "President's" and "Service Excellence" awards on proud display. These awards are handed out by the factory, not the consumer, and represent nothing more than acknowledgment of the amount of business generated in a particular month or year, or recognition for having fewer than a certain number of vehicle comebacks reported back to the factory within a certain period of time. I wouldn't be wowed by their gleam if I were you. Again, look for awards that came from within the community itself.

In a similar vein, local newspapers do quite a business selling advertising space to local merchants on the make for more business. Often these ads are couched as "stories" about the merchant, replete with the owner's photo and a blurb about how Tom of Tommy's Econo-Tune is a straight-talking guy with real concern for his customers' cars. I used to see ten or more of those "offers" per year. Be advised that not all "honors" are the genuine article.

11. National Institute for Automotive Service Excellence (ASE)

The ASE provides paper-and-pencil testing on a wide range of automotive subspecialties and certifies those who pass. This program is strictly voluntary. Because it is a paper-and-pencil test akin to the Scholastic Aptitude Test (SAT), it tests primarily for verbal and cognitive abilities, not mechanical aptitude. Therefore, it is not necessarily a good indicator of mechanical skill, although those mechanics who have earned a number of different certifications are more likely to be better trained, and more capable of communicating with you about your car.

Be advised that a shop can display the ASE sign even if there is only one mechanic on the premises who has become certified. Note too that there are numerous tests covering virtually every aspect of the automobile and the automotive service industry. An individual could conceivably fail seven tests while passing one, and still qualify his shop for the proud display of that sign.

Fig. 1–3. Logo, National Institute for Automotive Service Excellence.

"To become certified by the National Institute for Automotive Service Excellence (ASE), a mechanic must have worked in the automotive service field for at least two years and pass at least one of the tests offered by ASE.

"Certification is not easy—about one-third of the tests taken are failed—but more than 300,000 technicians nationwide are currently certified. Technicians who become certified in all eight categories receive "master technician" certification, and there are nearly 100,000 certified master technicians. To remain certified, technicians must retest and pass exam(s) every five years."

"Although you would expect shops that employ certified technicians to rate higher than shops without them on our survey's "doing work properly" question, this is not the case. One explanation is that ASE certification ensures the competence of only individual mechanics. Competence of one or a few mechanics in one or more service specialties does not guarantee competence of a shop's other mechanics, much less their diligence and honesty. And certification of some mechanics doesn't ensure good communication within the shop or between the shop and its customers.

"Despite the absence of positive correlation between certification and customer satisfaction, we are convinced that the ASE program is a well-conceived, well-managed effort, and advise you to ask that a certified mechanic work on your car. Request a mechanic who is certified in the particular specialty—say, engine repair—you need; certification in brake work, for example, says little about ability to make major engine repairs.[28]

---

28 checkbook.org Auto Repair Shop Articles; *How to Find a Good Auto Repair Shop.* . © Center for the Study of Services.

1

I was a certified master tech, which meant, presumably, that the ASE considered me qualified to rebuild both engines and transmissions. The fact was, however, that since I only saw perhaps one engine every six months that required an overhaul, and even fewer transmissions, I was never comfortable tearing into them. That's what specialists are for! One other point should be mentioned here: Not one of the dealerships that I ever worked in – and certainly not my shop – had the specialized equipment that our machinist and trans gurus possessed.

In all the years I was in business, I never had the opportunity to hire a mechanic who bothered with ASE certification. (I believe that most of them head for the dealerships.) In spite of that, at least four of my mechanics were very bright, intuitive, highly skilled craftsmen who found tremendous satisfaction in producing excellence. None of them had anything more than a high-school diploma – they weren't book-smart – but they were honest. More on that later . . .

Chapter 2. Scheduled Servicing

2

Notes

Scheduled Servicing

1. Introduction

Twenty-first century cars require significantly less maintenance than their predecessors. Engine "tuning" is basically a thing of the past. Most manufacturers now recommend using synthetic oils exclusively, which means less engine wear and longer intervals between oil changes. So typical service cycles have gone from once every six months or 6,000 miles to once every 7,500–10,000 miles.

One thing that hasn't changed, however, is that virtually all of the manufacturers distinguish between normal and severe driving conditions, with shorter service intervals for the latter. Most people consider their driving to be normal, and in most cases, they're correct. But all that short duration, banging around city streets stuff and frequent, hot summer trips can put many of us squarely into the "severe" category for at least certain portions of the year. Typically, the term "severe" includes cars driven:

- In short hops of five miles or less
- For ten miles or less in temperatures at or below freezing
- Extensively at temperatures in excess of 90°F
- Cars that live at the beach, in dusty terrain, or in the mountains
- Cars that tow trailers, and
- Taxis and delivery vehicles

"Normal" is defined as anything else.

The short commute qualifies as severe because the engine barely warms up before it's shut down again, brake usage is greatly increased, and all that acceleration and deceleration is hard on automatic transmission clutch packs and transmission fluid. High speed driving does not qualify as severe unless it occurs during hot summer months.

Most manufacturers recommend an oil change once every six months or 7,500 miles, whichever occurs first. Pure synthetic oils maintain their lubricative properties for significantly longer, and across a significantly wider temperature range. Consequently, pure synthetics need to be changed less frequently, typically once every 7,500–10,000 miles. Follow your manufacturer's recommendation. **Replace the oil filter with every oil change!**

Speaking of filters . . . Most maintenance schedules call for air filter replacement once every 30,000 miles. That's not often enough, even for normal driving. Spring and summer offer dirt and dead bugs, fall contributes leaves, and warm engines in winter provide many an enterprising rodent with a nice nesting spot. Check both the air and cabin filters at least once a year or every 15,000 miles.

Creating a service schedule that can encompass both the range and kind of miles that people drive in any given year is a very difficult, perhaps impossible, task. Yet those issues – what servicing should be performed on your car and when – confront every mechanic, service writer, and shop owner when you drop your car off for servicing. This is especially true with low- and high-mileage cars, and for vehicles that face a lot of hard miles.

2

So keep the following in mind:

❑ Most manufacturers alternate their maintenance procedures between one major and one minor service, each performed annually or at set mileage intervals, determined by whichever occurs first. As previously mentioned, mileage intervals have increased over the years: Most cars of the late 1990's had major servicing intervals at 12,000 or 15,000 miles; current models typically use 15,000 or 20,000 mile intervals. There are two principal reasons for that:

  • Cars are just being built better these days! Computerized control systems are far more consistent and reliable than the mechanical and vacuum control systems of the past.

  • Most manufacturers are recommending synthetic motor oils exclusively. Because these oils lubricate better, they spin up quicker in cold weather, resist thermal breakdown far longer, and seem to stay cleaner than conventional oils. So oil change intervals have grown longer.

❑ Pull out your owner's manual and turn to the section on scheduled maintenance. Most of you will find several pages of tables replete with footnotes, asterisks, and references to other pages. (Sounds a lot like what you're gonna get here!) **To avoid the confusion generated by both time and mileage intervals, from here on I'll be using the terms** *major service* **and** *minor service* **to define the one-year and six-month services respectively.** Table 2-1 (p. 34) identifies the rhythm within those services. (In preparation for writing this chapter I studied over forty different owners' manuals / schedules for service booklets spanning the years 1998 – 2019).

Table 2-2 (p. 35) and Table 2-3 Part 1 (p. 37) present an approximation of their content with one caveat: The first two minor services and the first major service are my own take on what's actually needed at that early point in a car's life.

❑ If you put significantly less than 15,000 miles on your car per year, be advised: Over-maintaining a low-mileage car is neither necessary nor cost-effective. Most shops – dealers and independents alike – will be inclined to sell you a complete six-month service even if you have only 4,000 miles on your car, then suggest a one-year service when you come back for your second visit at 9,000 miles, and follow that up with another six-month service at 14K. That's certainly excessive, but not necessarily a rip-off: If you come into the shop professing complete ignorance of what your car actually needs, then you are leaving it to the shop to assume total responsibility. The mechanic really has no choice but to go bumper-to-bumper. **Simply being more specific in your requests for service can save you lots of money.**

❑ In its early life, your car shouldn't need more than a basic levels check every 2,000 – 3,000 miles–you can do this yourself in four to five minutes–with an oil change, complete level checks, and a tire rotation at the minor service interval.

> Until you know how fast your car consumes oil, check the oil level every fourth or fifth fill-up, §1A, p. 93 and §1E, p. 97.

❑ In my opinion, the first major service need not be performed until you actually hit the 15,000 or 20,000 mile interval unless you are experiencing a problem that concerns you. Naturally, that assumes that you're monitoring your fluid levels and tire pressures.

The less involved you are in caring for your car, the more critical it is that you follow the manufacturer's schedule – time or mileage, whichever occurs first.

❑ As the car's odometer moves past 30,000, the early minor service oil change with basic levels should be supplanted by the more complete minor service.

❑ Those of you who log 30,000 miles per year should have an oil change with level checks every 7,500 to 10,000 miles, a brake inspection every 15,000 to 20,000 miles, and an annual service at least once every 30,000 miles. Highway miles are usually kind to a car. Naturally, if your mileage or performance drops off, you'll need to go in sooner for service.

❑ The frequency of routine servicing should increase after 75K.

2. The Early Services

Tables 2-2 and 2-3 present a leaner look to the first three services (minor-major-minor) than either your schedule for service or the dealership's "menu". Both of those will list inspections and procedures that are totally unnecessary, won't be performed, and are listed simply to provide justification for why they're charging you so much.

I can assure you that when you go to a dealership for servicing during the first 30 thousand miles of your car's life, their mechanic's perspective will be that the likelihood of it needing cooling system or exhaust work, door hinge or wiper pivot lubrication are close to zero. Naturally, if the coolant overflow bottle is empty or the undercarriage took a hit, he'll dig in a bit deeper. But as I mentioned in Chapter 1, the less a mechanic touches in those early years, the lower his chances of screwing something up and the quicker he can get on to his next job – critically important if he's working on commission.

The first two minor services are basically no more than a glorified oil change and tire rotation. Although inspections of hoses and belts, a brake check and 15 other items might be listed on your dealer's "menu", I can assure you that your car will be in and out in 25 minutes, usually with a bill for 1½–2 hours labor. While these inspections gain in significance with increasing mileage, they are largely irrelevant to your car's health while it's still a baby.

Perhaps the biggest factor to consider when deciding on how much servicing you want to pay for is whether you are willing to assume responsibility for appropriate, timely scheduling of the important work. Since a cooling system flush and transmission servicing are now scheduled for long past the end of most warranty periods, you'll need to stay on top of those levels regardless. Chapter 6 covers all of that in detail.

Speaking of warranties, if you can document routine oil changes and anything close to routine servicing, then you'll have no problems with the dealer should you encounter something ugly. In fact, I never saw a dealership ask for any documentation. The only cases they contest are engines with no oil in them, or engines full of sludge (due to never changing the oil). But make sure that you keep your receipts! You'll need them to refresh your memory when you begin to build a list of the work you've authorized, *§*7, p.45.

2

Repetitions Within Common Service Intervals
Major services typically repeat annually every 15,000 / 30,000 / 45,000 / 60,000 / 75,000 / 90,000 miles (etc.), or   20,000 / 40,000 / 60,000 / 80,000 / 100,000 miles (etc.)
Minor services typically repeat midway between the majors, or every 7,500 / 22,500 / 37,500 / 52,500 / 67,500 miles (etc.), or   10,000 / 30,000 / 50,000 / 70,000 / 90,000 miles (etc.)
The second (30K or 40K), fourth (60K or 80K) and sixth (90K or 120K) major services typically include more procedures than the first, third and fifth year services. See Table 2-2 and your owner's manual for further enlightenment.

Table 2 – 1

The shop that will be doing the work factors into this as well. If you know them well, trust them, and want to simplify your life, then you might find it advantageous to just turn your car's servicing over to them. Most of my customers made it pretty clear that they were far more comfortable entrusting their car's safety and reliability to us rather than to themselves, primarily because they felt that they didn't know enough to assume that role. But **getting your car serviced properly doesn't really depend on how much you know. A willingness to do your homework and to schedule the car when it's due are far more important.** Again, it all comes back to how much servicing you want to pay for.

3. The First Two Minor Services

As mentioned above, these services are really no more than a glorified oil change and tire rotation. But to prepare for the inevitable, have a set of corrosion rings installed at the 7,500 mile service while the battery is still pristine. Or do it yourself, §A3, pp. 224-228.

Speaking of prevention, if you're at the dealer, ask if there are any warranty campaigns currently under way that are relevant to your car. If so, you'd like them addressed while the car is in their shop. If you're not, check www.autosafety.org or www.nhtsa.gov at some point. The dealer is required to repair any and all warranty issues whether you spend a dime there or not.

Of the five dealerships I worked in, only one used a first-tier motor oil. The other four all used the same brand – widely known within the industry as the cheapest of the major producers. That may have changed somewhat now that so many manufacturers are specifying synthetic oils, but I tend to doubt it. **Wherever you choose to go for service, ask what brand of oil they intend to use, its API grade** (SN Plus is the current standard – 2023), **and its weight** (5W-30, 10W-40). Then take a moment to double check that the weight of the oil they'll be using is among those recommended in your owner's manual. As for

2

Minor Service Recommendations

Every 1,500 to 3,000 Miles (Owner's Responsibility)

Basic level checks   Tire pressures

The First Two Minor Services

Check OBD-II for codes[1]

Oil change with filter   Complete level checks
Tire inspection and rotation   Install battery corrosion rings (once)[2]
Issues specific to individual models[3]

Years Three, Four, and Following

Review of prior receipts   Check OBD-II for codes

Oil change with filter   Complete level checks
Inspect for oil leaks   Inspect tires, possible rotation
Inspect for coolant leaks   Brake inspection and adjustment
Inspect coolant hoses   Inspect for torn driveshaft boots (FWD)
Inspect coolant condition   Inspect the exhaust system
Inspect drive belts, tension as needed Inspect lights and switches
Issues specific to individual models   Road test

Table 2–2

1 OBD stands for On Board Diagnostics. See $6, p. 200 and $1, p. 211.
2 During the first minor service or shortly thereafter, then as needed, $3C, pp. 102-105.
3 Such as recycling the service interval warning light . . . Reference your owner's manual and either www.autosafety.org or www.nhtsa.gov for warranty issues.

brand names, that's up to you. If you're not satisfied with what they're offering, bring your own oil the next time around, but make sure to **save your receipt for that and staple it to your repair order!**

If you've brought your own oil and filter, leave them on the front seat and make sure the service writer notes that on your ticket. If you have wheel locks, leave their "key" in plain view. **Check that it's been returned before driving away.**

How can you tell what's been done? The truth is, on this service there really isn't much to notice unless you've taken the time to mark your two rear tires (to verify that they've been rotated). Devious? Yes. Paranoid? That too! But if you want assurance regarding their thoroughness or lack thereof, *you* need to do the homework. **Don't forget to jot down your mileage when you drop off the car. That way, you'll know whether they bothered to road test it. Make sure to check the oil level after you pick it up.** Engines can be underfilled or overfilled, and they can leak. See $3, p. 129.

2

4. The First Major Service

> Section 3, p. 34 contains information pertinent to every service – oil change issues, warranty repairs, and procedures for drop-off... Read it!

For most of you, this service marks the first time that anyone takes more than a superficial look at your car. A road test, brake inspection, an air filter and OBD-II[1] code inspection, plus a set of wiper blades (or wiper inserts) – these procedures commonly separate this service from whatever has gone before. The first major service is by far the simplest of the annual work-ups – everything is still clean and rust-free, and the engine is still too young to be bothered by coolant or oil leaks.[2] The drive belts probably won't need tensioning for another 15,000 to 30,000 miles; the coolant hoses should be fine for at least another 50,000 to 75,000 miles. So if there's not going to be much wrong with it, why bother getting the service at all? Four reasons:

❏ To protect your warranty – and your investment – by having a thorough levels check performed. Odds are that your windshield washer bottle will be the only reservoir that actually needs to be topped off at this point in your car's life, but there are two other levels of potential significance at this tender age: oil level and tire pressures.

Since looking at the other reservoirs takes so little time, those few seconds offer the mechanic assurance that there are no problems lurking on the horizon. It's highly unlikely that your battery fluid levels and spare tire inflation will be looked at. That's great with regard to your battery (*S*?, p.???). Your spare? Maybe ask them to check it...

❏ Speaking of tire pressures – Topping off the tires can extend their life and improve your mileage. The same can be said of air filter replacement.

❏ As for timely oil changes, the benefits should be obvious. But before you head in for service, check the frequency that your manufacturer recommends: Many of you will be surprised to learn that most are calling for a 10,000 mile interval these days – as long as you're using a full synthetic oil.

Other parts of the service that can be of considerable benefit include inspection for tire wear, tire rotation, wiper insert or blade replacement, and inspection of lights and switches.

❏ The amount of brake lining remaining at *each* wheel should be examined once a year or every 15,000 – 20,000 miles (whichever occurs first). That's very easy to accomplish if your tires are being rotated. The odds of your car needing brake pad replacement at this point are close to zero – unless you live on the side of a cliff. The amount of remaining brake lining – its depth expressed in inches (5/32″) or as a percentage (40% remaining) – should be noted on the customer's repair order to benefit all concerned.

To inspect rear *drum* brakes, the drums have to be removed, Fig. 12-5, p.258. While they're off, the brakes should be cleaned and perhaps adjusted (not likely at this mileage unless the car was driven extensively with the parking brake on). Adjustment will:

---

1 On Board Diagnostics: Every car built since the early 1980's has some engine/emissions monitoring capabilities. These became significantly more sophisticated in 1996 when the OBD II standard was implemented. Because the Check Engine Light only comes on if there's a problem with the system, odds are that it won't be checked. See *S*6, p.200 and *S*1, p.211 for more.

2 With the exception of a leaking oil drain plug or oil filter following servicing...

2

Major Service Recommendations

The First Major Service

Check OBD-II for codes

Oil change with filter
Inspect for oil leaks
Inspect for coolant leaks
Complete level checks
Inspect/adjust drive belt tension
Windshield wiper service
Inspect lights and switches
Issues specific to individual models[1]

Replace air filter
Replace cabin air filter
Tire inspection and rotation
Inspect for torn drivehaft boots (FWD)
Inspect the exhaust system
Brake inspection (percentages)
Lube door hinges, hood latch
Road test

Years Two and Following

Review of prior receipts[2]

Oil change with filter
Inspect for oil leaks[3]
Inspect for coolant leaks[4]
Check radiator coolant level (ATF to -??)
Complete level checks
Inspect/adjust drive belt tension[8]
Windshield wiper service
Inspect lights and switches
Issues specific to individual models[1]

Check OBD-II for codes

Replace air filter
Replace cabin air filter
Tire inspection and rotation[5]
Inspect coolant hoses[6]
Inspect for torn drivehaft boots (FWD)[7]
Inspect the exhaust system
Brake inspection (percentages)
Complete lubrication
Road test

Table 2-3, Part 1

1 Such as recycling the service interval warning light . . . Reference your owner's manual and either www.autosafety.org or www.nhtsa.gov for warranty issues.

2 You will never find this "procedure" in your owners' manual, but for your pocketbook, it ranks among the most important.

3 Oil leakage rarely comes into play until the car rolls past 60,000 miles–with one exception: a botched oil change– but because the consequences of an unattended leak can be so severe, this inspection is crucial. The term oil leakage encompasses the engine, transmission, drivetrain, and power steering components.

4 The coolant level needs to be checked every two to three months on older cars, §1B, p.95 and §3, p.100. Coolant hoses typically last well past 100,000 miles, but seepage can develop past hose clamps due to engine vibration.

5 Front tires always wear faster than rears due to the lateral forces of cornering–that's why routine tire rotation is so important. But as a practical matter, most of us end up with two tires that are in better shape than the other pair, with none of them sufficiently worn to require replacement. So your best tires get moved to the front where they will remain until replacement.

6 Coolant hoses usually don't need even a glimmer of attention until the car has 80-100,000 miles on it – assuming that the coolant level remains constant through routine inspections.

7 Front-wheel drive only

8 Certain drive belts can need retensioning as early as a few months after their installation; others may not need attention for three or more years. Serpentine belts can start wearing out by 60,000 miles (+/-).

2

Major Service Recommendations

You'll need to consult your owners' manual *Schedule of Service* for their recommendations on when the folllowing services should be performed.

Engine performance procedures:[1]

Spark plugs – Clean the throttle body prior to spark plug replacement.
PCV valve
Fuel Filter

Transmission service	Brake fluid flush[2]
Drive belt replacement	Timing belt replacement

Flush cooling system, replace coolant, replace thermostat[3]
Inspect front end (steering) components[4]

Issues specific to individual models	Road test

Table 2-3, Part 2

1 Engine tuning is essentially a 20th century concept. Recent models require little more than an air filter and occasional throttle body lubrication (*S*2A, p. 188). Spark plug and fuel filter replacement typically occur at 60,000 miles or every four years. Consult your owner's manual carefully to evaluate the manufacturer's recommendations.

2 The most cost-effective way to accomplish this is to have it done with either front or rear brake work.

3 *Always* check what type of coolant your factory recommends, and *always* use a factory replacement thermostat.

4 Typically not necessary during the first 60,000 miles unless the car pulls right or left (*S*3, p. 260 and *S*3B, p. 321), clunks either when turning or going over bumps, has unusual or severe tire wear, or you've been driving over things.

- Improve brake pedal response (by raising the height at which the brake pedal starts to "bite")
- Reduce the rear brakes' potential for making noise (by removing the bulk of the brake dust).
- Tighten the parking brake (reducing the number of clicks between fully off and fully engaged)

If the parking brake height is unchanged, it may be that your brakes were in proper adjustment when the car came in – four to six clicks is about right. Having less than three clicks can indicate overtightening and consequent binding to the rear brakes; check that the car can still roll on a slight incline, *S*9, p. 281. Overtightening causes more rapid brake wear and reduced gas mileage.

Cars running *disc* brakes on the rear may or may not have an independently adjustable handbrake, Fig. 12-6, p. 262. On those that do, it might get adjusted as part of the inspection. This adjustment will not affect brake pedal response (height). If your disc brakes squeak, see *S*4B, p. 264.

2

If this service is so much simpler than later ones, why should it cost as much as a three or five year service? Clearly, the answer is that it shouldn't. But there are three reasons why they usually do:

- Most car owners expect the shop to go bumper-to-bumper regardless of the car's mileage.

- It eliminates one of the many complexities in billing by averaging the amount of time to service both low- and high-mileage cars: While the mechanic might save 20 to 40 minutes on this one, he will probably have to spend every bit of that time on the next car that comes through the bay – the one with 100,000 miles on it.

  Does that seem fair? How about reasonable? No and no, but that's how the flat-rate manuals read, and in a commission shop, that's all that really matters.

To save money on this service, you first need to decide how much of it you want to do yourself. Limit the areas of responsibility that you entrust to the shop by clearly defining *on paper* (*S*3, p.54) what you'd like them to address. Viewed from the perspective of the manufacturer's warranty, relevant items include the oil change, a complete levels inspection, any engine issues specifically mentioned in the owner's manual (few, if any, at this age), and any warranty repairs that might have surfaced since you purchased the car. Your owner's manual must be your primary reference.

Another way to limit the size of your bill is to divvy up your list between the dealer and an independent shop. Let the dealer handle any relevant warranty issues (*S*2B, p.06), and have your local tire store or independent do the oil change, levels, tire rotation and brake inspection.

Don't be shy about supplying your own parts, either! *S*1, p.67.

In conclusion, if you intend to cut corners on any of these services, do so on the early ones, but be smart about it. A car is a sophisticated piece of equipment; in all probability, one that's nowhere near paid for. Although it's important to maintain the warranty, it is even more important that you maintain your investment in the machine. There can be a sublime feel to those years following payoff on a car loan. And while it always ends with turning "Old Paint" out to pasture, why hasten that eventuality by being neglectful along the way?

5. Later Minor Services

> The information in this chapter is cumulative. Section 3, p.34 contains information pertinent to every service – oil change issues, warranty repairs, and procedures for drop-off. Section 4, p.36 covers performance and brake issues among others, plus some thoughts on saving money. Read them!

By the second or third year, this service should no longer be considered a glorified oil change, but rather insurance against neglect. Those miles between 40,000 and 120,000 are the ones that will make or break your checking account, and assuming you're the original

2

owner, you will have to overcome the complacency engendered by owning a reliable car. Most of you have been lulled to sleep; it's time to start paying attention to the possibility of an oil leak, brakes wearing thin, etc.

Under normal driving conditions, the amount of brake lining that remains at each wheel should be examined once a year or every 15,000-20,000 miles (whichever occurs first). But if your driving qualifies as severe, then as the car gets older you might need to add a brake inspection to every service.

While the list of inspections can certainly fill up a page, in real time they don't take more than a half hour to perform. On the other hand, level checks, done properly, do take some time – perhaps another half hour. Add another half hour for brake inspection, five minutes more for lubrication. Add in the oil change, a road test, and the review of prior receipts, and your bill should probably reflect 1.5 hours labor.

The higher your mileage, the more important it is that the mechanic road test your car. So jot down the mileage when you drop it off. If that "road test" encompassed one or two tenths of a mile, speak with the service manager about it, and ask that he or the shop foreman road test it with you.

6. Later Major Services

> The information in this chapter is cumulative. Section 3, p. 34 contains information pertinent to every service – oil change issues, warranty repairs, and procedures for drop-off. Section 4, p. 36 covers performance and brake issues among others, plus some thoughts on saving money. Read them!

Annual service schedules always have some variations from one year to the next, regardless of the manufacturer. And because manufacturers use different mileage intervals to create their schedules and place even the most common service requirements into different years, it's impossible to create a neat little list like the ones above to encompass that variety. Which brings us right back to this chapter's introduction:

> Yet those issues – what servicing should be performed on your car and when – confront every mechanic, service writer, and shop owner when you drop your car off for servicing. This is especially true with low- and high-mileage cars, and for vehicles that face a lot of hard miles.

Table 2-3, Part 2, p. 38 lists service procedures that are typically included in higher-mileage major service schedules. It's up to you to correlate those procedures with your manufacturer's maintenance schedule.

- ❑ You will never find the procedure "Review of prior receipts" in your owners' manual, but for your pocketbook, it ranks among the most important. See §7, p. 45.

- ❑ The engine's air filter should get a hard look at least once a year. Most manufacturers call for fuel filter and spark plug replacement at 60,000 miles. The throttle body should be cleaned and lubed every year or two to prevent a sticky throttle (gas pedal).

If your engine has been having "issues"–rough idle, starting problems, poor mileage–read up before you make your appointment. You might just need a few tanks of better fuel (§5D, p.234) or a can of fuel injector cleaner. Purchased from a parts jobber, that can will probably cost half of what you'd pay at the shop. For more on injector cleaning, see §12, p.248.

Far less likely, but certainly possible, is that your problem is being caused by a faulty sensor or actuator. Check the following websites–www.autosafety.org and www.nhtsa. gov–for any ongoing warranty campaign. (§2B, p.06) And this would be a perfect time to check out that new scanner of yours! (§6, p.200)

❏ Most automatic transmissions require servicing every few years. But if you live in the mountains, haul a trailer, or spend your entire life in rush-hour traffic, then you might need annual servicing. If your car has a dipstick, then it's easy to determine: You can see if the fluid is dirty, you can smell if it's burned, §1A, p.357. If either is true, it's past time to get it serviced. If, on the other hand, your transission has no dipstick (arrrgh!), then you'll need to follow your manufacturer's recommendations for severe driving conditions.See §7B, p.377.

To some, servicing the transmission can mean nothing more than draining the old fluid and refilling with new. And in a perfect world–a low-mileage car with relatively clean fluid–that might be sufficient at 30- to 40-thousand miles. But over time, crud builds up that can start to clog the filter. You're in luck if that filter is similar to the engine's external filter: Just spin off the old and spin on the new!

Back in the day, servicing involved draining the transmission fluid, removing the transmission pan and the internal filter, cleaning and inspecting both, replacing the filter (if damaged) and it's gasket/seal (if broken or crushed), followed by reassembly and fluid refill. A thorough leak check and road test would follow.

The only problem with that type of trans is that the fluid contained within the torque converter remains in place. As the engine starts, your pristine new fluid mixes with the two or three quarts of original fluid that remained behind. That's OK as long as the owner is getting routine, timely servicing; not so much if that fluid is nasty and burned.

These days, better-equipped shops can replace the fluid by hooking up a flushing station to the transmission cooler lines, pumping in new fluid while displacing the old. Those machines cost $400-$600. But there are machines that cost $4,000 and up that will also flush the trans with cleaning solvents on their way to replacing the fluid. For most of us, that first alternative will work just fine.

Many manufacturers specify synthetic transmission fluid these days; consult the specifications page in your owner's manual. Verify that the shop has the proper fluid on hand.

❏ Antifreeze, like everything else, has improved dramatically over the years. Flushing the cooling system and replacing coolant used to be scheduled every three to four years. Current models require this work every six years or so. Follow your manufacturer's recommendation; don't get sucked into paying for this work more frequently than necessary.

2

As with every other fluid replacement we've discussed, you need to ensure that the shop you've chosen has the correct type of antifreeze on hand, or supply your own.

I always considered that replacing the thermostat made sense whenever the cooling system was flushed–insurance against overheating for the duration of the next cycle. **Overheating an engine is seriously expensive at best, catastrophic at worst.** This is another of those instances where I would always insist that the *factory's* original equipment parts (thermostat plus gasket) be used.

❑ Every manufacturer recommends flushing the brake system every few years. Two reasons for that:

- Non-synthetic brake fluids pull moisture right out of humid air. Water plus steel yields rust, and rust creates expensive problems–particularly within the ABS[3] modulator unit.

- Brake fluid doesn't cycle through the system; what resides in the calipers when you purchase the car basically remains right where it is until the fluid is changed. Over time, the temperatures that develop within the calipers can degrade brake fluid, resulting in sludge.

The most cost-effective time to accomplish flushing the brake system is when either the front or rear brake pads/linings are being replaced–the wheels are already off, the brakes have already been inspected, and that particular set of brakes are being disassembled anyway. See *§*5, pp. 71-73.

Check the specs on your car. If it uses synthetic brake fluid, verify that your shop keeps it in stock; otherwise, bring your own.

❑ Engine drive belts stretch over time, and as they do, the slippage that develops creates friction, glazing and cracking. The end result is a noisy belt. Drive belts seldom break any more unless the owner is particularly abusive or clueless. That being said, all of the annual services include drive belt inspection.

A slipping (noisy) belt does not necessarily require replacement; simply retensioning it may be all that's necessary, at least for the time being.

❑ The timing belt typically gets its first look between 60 and 90 thousand miles. That doesn't necessarily mean that it needs to be replaced at that point, but it will require inspection every year following until it is. A worn timing belt won't alert you to its demise by squealing its displeasure:

Because its purpose is to keep the engine's crankshaft spinning in lock-step with the camshaft(s), it employs a series of notched teeth to lock the two together, Fig. 2–1. If even two or three teeth are torn off, the engine will quit and won't restart until that belt is replaced. The degeneration is silent, quick, and in some cases catastrophic.

The timing belts on many engines can endure longer lives– significantly past the interval recommended by the factory– simply by retensioning the belt at that first inspection. That's because the oscillations that develop in a loose belt put far more stress on its teeth than one that is properly tensioned. The only question is, how much extra will it cost to gain access to the tensioner? If too much disassembly is required, then replace the belt in accordance with factory recommendations.

---

3 Antilock Braking System

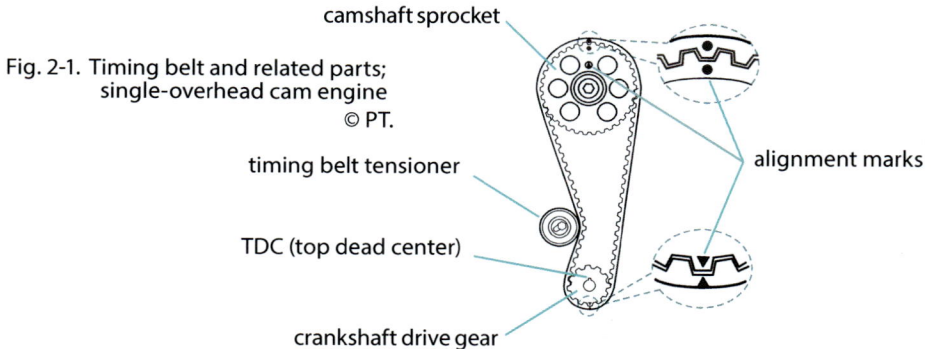

Fig. 2-1. Timing belt and related parts;
single-overhead cam engine
© PT.

camshaft sprocket

alignment marks

timing belt tensioner

TDC (top dead center)

crankshaft drive gear

Before you read the next paragraph, please note that there is a distinct difference between a timing belt and a timing chain . . . One is fiber, one is steel . . . The first will disintegrate silently; a loose chain will rattle–sort of a tinkling sound–right at the front of the engine, giving you ample time to address the problem before disaster strikes. How can you know for certain what you own? Ask the dealership's shop foreman or consult an online resource such as AllDatadiy.com.

> If you're going to get involved on any level with maintenance/repairing your car, you'll need access to the factory's shop manual. Your owner's manual should provide an order form for the book; AllDatadiy.com is a resource on the internet that presents the identical information.

Believe it or not, these days many engines are designed with an overlap to the space through which both the valves and pistons must travel. The result is that if a timing belt is connecting the crankshaft and camshaft(s), the upper end can stop moving with no warning. Any valve that's open – there will always be at least one – will be smacked by its corresponding piston as it moves upward within the still-spinning lower end. As the two collide, the piston will either bend the valve, end up with a hole punched through its upper surface, or both; creating a repair bill that will take your breath away.

The engineering staffs who designed these engines will point to the increased compression that these configurations permit. That coupled with variable valve timing (§1C, p.185) yields significantly increased fuel economy and reduced emissions.

The only sensible course of action with these engines is to replace the timing belt when the factory recommends. Fortunately, many of the engines that were once built with belts have changed back to timing chains, but in many cases, those chains are now single-link–like a bicycle chain–rather than the old-school double-linked versions of yesteryear. All admirable results as long as owners keep their engines full of clean oil! And if your manufacturer calls for synthetic oil, by all means use it! Sludge buildup within an engine can plug the timing belt tensioner's oil feed hole, reulting in that tinkling described above–one step away from a broken chain . . .

2

A belt that traces a squashed oval (Fig. 2–1) will always last longer than a serpentine belt that must double back on itself as it wraps around a collection of gears (Fig. 2–2). Naturally, the condition of the belt will dictate its future: A zillion hairline cracks in the backside of a timing belt should be of far less concern than deeper cracking around the teeth themselves. If that's what you're seeing, replace the belt *now*. In any event, the safe money is on replacement by 90,000 to 100,000 miles.

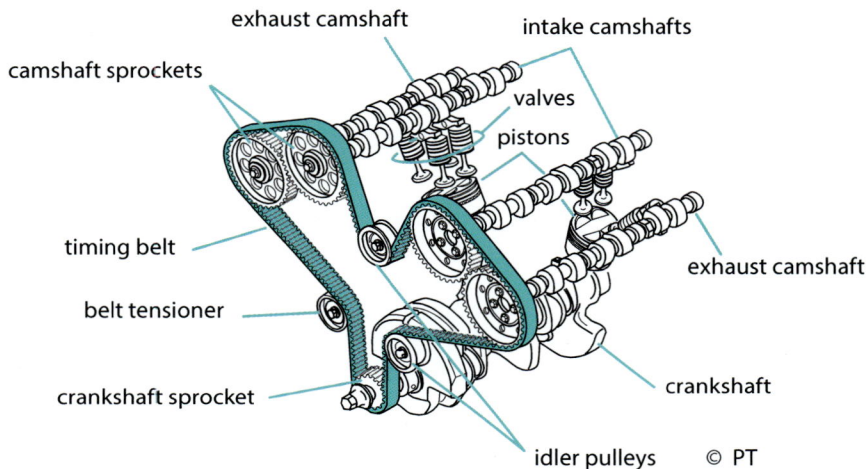

Fig. 2–2. The timing belt configuration on a six-cylinder (V-6) engine sporting four camshafts and four valves per cylinder. (For the sake of clarity, only two pistons – one in each bank – with accompanying valves are shown.) Note the semicircular indentations in the upper face of each piston. These provide sufficient clearance to prevent the disasters described above.

I've been saying all along that the more specific you are with your service requests, the less your servicing will cost. I stand by that statement absolutely. But if you've found a mechanic with extensive experience in your particular model and you've worked with him enough to trust him, then give him the authority to make a wide-ranging search – particularly with cars rolling past five or six years, or 80,000 miles. Don't tie his hands before he has a chance to look the car over. Make sure that you provide him *in writing* any relevant information that you have at your disposal (make it brief – *§*3, p. 54), as well as any limits that would be dictated by common sense. For example: "Please skip the brake inspection. You folks checked them 3,000 miles ago; 35% remained up front, 45% were left in the rear. I'll get them rechecked in another 6,000 miles." Or, "I've included the following list of glitches. I'd like to have some of them repaired if the rest of the car checks out O.K. But please don't spend any time on them yet, I'm only including them here for the sake of completeness."

> **The older the car, the more important the major services become.**

7. Review of Prior Receipts

Imagine for a moment that you purchased one of those extended warranties that dealers love to sell, and at 52,000 miles your engine blows up. You locate the warranty, make the call, and find that they have absolutely no liability unless you can document routine oil changes throughout the life of the car. This happens more often than you might think.

2

Another example: You've been taking your car to the dealer faithfully every 7,500 miles, and at 90,000 miles you get the call: "The fluid in your transmission looks like mud." Without looking back through your records, how would you know that you already paid for a transmission service in their shop at 75K? How far do you think you'll get – requesting that they service it free of charge – without being able to fax them your documentation?

Now let's take the far more likely scenario: Your car is coming up on its six-month service interval; you know it needs an oil change, but you're unsure about the rest. You have two choices:

❑   You can let the shop determine the appropriate level of repair – the blind trust gambit.

As a shop owner, I found that trust to be gratifying but problematic: The customer is placing the responsibility for his or her safety and the car's reliability squarely onto the shop's shoulders, but implicit in that trust is the assumption that the shop's charges will be reasonable. How can the mechanic possibly know how deeply to search for underlying problems? If the servicing is limited to an oil change and basic levels, that bald right rear tire could be overlooked, but on this one day the owner would be happy with the bill. On the other hand, if time is spent looking for a possible flat spare tire or for a coolant hose that's about to blow, and no problems surface, the uninformed customer might just consider the shop guilty of highway robbery when an additional $15 for complete levels is tacked onto the bill. The only effective way to protect both the customer and the shop's reputation is to review the car's repair orders and to follow that up with a few carefully selected inspections / recommendations designed to protect both parties.

Customers who come in professing their profound ignorance of all things mechanical usually end up paying larger bills, not necessarily because the shop is a haven for crooks, but because their staff must spend time figuring out what level of servicing is appropriate, or – more likely – because the safest approach to protecting both parties is to perform a complete service.

❑   Or you can choose to spend the time necessary to review old receipts and the appropriate sections of this book, then produce a short note detailing your car's needs.

To review your receipts, start by placing the most recent three years' worth of tickets into chronological order. Note your present mileage. Then write down all relevant work as detailed in Table 2–4, p.46. Add to this list any recommendations that were made over the past few repair orders, include previously-noted brake percentages (amount of lining remaining). You should now be able to evaluate for yourself exactly what work is required.

Review of Prior Receipts

Type of work performed	Mileage at the time	Current mileage	Suggested interval
Last oil change			6,000 to 10,000 miles
Basic level checks			2,000 to 4,000 miles
Complete level checks			Twice yearly
Under-chassis inspection			Twice yearly
Brake inspection			Once/year or every 15,000 miles
Air filter replacement			Once/year or every 30,000 miles
Engine performance, maintenance			30,000 miles
Transmission service			Consult your owner's manual
Flush cooling & brake system			Consult your owner's manual
Timing belt replacement			Consult your owner's manual

Table 2–4

5 Of the nine inspections that follow, you'll note that more than half of them aren't scheduled until two to four years down the road. That's not to say that those aspects of the car are bulletproof; only that the likelihood of a problem developing are so slight that you needn't pay someone to look at them unless you've heard strange noises, one of your instrument panel gauges is reading abnormally, or you've driven over or into something worrisome.

Any competent mechanic will notice if there's a problem with any one of these items while he's doing the rest of your service. One exception–the steering components–but unless you've smacked something *hard*, the odds of low-mileage problems are vanishingly small.

6 Checking the coolant level *within the coolant overflow bottle* is part of every basic levels check. If that's low, it should trigger the two following inspections regardless of mileage. The coolant level *within the radiator* is typically not checked at the six-month interval because the minor service is too short to permit opening the radiator cap safely. Do it yourself some morning before starting the car. For more on this, see §3, p. 151.

7 Just an air filter during the first two years.

8 Chassis lubrication is largely a thing of the past. Most undercarriage components are sealed units. What's left? Door hinges, hood and trunk hinges, the latches for each, the wiper arm pivots, convertible top hinges, things that squeak enough to drive you nuts–you get the picture.

9 The most cost-effective time to accomplish this is when either the front or rear brake pads/linings are being replaced–the wheels are already off, the brakes have already been inspected, and that particular set of brakes are being disassembled anyway. See §5, p. 71..

Compare your manufacturer's maintenance schedule with Table 2–5.

Mileage intervals vary, the year that some procedures are called for will vary, some cars require specific procedures that are not listed here.

Another Way of Looking At It

A = Major Service – Once per Year or Every 15,000 – 20,000 Miles
I = Minor Service – Halfway Between Each Major Service

	.5	1	1.5	2	2.5	3	3.5	4
Year / Six Month		(1)	(1.5)	(2)	(2.5)	(3)	(3.5)	(4)
Major (A) / Minor (I)	I	A	I	A	I	A	I	A
Basic level checks	Every 2,000 – 3,000 miles							
Complete level checks		☐		☐		☐		☐
Oil change with filter	Every 7,500 – 10,000 miles							
Review prior receipts		☐	☐	☐	☐	☐	☐	☐
Inspect for oil leaks [5]		☐	☐	☐	☐	☐	☐	☐
Inspect coolant level & condition [6]		☐		☐		☐		☐
Inspect for coolant leaks		☐		☐		☐	☐	☐
Inspection of coolant hoses						☐		☐
Inspect drive belts				☐		☐		☐
Inspect driveshaft boots (FWD)								☐
Inspect exhaust system				☐		☐		☐
Inspect lights & switches		☐	☐	☐	☐	☐	☐	☐
Inspect steering components								☐
Engine performance, [7] maintenance		☐		☐		☐		☐
Windshield wiper service		☐		☐		☐		☐
Brake inspection		☐		☐		☐		☐
Tire inspection, (possible) rotation	☐	☐	☐	☐	☐	☐	☐	☐
Complete lubrication [8]		☐		☐		☐		☐
Automatic transmission servicing								☐
Flush cooling system & Replace thermostat								☐
Flush brake system [9]								☐
Timing belt inspection								☐
Install battery corrosion rings	☐							☐

Table 2 – 5

Page References to Specific Service Procedures

Table 2–6

Chapter 3. The Days Before Drop-Off

3

Notes

The Days Before Drop-Off

1. Advance Preparation

Put a notepad in your car so that as all those little annoyances occur, you can write them down. Don't forget the big ones!

Consult the maintenance schedule in your owner's manual. Compare that to Chapter 2. Check your prior receipts to ensure that you won't be asking for work that was recently performed.

3

If you anticipate that the work will include either brake work or an annual or semi-annual service, get some accurate mileage readings to allow for before and after comparisons. (The purpose of checking your mileage in advance of brake work is to verify that none of the wheels ends up binding as a result of the repair.) For information on mileage calculations, see §5A, p. 233.

Make your appointment well in advance, and if possible, choose a day that will permit you to wait with the car – at least until it's been checked out and repairs have been approved. The shop needs to know in advance that you will be waiting. If your schedule won't permit you to wait, be certain to pick a day during which you can be by the phone all day. It will be to your advantage to arrive at the shop right at the start of the business day.

Mondays typically offer the lightest schedule because there's no carry-over from previous days, but that pesky human element – those "First-Day-Back-to-Work Blues" so prevalent to the trades – makes Tuesday, Wednesday, or Thursday the better choice. Fridays are the pits no matter what time of year: I'm not superstitious, but bolts do seem to break entirely too often late that day, and parts deliveries are atrocious.

Auto repair is a seasonal business. The busiest months are June, July, and August; the phone lines start burning again in November and December. The summer customer is usually seeking assurance right before heading off on vacation. As for winter checkovers, the first few frosts start a stampede; get a jump on the herd by getting your servicing in October or early November. Except for starting problems, January through April tend to be absolutely dead; people are too wrapped up with Christmas bills and Uncle Sam. September and October are almost as slow – the weather is just too nice to fret over car repair. Where a shop might see 18 to 25 cars daily during the dog days of summer, there may be no more than 8 to 10 cars in the merry month of March. Take advantage of those lulls in the business cycle. You'll get much better service and (hopefully) more attention to detail. If you don't get it then, you're in the wrong place, because you sure won't get it in July!

I can't emphasize strongly enough the value of making your appointment on a wait basis. Assuming that you arrive as the shop opens, your car will be among the first ones in, and while you're waiting, you'll have a chance to develop a feel for the quality and integrity of the place, simply by eavesdropping on the conversations of those around you. Early morning is the busiest and often the most stressful time of day in a shop; it provides an excellent opportunity to observe how these people operate. Once your car has been checked over, you will be there to verify first-hand exactly what the mechanic has found, to make an appraisal of the mechanic himself, to discuss priorities if budgeting is necessary, and finally, to authorize whatever's appropriate. Depending on the length of time needed to complete the repairs, you will then have the freedom to stay or go. If you leave, be sure that they have your phone number, and be available in case they need to call.

3

If you've never been to this shop before, you'll want to assess their ground rules before you decide to play. Obviously price is an issue, but it's nowhere near as important as not having to return a second or third time. The following list covers most of the relevant issues; it starts off with the easy questions, then wades into the swamp. By moving gently into this interrogation, you stand a chance of projecting a personable, yet well-prepared image of yourself. (Those of you in need of sensitivity training–and those convinced that its pursuit would be a waste of your valuable time–should perhaps read *§2* through *§7, pp.77-84* before initiating this phone call!)

❑ What are their hours? Do they have evening drop-offs? Can you pick up your car after they close? (Shops that operate under the fear that their customers might stiff them will never release a car without seeing their money first. Those shops that provide the option of a pick-up after hours signal that they have no reason to be concerned with your honesty. Trust, after all, is a two-way street.)

❑ Do they have access to public transportation? How far away is it? Do they offer courtesy transportation, both back and forth?

❑ Do they take major credit cards? How about personal checks? (I've had the misfortune to work (briefly) in two dealerships where a personal check was not acceptable as payment. Both of those companies had such a large number of people stopping payment on their checks that they had no recourse. If a shop won't take a check, keep looking!)

❑ Let them know that you will be waiting on the car and will want to see what they find before authorizing any repairs. If the individual that you're speaking with indicates that insurance regulations prohibit customers from entering their shop, tell him you know better. (Signs stating that insurance regulations prohibit customers from entering the shop are intended to keep unsupervised folk from entering areas where they might injure themselves or others – not to prevent them from entering the garage in the company of shop personnel for the purpose of a "show and tell.") If he won't budge, find another shop.

❑ Do they use OEM parts exclusively? Are they willing to use factory parts on request? What brand of oil do they use? Do they stock the appropriate grade and viscosity for your car? Or should you bring your own? (Be prepared! Consult your owner's manual before you start asking these questions.)

❑ If you'll be getting a transmission service, find out what kind of equipment they have and what type of fluid they'll be using. See *§1B, p. 359* for more information.

❑ If either the cooling system or the brake system will be flushed, find out what type of fluid they'll be using. See *§5 through §7, pp.153-156* (coolant), your owners' manual for the type of brake fluid required.

❑ If you're looking for a major or minor service, ask what's included in their package. If you have items on your list that aren't included on theirs, ask why.

What about the flip side of this question? Let's say that your owner's manual lists just four items in the 10K service, but the shop intends to charge you two hours labor for going over the entire car. Do you pick up the phone book and start over? Not necessarily. Remember that most people who bring their cars in for servicing are there seeking assurance that their car will carry them safely and hassle-free for the next full year. Perhaps shop policy has evolved toward the more encompassing bumper-to-bumper investigation to protect the interests of both parties. The service writer certainly has no means to evaluate your car over the phone, although he's sure getting a handle on you! Redirect the conversation back to your list, asking for a quote à la carte if need be.

3

If all of this seems reminiscent of the Great Inquisition, you might seek comfort in the knowledge that you are now addressing issues of substance rather than passively submitting to boundless fears of getting screwed.

You can get at some of this information using less interrogatory methods by limiting your first visit to something cheap, such as an oil change, or by asking the chap who referred you if you could take a peek at his prior bills.

In almost all cases, the service writer who writes the repair order will be your sole line of communication to the shop, so if you like the person with whom you're speaking, get their name! If not, perhaps call back in search of another. It's your responsibility to improve your odds, not theirs.

It's also your responsibility to avoid creating a negative image for yourself. You might try something like this: Acknowledge that you must sound like the very definition of trouble, but that your Dad/sister/son is a mechanic back in Pocatello and insists that you do this for your own protection . . . Or you can just blame me! Doing so might take off some of the heat while still ensuring that you get the information you need.

2. The Weekend Before

Take some time to look over your car, make note of any issues as yet unaddressed, and otherwise just commune with the thing. Check your oil. Is it low? How dirty is it? Ditto for automatic transmissions (whether you intend to get it serviced or not). How about basic fluid levels? Anything need to be topped off? Is there corrosion on your battery terminals? Any bulbs burned out?

Also, regardless of whether you expect any brake work to be performed, road test your car following the guidelines of §2 and following, starting on page pp.259. You never know what your mechanic might find: An attentive road test performed before his work allows you a much more meaningful road test following. Along those same lines, incorporate the road test for tires and alignment, §3, p.320.

Empty your trunk of gardening tools, clothes to be dropped off for the homeless, etc. Accessing the spare tire or repairing taillights becomes a much greater challenge if the trunk has to be unloaded first.

3. Your Note

At the top left, list your name, address, and phone numbers – work, home, and cell. On the top right, list the year and make of the car, its color, tag number, and mileage. Providing them with all of your phone numbers may be irrelevant today, but odds are that if you use this shop again, those numbers will come in handy for both of you on some future visit. As for identifying your car, suffice it to say that on a busy day, those who don't take the time to do so could end up at the bottom of the routing chart.

3

Set down a description of what you want, clearly and concisely. If it's standard servicing, write that at the top of the page – not halfway down – then list any additional issues or questions below that.

What you need to say (or not say) in your note depends on the type of shop you're in and the type of work you are requesting. There's very little ambiguity to an exhaust leak or a broken taillight, but a major or minor service can mean very different things to a dealer and a gas station. The dealer and most independents that specialize in your brand know exactly what's included in the manufacturers' maintenance schedules. For them, the following would be sufficient:

> It's due for its 60K service. The spark plugs and fuel filter are all original. The transmission has not yet been serviced. The timing belt is due for an inspection.

General purpose shops might need a bit more specificity:

> It's due for its 60K service. I put the maintenance schedule on the passenger's seat. Please do me the favor of looking it over. The spark plugs and fuel filter are both original. The transmission has not yet been serviced. The timing belt is due for an inspection.

Don't itemize the entire list of operations detailed in your owner's manual: Doing so not only makes the letter unmanageably long, but also may appear to insult the mechanic's intelligence. Just leave the schedule on the passenger's seat, open to the correct page, secured with a paper clip.

If you're coming in for a specific problem rather than for servicing, **describe its symptoms: Don't try your hand at diagnosis.** Tables 3–1 through 3–3, pp. 56-63 should help.

The more effort you put into the note, the better your results will be. If your note is clearly written, well-organized and succinct, the odds are strongly in your favor that the service writer will simply hand your note to the mechanic. The reverse is also true: If you come in with scattered notes or random thoughts vaguely expressed, then you will have to rely on the service writer to organize and translate your ideas for you.

To identify the positioning of something on your car, use the terms driver's side and passenger's side rather than left and right. (While one owner might assume that left is always viewed from the driver's seat, another will use the term standing at his front bumper, facing the engine compartment.)

Be sure to request the following, where applicable:

- For oil changes, ask that the oil's brand name, grade, and viscosity (§1 and §2, pp.125-129) be listed on the repair order, as well as the brand name of the oil filter. If that seems provocative, explain that you once had to buy a replacement engine as the direct result of a defective oil filter. Or simply bring your own.

- For winterization inspections, checkovers for summer trips, annual services, or cooling system flushing or repairs, request that the antifreeze protection level within the radiator (§3, p.151) be noted on the ticket. If the antifreeze is to be replaced, verify that they have the correct type on hand and that it be listed on the repair order.

- For transmission servicing or differential draining and refilling, the type of fluid used and its viscosity (§1A through §1C, pp. 357-360) should be written on the repair order.

- For brake inspections, request that the percentage of brake lining remaining – both front and rear – be written on the repair order, as well as the condition of any other brake parts that are on their way out (§4, pp. 36-38). If the brake fluid will be flushed, verify that they have the correct type on hand (consult your owners' manual) and that it be listed on the repair order.

3

If you are still under warranty, you might want to request that the repair order list the brand name and part number of all parts used. If they're not set up for that, ask that the mechanic toss all of the boxes and bags that your new parts came in into the trunk.

The following statement or something very close to it should conclude every service request that you write:

> Following your road test and inspection of the car, please notify me of your findings. I need to know what the work will cost before I can authorize the repairs. I understand that it will take some time to inspect my car, and I'll be happy to pay for any reasonable charges associated with that estimate. If you expect these charges to run more than the cost of an hour's labor, please let me know before you begin.
>
> I'm also interested in any other problems that your mechanic might notice as he investigates the issues that I've set down, particularly those that in his judgment are more critical to the safety and well-being of the car than the ones I've listed.
>
> If anything out of the ordinary comes up during the course of your work that will significantly increase the cost of the authorized repairs, please don't hesitate to call me.
>
> Thanks very much for your help,

Make a photocopy of your note and give them the original.

Include copies of the past three years' repair orders, you keep the originals.

3

Table 3–1. Describing Your Car's Symptoms

❑ **What made you notice the problem? Something you:**

- Hear
- Feel
- Smell
- See

❑ **Noises:**

- Loud – Moderate – Faint
- High-pitched – Low-pitched
- Rattling
- Buzzing
- Screeching
- Scratching
- Rubbing
- Whining
- Groaning
- Moaning

❑ **Vibration:**

- Strong
- Light
- Moderate

❑ **Where do you think the noise (vibration, problem) is located?**

- Ahead of you
- Engine compartment
- Behind you
- Passenger's compartment
- Below you
- Dash panel
- Above you
- Trunk
- Driver's side
- Wheels or tires
- Passenger's side
- Brakes

❑ **Smell:**

- Gasoline (*§2A, p.213*)
- Exhaust fumes (*§3, p.447*)
- Oil (*§5, pp.132-137*)
- Rotten eggs (*§6C3, p.239*)
- Mold or mildew (*§2F, p.396*)
- Burning smells (*§5, pp.132-137 & §6A, p.268*)
- Melting plastic (*§3A, p.449*)

❑ **Is the problem:**

- Constant or Intermittent
- Repeatable or Unpredictable

Part 1 of 2

Table 3–1. Describing Your Car's Symptoms

❑ **If the problem is intermittent, is there a pattern to it?**

- Only when the engine is cold
- A certain number of miles or minutes past start-up
- During warm-up
- Only when fully warmed
- Whenever it overheats (EGAD!) (§8B, p.158 and §13, p.163)
- Only in wet weather?   (rain – snow and ice – high humidity)
- What about air temperature?   (cold – hot – irrelevant)
- Does it change at different speeds?

3

❑ **If you know how to produce it,**
        describe how to do it in the following terms:

• Sitting at idle	• Accelerating	• Decelerating
• Maintaining a constant speed	• What speed?	
• Within a certain speed range	• What range?	
• Within a certain RPM range	• What range?	
• Only in a certain gear	• Which gear?	
• At any RPM	• At any speed	• In any gear
• Over a smooth road	• Over a bumpy road	• Only over speed bumps
• Going uphill	• Going downhill	
• Sharp turns – left or right?	• Gradual cornering – left or right?	
• Accelerating hard while cornering	• When braking while cornering	
• Only during light braking	• Only during hard braking	

❑ **Can you duplicate the problem with the car sitting still,**
        simply by revving the engine?

❑ **Does the problem change when you shift gears?**

- Upshifts or downshifts?   • Which gears?

Part 2 of 2

Table 3–1. Describing Your Car's Problems / 57

If you're heading to the shop for a specific problem rather than servicing, describe its symptoms: Don't try your hand at diagnosis.

The following Tables are not intended to contradict that statement; rather, they are intended to assist you in determining just how critical a particular noise might be.

The absence of a 💣 next to a particular noise does not mean that the problem underlying the noise will not create damage. Any new sound can indicate a problem, and any problem left unaddressed will degenerate. The longer a condition is allowed to fester, the more serious the problem can become.

Having said that, axle bearings, wheel bearings, and CV joints are not repairable items. Typically, they last for quite some time before they freeze up (stop turning). The noises they create simply get louder and more disturbing with time. They should always be replaced before that point.

Table 3–2. Starting, Stalling, & Engine Performance Problems

❑ **Starting problems** (*§2* through *§4* pp.211-232)
- Cranks but won't start   • Won't crank (turn over)   • Absolute silence

❑ **Stalling problems** (*§10*, p.246)
- Stalling cold   • Stalling hot
- Stalling during warm-up (in transition between cold and hot)

❑ **Engine performance issues** (*§6* through *§9*, pp.238-246)
- Bucking   • Surging   • Power loss
- Misfiring (Missing an occasional beat? Or a whole string of them?)
- Hesitating under acceleration

❑ **Are any of the dash warning lights / gauges involved?**
- Which one(s)?
- What do they indicate?
- Does the problem occur only if one of the accessories is turned on?
- Which one? More than one?

3

Table 3–3. Noises

CAUTION: The 💣 symbol indicates immediate potential for damage to the vehicle or danger to passengers.[1]

❑ **The Sound of Boiling Water** 💣
Check your temperature gauge **NOW**, pull off the road immediately, turn off the engine, then turn to §1, p. 147.

❑ **Noises Heard at Idle** (engine on, standing still)
- **Growling from the engine compartment:** Usually a bad bearing in one of the components being driven off the drive belt(s).
Likely candidates include the water pump, alternator, power steering pump, air conditioning compressor, or the idler pulley (which tensions the drive belt) for either the power steering pump or AC compressor. The timing belt tensioner is another possibility. (§18, p. 175)
- **High-pitched humming or whining sound from the engine:** Same possibilities as above, usually the alternator. (§18, p. 175)
- **Low-pitched grumbling or growling sound from the engine** (as if rocks were being chewed up in a grinder): Same possibilities as above, usually the water pump 💣 (§18, p. 175) or an idler bearing. 💣 (§18, p. 175)
- **Tick-tick-tick- coming from the engine:**
  - You might be very low on oil. 💣 (§1E, p. 97)
  - A collapsed hydraulic lifter, a worn camshaft, a valve in need of adjustment, or (§B1, p. 237).
- **An electrical kind of ticking** (otherwise known as arcing): Bad coil or coil to spark plug connection (§2B, p. 214)

Part 1 of 5

1   See Note, previous page.

Table 3–3. Noises / 59

3

Table 3–3. Noises

CAUTION: The ☙ symbol indicates immediate potential for damage to the vehicle or danger to passengers.[2]

❑ **Noises Heard Only When the Engine Is Accelerated, Stationary**[3]

- **Rattling or buzzing sounds, either from beneath the car or within the engine compartment:** Loose or rusted exhaust system heat shields, loose or rusted exhaust pipe hangers, loose or rusted baffles internal to the muffler. (§3C, p. 451)

- **Screeching:** Almost always associated with a loose or oil-soaked drive belt. Another possibility: a "frozen" bearing (one which has seized and can no longer turn). ☙ (§18, p.175)

- **Vibrational noises related to specific engine speeds (specific RPM range):** The exhaust components mentioned above; loose dash panels, trim panels, etc.

- **A disturbingly loud whap-whap-whap coming from the engine:** Either a rod knock from an engine that's coming unglued, ☙ (§6, p.138) or a piece of carbon that's become stuck to an exhaust valve. (§9D, p.244)

❑ **Noises Heard Only When the Engine Is Decelerated, Stationary**[2]

- **Rattling or buzzing sounds, either from beneath the car or within the engine compartment:** Loose or rusted exhaust system heat shields, loose or rusted exhaust pipe hangers, loose or rusted baffles internal to the muffler. (§3C, p. 451)

- **Vibrational noises related to specific engine speeds (specific RPM range):** The exhaust components mentioned above; loose dash panels, trim panels, etc.

Part 2 of 5

---

2 See Note, top of p. 60 .

3 This category is limited to noises that can be duplicated in motionless cars, transmission set in Neutral or Park, with the engine being revved from idle to 3,000 or 4,000 RPM, then decelerated. These noises are typically first noticed while driving the car under normal circumstances.

3

Table 3–3. Noises

CAUTION: The 💣 symbol indicates immediate potential for damage to the vehicle or danger to passengers.[4]

❑ **Noises Heard Only in Motion**

- **Growling or rumbling from any wheel, usually at speeds over 30 mph, more pronounced at higher speeds:** A wheel bearing or axle bearing. 💣 (*§*11, p.342)

- **Wua-wua-wua-wua or wo-wo-wo-wo sound from the rear on a rear-wheel drive car, usually at speeds over 30 mph, more pronounced at higher speeds:** An axle bearing. (*§*11, p.342 and *§*11A2, p.345)

- **Whining from the rear:** Either a rear-wheel bearing on a front-wheel drive car (*§*11A1, p.344), or differential wear on a rear-wheel drive car. (*§*3B2, p.371)

Badly chopped tires and tires with ply separations can be just as noisy as a bad axle bearing or wheel bearing, but with a somewhat softer sound:

- **Galumph – galumph in sync with road speed, typically at low speed:** Ply separation in one of the tires. 💣 (*§*3A, p.320 and *§*4A, p.322)

- **B-d-d-d- increasing in volume with increasing speed:** Lumpy tire (badly chopped) or ply separation. (*§*3, p.320)

- **Screeching when cornering:** Low tire pressures or worn tires. 💣 (*§*1 and *§*2, pp.317-320, *§*1A, p.94) 💣

- **High-pitched screeching sound from a wheel:** Usually a warning that the disc brake pads are wearing thin, produced by a depth warning indicator cutting into the rotor. (*§*4A and *§*4B, pp.263-264) 💣

- **Rattling:** Loose stuff in the trunk, usually the jack, or a loose or rusted exhaust system component. (*§*3B, p.450)

- **Hundreds of little popping sounds at low to moderate throttle while climbing a hill:** Engine ping. Get some high-octane premium gas and a can of decarbonizing fuel system cleaner. (*§*12, p.247) If the problem persists, check the OBD system for codes.

Part 3 of 5

4   See Note, top of p. 60.

Table 3–3. Noises / 61

Table 3–3. Noises

CAUTION: 💣 The symbol indicates immediate potential for damage to the vehicle or danger to passengers.[5]

❑ **Noises Heard Only When Applying the Brakes**
- **High-pitched screeching sound:** A warning that the disc brake pads are wearing thin, produced by a depth warning indicator cutting into the brake rotor. 💣 (*§4, p.263*)
- **Screech or squeal:** Brake pads or linings of the wrong composition (too hard), or a problem with hardware. (*§4B, p.264*)
- **Grinding, crunching:** Spent brake linings running metal-to-metal. 💣 (*§4A, p.263*). Oil-soaked brake shoes (*§4D, p.266*). 💣

❑ **Noises That Go Away Only When Applying the Brakes**
- **High-pitched screeching sound:** A warning that the disc brake pads are wearing thin, produced by a depth warning indicator cutting into the brake rotor. 💣 (*§4B, p.264*)
- **Groan or moan:** A brake caliper that's binding, yielding extreme temperatures. 💣 (*§4C, p.266*)

❑ **Noises Heard Only When Cornering**
- **Squealing:** Worn tires or low air pressure. 💣 (*§1 and §2, pp.317-320, §1A, p.94*)
- **Rapid clicking when powering through a corner:**
  - Worn CV joint, or
  - Torn CV boot yielding loss of lubrication within the joint 💣 (*§4, p.372*)
- **Groaning when turning the wheel, first thing in the morning:** Low fluid level in the power steering system. (*§13B, p.349*)

Part 4 of 5

---

5   See Note, top of p. 60.

Table 3–3. Noises

CAUTION: The 💣 symbol indicates immediate potential for damage to the vehicle or danger to passengers.[6]

❑ **Noises Heard Only Over Bumps**
- **Creaks when going over a speed bump:** Chassis in need of lubrication, worn suspension components (usually bushings). (p.324)
- **Clunk going through potholes at speed:**
  – Blown strut or shock. 💣 (§12, p.345)
  – Worn suspension component. 💣 (p. 324)
- **Tock, or tock-tock going over bumps:** Blown strut or shock. 💣 (§12, p.345)
- **Chatter over rough road surface:**
  – Worn suspension components (higher-mileage). 💣 (§12, p.345)
  – Worn steering components (higher-mileage). 💣 (§13, p.348)
- **Rattling:**
  – Loose or rusted exhaust system heat components. (§ 3C, 451)
  – Loose stuff in trunk, usually the jack..

❑ **Squeaks From the Interior**
- Loose or rubbing dash panels: Try snugging all relevant dash panel retaining screws. Applying silicone spray using a pinpoint sprayer into the seams between panels can help as well. The trusty matchbook cover, properly folded, works wonders for some.

❑ **Rumbling, Growling**
- A bearing somewhere: Pin down its location and the circumstances of its occurrence. (§18, p.175 and §11, p.342)

Part 5 of 5

3

---

6   See Note, top of p. 60.

Table 3–3. Noises / 63

4. Dream Vacations

❑ When preparing for your vacation, always plan to have your car serviced at least one week – preferably three weeks – in advance of your departure to allow sufficient time to acquire parts (if some critical part goes on back-order) and to give you time to evaluate the work performed. Those who procrastinate till the last few days put themselves at risk in a number of ways:

- The shop you prefer may be booked into September, leaving you with the unenviable quandary of where to go for service on short notice, or whether to go at all.

- You'll be more vulnerable to the BIG SELL. For example, if you don't have sufficient time to evaluate whether the less expensive of two brake jobs will be adequate, you leave yourself no choice but to opt for the more conservative, and often considerably more expensive, alternative.

- If your shop does manage to squeeze you in on the day before you leave, then screws up either because they're too busy or because the kid they've hired to handle the overload forgets to tighten your oil drain plug . . . Well, then you might find yourself sitting at home the next morning, or 150 miles away, fully loaded, at the mercy of some robber baron who has set up shop along the highway.

❑ In the heat of summer, try to schedule your car into the shop first thing Monday morning, before the pressures of the week start to build. (I'm not suggesting that you call Monday morning or descend unannounced. I'm suggesting that your car be at the shop with a pre-scheduled appointment at the start of their business day.)

Chapter 4. Dollars and Sense

4

Notes

Dollars and Sense

1. Penny-Wise, Pound-Foolish?

Those who focus on the price of a certain job rather than the reliability of the work performed are doomed to suffer the consequences. By looking for the best "deal" on labor, people often end up selecting a garage that employs workers with no more than rudimentary skills.

Shops need to bill enough to make ends meet. Lowering labor prices to attract more business can certainly help increase revenue, but as I mentioned above, it ultimately has a chilling effect on the quality of the staff. Now consider what lower labor revenues do to parts prices: If a shop can't pay its expenses from one column, it will most certainly try to make up the difference in another. The following prices came off receipts for parts I purchased for a 2014 Subaru Outback (2017).

	Regional Repair Chain (aftermarket parts)	Subaru Dealer (OE parts)	NAPAonline.com (OEM parts)
Air filter	$27.20	$21.95	$13.49
Fuel filter	$32.92	$22.80	$24.49
Oil filter	$ 9.62	$ 6.65	$ 6.49

The difference in pricing speaks for itself. But when you compare the filtration material used in the three air filters–tufted fibers in the OE parts, rough cardboard in the other, you begin to appreciate just how significant those differences are. Poor filtration produces internal engine wear, hastening increased oil consumption, pollution, and performance losses.

It is impossible to compare price without considering several hidden variables, one of which is the quality of the parts being used.

2. Estimates vs. the Actual Bill

Estimating is not a precise science. One or two unexpected parts, shop materials, and that extra bit of labor to align this or adjust that always seems to creep in to run up the price. Computerized billing has helped somewhat, particularly with routine service work performed at standard service intervals. But many jobs are, by their very nature, open-ended – at least until the car has been disassembled sufficiently to enable the mechanic to see the whole picture. That being said, an estimate that starts out at $200 and ends up as a bill of $400 is, in my opinion, definitely not within the forgivable realm unless:

- the owner has been forewarned of the above possibility prior to authorizing the work,
- **has subsequently been notified** of the necessity of the additional work, and
- has approved the overage.

There is one category of exceptions to the above statement, invariably unpleasant for both the shop and the customer, but legitimate and unavoidable: Bolts do break, threads occasionally do strip (particularly in aluminum), problems do arise that could not be predicted. Statistically, this category is quite rare, particularly in competent shops.

Estimates that skyrocket are much more likely to be examples of an all-too-frequently-employed tactic known as the "low-ball" or "come-along" quote: Newspapers and mailers are filled with them – ads for "safety inspections", brake pad replacement, mufflers. The fine print certainly seems reasonable; the price, often irresistible . . .

During my years of repair shop ownership, one of my competitors would mail out an advertisement every month or so that offered various "specials", one of which was for timing belt replacement. This "special" included the cost of the timing belt plus the labor to put it in – nothing more. I suppose that in a court of law, that ad would be considered perfectly legitimate – an accurate reflection of cost. But in the real world, once the customer's engine was lying in pieces across some mechanic's bench, the price would climb exponentially as "problems" became "apparent":

❑ The timing belt is located behind the other drive belts, separated from them by the timing belt cover. To gain access, every one of them must be removed. More often than not, those drive belts are just as old as the timing belt. Shouldn't the cost of new drive belts at least be mentioned as a possibility in the original estimate? And since these drive belts not only have to come off, but also have to go back on to complete the timing belt job, wouldn't you think that there shouldn't be any extra labor charges for installing new ones? In that particular dealership, the answer to both questions was a resounding no . . . Anything above and beyond the timing belt incurred extra labor charges.

❑ The camshaft and crankshaft seals are positioned directly behind the timing belt drive gear and sprocket (Fig.2-1., p.43). To my mind, unless the owner intends to sell the car within the next year or two, it makes a lot more sense to spend an additional half-hour's labor to replace those seals while they're so easily accessed, rather than tearing the whole thing down again after the engine starts leaking enough oil to become obvious to the owner. (By that time, of course, the new belt will be soaked in oil.) After all, any engine that has spun through the countless millions of revolutions that led to 60 or 90K has a worn set of seals. The odds of their making it until the timing belt needs to be replaced again are nil.

❑ And then there's the timing belt tensioner. While not as likely to fail as the drive belts and seals – at least the first time around – the act of increasing the tension on a worn bearing will ultimately result in bearing noise . . . down there inside the engine, right behind all those other bearings (water pump, alternator, power steering pump, air conditioning compressor). Harder to diagnose – or easier to misdiagnose – raising the possibility of yet another blown quote.

We used to mention all of the above in our initial quote so that the customer could only be pleasantly surprised if we found it unnecessary to replace the tensioner or drive belts. Naturally, we priced ourselves out of a number of timing belt jobs, but educating our customers produced both loyalty and referrals.

> Beware the low-ball – your car will be in pieces before you even get the call about the need for "additional" work.

There are only two ways to avoid being skewered by this tactic: Either educate yourself in the mechanical workings of your car so that you can ask the right questions before your car is in their hands, or develop a working relationship with a shop that lays out all your options in advance.

> It is impossible to compare price without knowing specifically what is included in a particular quote.

The written estimate is a very useful device for keeping estimates and final bills in line. Not only does it give you a clear picture of how your money will be spent, but it also puts the shop on notice that you are not one to be taken lightly. But the written estimate does have its downside:

4

- ❏ The very fact that you are requesting an estimate in writing injects a note of distrust. Although that might be understandable – even reasonable – the first time you walk through a shop door, it gets downright insulting if used to excess. If you can't trust the shop you're in, start looking for another!

- ❏ Most jurisdictions allow a 10% cushion to cover the little extras (shop materials, disposal of hazardous wastes, etc.), and acknowledge the shop owner's right to pick up the phone whenever some unforeseen problem requires raising an estimate. This gives the low-baller an immediate out, regardless of your insistence on having the estimate in writing.

- ❏ In requiring a shop to stick close to its numbers, a mechanic's inclination to address any other issues that might arise can be greatly reduced. In the example above, you've been given a quote for a timing belt and the labor to put it in; you have it in writing. But your crankshaft seal is starting to leak. What's the shop going to do about it? Their perception of you is already set: A phone call raising further issues will likely produce bellyaching or worse. Why not just slam the belt in, then make a note on the ticket about your crank seal? That way, the shop can take you twice and feel justified that it couldn't have happened to a nicer guy! Cooperation and understanding – in an honest shop – will reduce this polarization.

What about the car that's been dropped off for standard servicing, but has a multitude of problems that the owner has no idea about? Even a reasonable estimate could climb considerably higher than the owner's expectations. What is the shop supposed to do if the owner can't be reached? Customers never enjoy surprises like that. We would generally "ice" the car until the owner resurfaced. But there can be extenuating circumstances: What if the owner absolutely, positively has to drive thirty miles home tonight? What if the car is literally dangerous in its present condition? Do yourself a favor:

> Whenever you take your car in for service, stick around while your mechanic examines your car and builds his estimate. Then you'll be able to go over your car's problems with the man who's going to fix it, to ask questions, to judge his character and skills. Then, and only then, should you leave for work.

4

3. Labor Guides

Mechanics who work on commission get paid a specific number of labor hours for a given job regardless of how long it takes to complete. How do shops arrive at those figures – their quotes for the number of hours to charge for a particular job? Virtually the entire industry relies on the figures set down in either the Chilton, Motor, or Mitchell Labor Guides. An internet subscription to one of their Pro websites provides not only "flat-rate" times for the replacement of virtually every part on almost every kind of car, but repair guides for these procedures as well. For the general purpose shop, these websites are invaluable, allowing them to provide a reasonably accurate quote on work they've never done before . . .[1]

Not everyone adheres to the numbers given, however. Most of the factories produce their own books that detail warranty times. These can be identical to, or quite different from, the hours billed to retail customers. Specialty shops often quote their work based on their own experience with a particular make or model. Collision estimates are computer-generated; I haven't a clue where their numbers come from.

Although any good mechanic can beat the "book" time on standard service work, a broken bolt, excessive rust, or previous butchery can soak up large chunks of time unaccounted for in any of the flat-rate manuals. These are the ugly problems of car repair – the unforeseen additional charges, source of many consumer complaints.

What about unbilled time? Dealership mechanics have to deal with some ugly warranty issues, most of which must be diagnosed "off the clock." Suppose that a fuel injection problem results from a poor electrical connection buried somewhere in the wiring harness. The factory pays for the repair but rarely for the time to diagnose, much less to locate the damned glitch! Dealership mechanics must usually "contribute" that time to the cause. Similarly, mechanics across the spectrum of car repair have their own demons to face: Cars that won't fix and won't go away regardless of how much unbilled time goes into them.

I don't mean to paint a picture implying that mechanics aren't making a decent living; obviously, the good ones do. But as long as the time charged for work performed comes out of a book rather than being based upon the actual time spent on a particular car, customers buying standard service work will be subsidizing those who drive in the nightmares.

4. Minimum Charges

The minimum charge is an aspect of billing that can often lead to grossly inflated costs. I suppose these were initially conceived to ensure that the shop could cover the costs of writing a repair order, possible road testing, etc. These things take time, and as a former business owner, I can understand the motivation of management in wanting to defray such costs. But the minimum charge has taken on a much more costly dimension in many shops. Rather than applying the concept strictly to those little nuisance jobs that take place out in front of the shop, many shops use it to define the smallest fee that they will charge for any

---

1   These resources are also available to the do-it-yourselfers among you: chilton.cengage.com, themotorbookstore.com, eautorepair.net and alldatadiy.com. Other resources can be found using the following search criteria: online factory diagnostic and repair information; automotive labor time guide free; car manuals online. My personal choice has always been alldatadiy.com because it adheres closely to the auto manufacturer's repair manual–albeit in database format.

type of work. Thus, a procedure that takes five minutes is billed as if it took ½ hour; with four of those on one ticket, the mechanic (working on commission) might earn two hours' labor for twenty minutes work. During my years in the dealerships, I saw a tremendous amount of customer resentment over minimum charges, and concluded that the amount of anger it generated far outweighed any financial benefit.

5. Double-Billing

The following section has more exceptions and qualifications than most. Misinterpretations are inevitable. And yet the underlying premise – that the pen is mightier than the wrench in extracting cash from your pocket – is critical to an understanding of how the greedy can hustle the uninformed. Perhaps it's enough to know that it happens. If you encounter a recommendation for two repairs in the same general area of your car, you will now know to question whether the procedures overlap.

I've seen countless repair orders brought in by new customers on which labor had been tacked on for the replacement/adjustment of certain parts that had to be removed to gain access to the primary problem – for example, additional labor charges to replace drive belts during water pump replacement. To my knowledge, every engine on this side of the galaxy requires drive belt removal to gain access to the water pump; that's part of the reason why the job costs what it does. Naturally, the price of the replacement belts will add to the bill, but additional labor should not be charged.

To discern whether you are paying for the same work twice, it is essential that you have a certain familiarity with the anatomy of your car – another reason for you to wait at the shop until you are presented with an estimate. That way, you can actually see what they're talking about.

Procedures that offer fertile ground for double-billing include:

❑ Any Annual Service: Since all of these involve multiple inspections, lubrications, and adjustments that span the entire car, just about any additional work can produce some overlapping labor charges.

Here's an example. Annual servicing typically includes a brake inspection and road test (Table 2-2, p.35). If it turns out that your car needs front brakes, the labor charged for their replacement is usually identical to the price you'd pay if you had made an appointment for the sole purpose of checking out a brake noise. In both cases, the mechanic will spend a good twenty to thirty minutes road testing, putting the car up on his lift, yanking the wheels, and performing the inspections. Clearly, the shop has the right to bill for this time. Equally clear: The annual service already includes time for doing these things.

Hypothetical question: If you went in for nothing other than a brake inspection and ended up buying front brakes, does the repair order have two distinct line items: one for inspection, and one for the brake work? Usually not. Most shops don't differentiate: road-testing, inspection, and repair are all rolled into one labor figure. Why? Because when a customer calls to inquire about the price of a certain job, he wants to hear one number, not some vague "It depends on what you ask for". And when he comes to pick up the car, he expects to see that number, not a bunch of add-ons – no matter how valid they might be. Real-world question: If you take your car in for annual servicing and in the process the shop uncovers a bunch of other issues that require attention, will you be paying overlapping labor charges?" The answer: Probably, but you are better off not asking the question . . . The fact is that annual servicing is more profitable than most repair work. The point is to be aware of it.

Other overlaps with annual servicing can involve battery servicing, aspects of engine tuning, oil change, lubrication and level checks, tire rotation, tire balancing, and drive belt tensioning (not replacement), to name a few.

Aside from the time involved in setting up the car on a lift, there are only four procedures that do not overlap with annual servicing in any significant way:

- Air conditioning work
- Exhaust system repair
- Wheel alignment
- Interior work –dash panels, seats, etc.

❑ If you are paying for a brake inspection, then a four-wheel tire rotation should not cost a penny. However, rotating five tires does take longer because of the time required to extricate the spare and secure its replacement. In any case, tire balancing will incur additional charges.

❑ The replacement of any one coolant hose, the thermostat, or the water pump should significantly reduce the cost to replace other hoses in the system, because the time to drain, refill, and perform a leak check need only be done once. (Remember the two dealerships that wouldn't accept a personal check? Each of them charged ½ hour per hose, regardless of the number changed – the minimum charge at work!

❑ Replacement of a front (rear) strut or strut cartridge should reduce the costs of front (rear) disc brake repair on that side of the car, because the brake caliper usually has to be removed to free the hydraulic brake line from its attachment to the strut. Likewise, disc brake work should reduce the cost of wheel bearing replacement and vice versa, assuming that both jobs involve the same quadrant of the car. Strut and wheel bearing replacement may or may not overlap, depending on design.

❑ Driveshaft removal (front-wheel drive) for rebooting or replacement should drop the cost of driveshaft seal replacement to about fifteen minutes. It should also slightly reduce the cost of front wheel bearing replacement.

❑ On most cars with manual transmission, clutch replacement requires that the transmission be removed, an ideal time for transmission front seal and engine rear main seal replacement. In addition, on rear-wheel drive cars, the transmission's rear seal becomes accessible; on front-wheel drives, the driveshaft seals become exposed. The clutch release cylinder (hydraulic systems) or clutch cable should also come cheap.

As for the automatics, the exact same seals become exposed if the transmission must come out.

❑ During any deep engine work, such as removing the engine's front cover(s) or cylinder head, many components become much more accessible. Therefore, the labor cost to replace or recondition them should be greatly reduced. The following parts or procedures suggest a range of repairs made possible by the disassembly. In looking over a bill (or estimate), I would expect to see one sizable chunk of labor for the removal and reinstallation (R&R) of the principal component(s). Additional labor charges for the following items should then range from no charge to nominal. These lists should not be considered exhaustive in any sense, nor necessarily applicable to your engine:

❑ Front cover removal: timing belt or chain, its tensioners and guides; front camshaft and crankshaft seals; the oil pump, its gasket or oil pump drive seal; water pump and related parts, and drive belts.

❑ Cylinder head removal: valve work including refacing or replacement, valve guide replacement, valve seating; cylinder head planing; replacement of fuel injector seals, intake or exhaust manifold gaskets, cooling system repairs of all sorts, drive belts, and oil change.

❑ Unless you are already paying for annual servicing, the following procedures do not overlap in any significant way with other work.

- Air conditioning work
- Battery servicing
- Drum brake repair
- Engine tuning, major or minor
- Exhaust system repair
- Lubrication work
- Oil change
- Rear axle bearing replacement
- Transmission servicing (automatics)
- Wheel alignment

6. Checks, Cash, or Credit Cards?

Some smaller shops provide discounts for cash; it never hurts to ask. Even those that don't usually appreciate those who consistently do business in cash. Personally, I've always thought that a hefty tip laid directly into the hands of the mechanic is far more effective than any amount of cash put into the business owner's coffers. However, if you've never worked with this particular mechanic, return with the tip after you know you're fully satisfied.

Using a check or credit card provides some cover if you feel the need to protect yourself against the unscrupulous or the inept. Unfortunately, neither alternative is perfect. A stop-payment order has be filed in time to beat your check back to the bank. T hat leaves you with no more than a day or two to realize that you've been had – and to react. The use of a credit card gives a bit more flexibility because there is no significant time constraint. But you must write a letter detailing both the problem and what you expect as settlement. Send it by certified mail, return receipt requested, *within thirty days*, to both the credit card company and the repair shop. The limitations of this approach are numerous:

- You must first try to resolve the problem with the shop.
- You may withhold no more than the amount of the disputed repair. Other work performed that day must be paid for.
- The cost of the repair in question must be over $50.
- These confrontations can drag on all the way to the courts. Although you won't be paying penalties and interest to the credit card company, you might be paying heavy dues on your tranquility quotient.
- There may well be other stipulations that I'm not aware of. Check one of your monthly statements or call the credit card company itself for more information.

4

Personally, I'd probably use a check because it enables a stop-payment: It's decisive, and it's in your hands. But since the shop owner could sue you, be sure to cover your backside with a letter (return receipt requested) and a smaller check for whatever repairs were not botched. Write this letter promptly and be reasonable in figuring the amount of your second check. Being fair will reduce the likelihood of a lawsuit. **If the shop prevails in court, you may be liable for court costs and additional service charges for the stop-payment process.**

One last thought: Credit cards reduce the shop's take by two to four percent – a significant loss to the small business owner.

Chapter 5. At the Shop

5

Notes

5

At the Shop

1. Before Climbing Out of Your Car

When you arrive, don't leave your car blocking a garage door or the middle of the service lane. Shops can be very busy first thing in the morning; the last thing a mechanic wants to see is a ton of steel blocking access to his bay – particularly if you have the keys!

Before leaving your car:

❑ Put your owner's manual on the front seat, opened to the specifications page, preferably with a paper clip to hold it open.

❑ Count the number of clicks your parking brake produces between fully off and fully engaged, and write that number down.

❑ Write down your mileage as well, and put both pieces of information in your pocket. (Many mechanics, myself included, take offense when customers push their trip counter back to zero: It implies a lack of trust. Because using that button can affect the work, I'd suggest writing down your mileage on a piece of paper that leaves with you, unnoticed.)

❑ If you have wheel locks, put their key in plain view.

2. Dealing with the Service Writer

Your note should be so complete that you need only flesh out a few details with the service writer:

❑ Where you'll be waiting (always preferable, at least until you know these folks), or how long it will take for you to get where you're going.

❑ Give the service writer your keys – both to the ignition and the trunk. (We can't replace a burned-out brake light bulb or check the inflation in your spare tire without one.)

❑ The Refresher – "Remember me? I'm the 'schnutz' who will only settle for OE or OEM replacement parts. I made a point of getting here early to give you as much time as possible to get them. By the way, thanks a lot for going to the trouble." (Because your phone call might have put you squarely into the Carl Consumer category, you want this conversation to be as lighthearted and pleasant as possible. A bit of self-deprecating humor never hurt anyone.)

Obviously, if you already know these folks use high-quality parts, then your note should be all that's necessary.

❑ Any if-then issues: "If you complete the inspections for the one-year service and find no problems that would increase the bill, then feel free to go ahead with the work; you won't need to call." But re-emphasize the point contained in your note that if the mechanic starts finding additional problems that need to be addressed sooner than later, you would really appreciate being advised of those issues *before* he starts wading into the actual repairs that you've authorized in your note. Getting slugged with a series of problems as yet unaddressed when you come in to pay for your $400 worth of repairs can be enough to make a grown man cry – particularly if those additional problems would cost another $700 to repair.

The more aggressive among you will probably feel compelled to close by saying that you don't like surprises. Personally, I find that aggression does get my attention, but in a negative, counterproductive way.

❑ Naturally, if you can convey to the service writer that you know how to open your hood, it will definitely work to your advantage, but be subtle about it. For example, "My coolant looks clear, but I'm not sure of its protection level; would you check it for me?" Another: "The drive belts on this engine are all original, but they look O.K. to me. Would you take a look at them and give me your opinion?"

❑ If yours is a problem car, one that's been through one or more different shops to no avail – or one that's been through this shop more than once – you should definitely point that out, and ask how your repair order (RO) will be handed out within the shop. There's no sense in your paying a rookie – or even a journeyman – to look over the car when you need the shop foreman. If your travails have encompassed other shops, be sure to bring copies of those repair records with you for their inspection.

❑ If the car is having cooling system work or "standard package" service work, verify that they have no intention of putting additive packages in your cooling system. There are many shops in this country that hustle additives: For every can that's added, the mechanic makes fifty cents or more. Water pump lubricant is oily, many of the offerings seem to create slime that plugs the inlet valve within the radiator cap, and in my opinion, it's unnecessary – water pump bearings are sealed.

Another opinion/fact: Radiator stop-leak is a product to be avoided at all costs – it can literally plug your cooling system from the inside out. If you ever see either of these products on your bill, particularly if you clearly stated your opposition to them less than twelve hours before, I'd suggest that you insist on a thorough flush of your cooling system at no charge. For more on this, see §8A, p. 158.

❑ Examine the repair order. If it contains vague or sweeping statements such as "fix rough idle" or "repair brakes," ask him to rewrite it as follows: "Investigate rough idle (brake noise) and call with estimate." Never sign a blank work order and never give a blanket authorization such as "put this car in tip-top shape" unless (1) you know these people very well, and (2) you set an upper limit on the cost of the repairs in your letter, and verbally reiterate it. Be very specific about what you're authorizing, and careful about what it will cost.

It's in the nature of humans to believe that their car really doesn't need that much work. After all, it's running so well! Most people believe as they walk in the door that their bill couldn't possibly be more than 200 bucks. To avoid the shock that comes with the realization of what things really cost, feel free to get some idea – a few basic figures – before you authorize much more than the initial inspection of your car. But don't stand at the guy's desk and badger him over the price of individual aspects of the job; in many senses these numbers are meaningless until the mechanic has actually examined your car. Giving your service writer the space he needs to compose a meaningful estimate is far better for both sides.

Once your conversation with the service writer is complete, do everyone the favor of retiring to their waiting area – or to your office if you must – and let the folks in the shop do their job without distractions. If questions arise, write them down and save them until the service writer comes to talk to you. He won't have any answers for you anyway until the mechanic has had a chance to look over your car.

When the service writer finally does call you in, listen attentively, ask to see why they are recommending this particular repair, then ask whatever questions remain. Actually looking over your car with them will give you the chance to evaluate their honesty and them the chance to prove it. To get at that question, ask them to prioritize the work – to give you a schedule of what needs doing, and when. "Which repairs are essential to the well-being of my car today, and which ones can I safely postpone?" Honest mechanics understand as well as any what it means to live within a budget and they certainly won't begrudge you that.

If you intend to put off some work, several questions need to be asked:

- Will the labor be the same if I put this off for a few paychecks, or will it cost more later because you're already in there today? How much more?

- Is there anything else that I've already authorized that is less important than these newly found problems?" By appealing to their guidance and judgment rather than by reacting to them as adversaries, you will get a lot more – and better – information.

If you do have the misfortune of being handed some outrageous quote, don't ever say that you'll need to get a second opinion. After all, they've got your car lying there in pieces before you! Instead, say "Geez, considering all that you've told me, I think I'd better give myself a few days to decide whether I want to sink that kind of money into this car, or whether I should replace the damned thing. Let me pay you for your time, and once I've made up my mind, I'll let you know. Could you write that stuff down for me? I'm sure I'll forget some of what you've said without a summary." Don't leave until you have those prices broken down line by line.

If you are not in the shop when you receive the dreaded follow-up call (detailing additional work), you are faced with a conundrum: Do you trust them and authorize the repair, or do you stand firm on your initial agreement? There's obviously no single answer, but you can always avoid insulting their credibility by explaining that you just don't have sufficient funds to complete the job in one trip. The only problem with this approach is that it doesn't answer the question of whether you actually need this additional work. Naturally, you can ask them to show you the problem when you pick up your car, but if it's not something visible (such as brakes), that option is remote. In all fairness, lift space is scarce; mechanics are entitled to complete their work, road test, and move on. The options of a second opinion or return visit are always open, but these will cost time or money, probably both. As I said before: **When you're talking car repair, the smart money is always on waiting around long enough to see, understand, and authorize.**

5

3. Shop Time

Good shops are busy places, and once a car has been torn down, that bay is tied up until either the customer gives authorization or they get tired of waiting for you. It can be a terrible waste of a mechanic's time to force him to button up your car because you've decided to go for a walk. So . . . Don't disappear while you're waiting!

The same holds true for those of you who have dropped your car off. Don't make an appointment on a day when you can't be reached! If a meeting comes up, be sure to have your cell phone with you–and turned on. Remember, you're trying to work with these people, not against them.

Avoid being a nuisance. Customers who badger a busy service writer with inconsequential chatter might just find that their bill reflects that error in judgment. Naturally, if some thought springs to mind that is truly relevant to the issue at hand, by all means let the service writer know about it. But save all those little questions for his conversation with you over findings and cost.

The same holds true for those who call hourly with the nervous, "Have you looked at my car yet?" One call is all it takes. They'll give you a call once your estimate has been prepared.

5

When it comes to authorization of the work, be decisive. Don't leave the shop twisting in the wind while you hem and haw over finances. Car repair never comes cheap; know what you can afford before you go in or be willing to use a credit card. Whenever a bay gets tied up for a half hour or more, standing idle for whatever cause, the man whose tools reside there – *your* mechanic right now– is losing money. Take the time you need to inspect your car's problem(s), to discuss your options, and total their cost. Then move on: Agonizing and foot-dragging are counterproductive.

If the decision to authorize must be made by your spouse, let the service writer know that up front – before he spends ten minutes providing you with the blow-by-blow. If the two of you have to arrive at this decision jointly, or if you are indecisive by nature, do the shop the favor of telling them that. Give them an idea of when you'll be calling back. That way, the service writer can let his mechanic know to button the thing up and run it out. You both win: You'll have sufficient time to reach a decision you can live with without costing both the service writer and mechanic the income they will lose while your car sits idly on that lift.

4. Opinions and Recommendations

❑ Customers who routinely take their cars to the same shop stand a much better chance of getting the service they need than those who jump all over the map looking for a deal here, a break there.

❑ Finding the right shop can be more than a one-step process, often involving the exchange of money just to find out what type of place you're in. Never start out your relationship with the "I want to form a relationship" line – that will grow naturally if you have been satisfied with the first one or two visits.

❑ A busy shop is not an appropriate spot for attempts to ingratiate yourself; the individual that you are facing is far more interested in your trust than your posturing. And most good shops are normally too busy to permit much in the way of the social graces anyway – too many cars to get out by 6:00 PM!

❑ Don't act like an expert. So often we'll be told to follow some list of the owner's devising – you want this and nothing else. There are two problems with that approach: First, your idea of what needs attention might be missing the mark; second, an impression is left that the owner can't trust the shop to suggest a course of action that's beneficial to both.

❑ Recounting all those horror stories from other shops you've been through might just label you as a problem customer. A simple statement such as "I've had my share of problems finding someone who can fix this damned thing" will do just fine. Wearing hostility on your sleeve is no way to gain a receptive ear.

❑ Once you've become comfortable with a shop, you might be lulled into believing that it's unnecessary to hang around waiting for that estimate. However, it's been my experience that shops can change drastically over the course of a year or two: Mechanics move on, service writers burn out. If it were my money, I'd make the time every time.

❑ There's no harm in asking for a specific mechanic when you drop off your car, but every shop has its own scheduling concerns that might preclude your request.

❑ Many mechanics are not comfortable working with the public. If they were, they probably would have found another niche for themselves, more closely aligned with people than with machines. The fact that they work on steel should serve as an obvious sign that they are more at ease manipulating their hands than their words. Don't lose sight of that fact should you decide to stand there and watch him work. That "up close and personal" approach may well be viewed by a mechanic as an obvious sign of mistrust on your part rather than the perfectly normal curiosity that you might consider it to be.

❑ Ask informal questions rather than giving the third degree. We in the trade have had more than our share of that. And don't assume because you are speaking with someone whose command of the English language is not as refined as your own, that you have the right to insult his intelligence or integrity. Although it's true that many mechanics were poor scholars, reaching the conclusion that they aren't bright could be more of a reflection on you than on them.

❑ Speed should not necessarily be equated with carelessness. Most great mechanics thrive on working fast – it's in their nature. Should you encounter a "wrench" who appears to have fur flying around him, give him the space he needs to focus on whatever he's doing. His intensity may be his best tool so long as others don't insist on breaking his concentration.

5

❑ And then there are the problem cars: If a car isn't fixing, the last thing in the world a mechanic wants to do is share his frustration – or listen to yours. And that service writer is skewered between the two of you like some perverse shish-kebab. Give these guys some room; they'll call when they've got some information for you. I understand that it's not always that easy; you could be feeling skewered yourself by some goon with his meter running. (All the more reason to ask the right questions before you set up that appointment!) But if you have any faith in these guys, remember that you have an important role to play in establishing this partnership: Trust and mutual respect are essential ingredients.

5. Reflections

I put off writing this section for many months, knowing how difficult it would be to confront the stereotypes of my profession without falling prey to a sense of sadness and a certain loss of objectivity. You see, the uniform, the dirt, and the noise and smells of the shop set me and my colleagues apart from the more "civilized" of our society. This view of mechanics as belonging to a lower caste has done tremendous damage both to the self-esteem and to the charity of many in my profession. The way people so often talk down to us, unabashedly declare their mistrust, or express their indignation over the cost of a repair often leaves me feeling cold, angry, or just beaten down. Little wonder that so many mechanics end up feeling hard-hearted!

If you want to know the one characteristic shared by every "good" customer of mine – the one trait that would lead me willingly into going the extra mile for that person – it is simply this: They approach the individuals within my shop as their equals. I am not speaking here of hollow flattery or ill-conceived attempts to ingratiate oneself. Rather, I am describing a fundamental respect for the skills of another, a communication of trust. Not one word need be spoken; in fact, these feelings are best addressed on a nonverbal level: They will shine through in any conversation that lacks fear or hostility.

Personally, I have no problem with a new customer projecting an air of nervousness or anxiety; most of these people have been ransacked more than once. Whenever I sense that a customer is consumed with fear, I work hard to overcome these suspicions through a brief course in mechanics. By addressing their questions, I do what I can to put them at ease. But not everybody can do that, and nobody's willing to do it all the time. As the relationship develops over the course of several visits, I feel that there should be a distinct change in the content of our communications, away from anxious testing toward a shared commitment to problem-solving. If mutual trust and respect are not growing, then either you are in the wrong shop or perhaps you are the one who is guilty of offensive behavior. Rehashing the same old cliches about the pervasiveness of ripoffs in the auto repair industry can dim your persona quickly in the eyes of an honest shopkeeper. He's heard it all entirely too many times before. Naturally, a bit of complaining is almost expected in this business, but past the first one or two visits, the customer should be transcending his knee-jerk distrust toward evaluating the shop strictly by the quality of their work.

No one deserves the label of butcher, low-life or thief unless or until he earns it. And if he does, then your shadow should never cross that threshold again. But what if the shop that you're about to enter offers high-quality service at a competitive price? Projecting a patronizing or contemptuous attitude might be truly counterproductive.

"Do unto others . . ." – The Golden Rule is just as valid in the arena of car repair as anyplace else. Those who would heap scorn on this industry and yet not take the time to look into the eyes of those they accuse are getting exactly what they deserve . . .

Taking the time to educate yourself in how best to get things done, how to evaluate various types of shops, and how to assume responsibility for your car's needs are all extremely important. But unless you actively pursue good service as a willing partner, letting go of your preconceptions regarding those who would voluntarily get that filthy for a living, then this book will have only brought you halfway. It's one thing to learn how to shun the butcher shop, it's quite another to learn that we're not all out to get you.

6. Doctors and Mechanics: A Few Similarities, A Few Differences

I don't want to belabor this analogy, but these two professions do share a few things in common, and certain of their differences are instructive.

First of all, there is no mechanic or doctor on this planet who is capable of fixing everything that comes through his door. But both customer and patient want to experience the comfort of knowing that they are in good hands. This need for assurance often creates an implicit consumer demand; namely, that the doctor and mechanic should know far more about the more puzzling, ephemeral problems of their trades than is actually possible. The statement "I don't know why . . ." injects a level of honesty into the equation that most people just don't want to hear. Yet there are certain problems in both health care and car repair that we (patient and doctor, customer and mechanic) just have to muddle through together. Lucky for the doctor, he has a way out: It is both customary and acceptable for him to refer you to other specialists and still get away with billing you! The mechanic who cannot fix the problem car on its first visit is usually viewed as either stupid or incompetent. If he were to refer you to another shop and hand you a bill, he'd be guilty of brazen greed as well – never mind how much time he's spent eliminating possibilities. Most consumers with a problem car usually start out believing that for the kind of money they're shelling out, they have the right to demand perfection. If they could only realize that certain problems really are unusually difficult! Sometimes a second opinion or just some cooperation – precise observation by the owner, perhaps? – might help to resolve the problem.

Clearly, there are distinct differences in how frustration is handled by the patient and car owner. How many of you have trained your rage upon a doctor, even when you know that mistakes have been made? About as often as you listen to opera, right? All those years of schooling and clinical training confer upon the doctor a far higher level of respect than a mechanic could ever hope to achieve. I can't imagine anyone in the automotive business who hasn't encountered his full measure of verbal abuse. And yet it takes every bit of five years apprenticeship to make a good mechanic, and another five to ten years to become a master technician. There are good mechanics in this country, men and women with both skill and an interest in the well-being of their clients, which leads me to the one fundamental similarity between mechanic and doctor: The good ones care; the bad ones have attitudes!

7. Picking Up Your Car

So much has been written by consumer advocates suggesting that you insist on seeing your old parts, then demand the freedom to road-test your car prior to paying the bill. Let's get serious now. Imagine how you would feel if you were an honest mechanic, his service writer, or the owner. I'm trying in this book to teach reasonable approaches to the very real problems of getting your car repaired. Some of you will indeed get burned the first time or two you venture out for service, even if you have read this book. But there are ways to deal with comebacks (*9, p. 86-89), and not one of them includes attacking the integrity of the very people whose support you will need to reach a satisfactory solution to the problem.

When you pick up your car, by all means feel free to ask questions: Don't leave until you understand what you've paid for. Look over your repair order to be sure that it includes:

- Brake percentages if you paid for a brake inspection
- Antifreeze protection level if you paid for a summer or winter checkover, or for a major or minor service.
- The brand of motor oil used, including its weight and grade
- If relevant, the type of transmission fluid/antifreeze/brake fluid used.

The proper time to ask that these bits of data make it onto your repair order is in the morning in your note to the shop, not at 6 PM with a line of people standing behind you. And speaking of lines, pocket calculators are really tacky – do the additions at home – later. If you authorized brake pad replacement over the phone, feel free to ask for them, but only if you told your service writer earlier in the day that you wanted them returned; otherwise, they'll probably be in the trash. In short, there's a proper way to extract the maximum amount of information, every bit of which you're entitled to; and then there's the heavy-handed, confrontational approach that is guaranteed to confirm for you everything that you've ever had to say about this profession!

If you are pleased with your service writer, be sure to say so – a little gratitude goes a long way in this business. If you are impressed with the mechanic who worked on your car, express that as well and get his name for the next time in.

Before driving off,

- Check your mileage.
- Make sure that your owner's manual is still on the seat.
- Those of you with wheel locks, be sure to check that the key is where you left it.

Chapters 6 through 18 contain in-depth procedures for inspecting your car once it's back in your hands. The "Short Course" below will get you started. Do yourself a favor and take the time to do it. Should you find that the work was well done, mail your mechanic a note of appreciation, possibly with a generous tip. I can assure you that it will be money well spent!

It's always worthwhile to spot-check the prices that you've been charged for parts by comparing them with comparable internet offerings from parts jobbers, dealers, and on-line suppliers of factory parts. Doing this is obviously not practical at the point of

5

authorization because (1) it drags out the time required to authorize the work and (2) you don't have any part numbers yet. But doing so after you have your bill in hand is definitely worth the effort. In the final analysis, a couple of bucks here or there is certainly not worth getting worked up about – it's the quality of the end product that really counts. But if it's clear that you've been hosed, revisiting the owner or service manager with the evidence could well result in a refund – or a bridge burned forever.

8. Evaluating Standard Service Work, The Short Course

If you were hoping that there would be some way to skip all that reading, this could be it – sort of like "Cliffs Notes" for the automotive service set:

❑ Check your oil level. Any puddles beneath the car? (*§*3, p.129)

❑ Check the coolant expansion tank, (*§*3, p.151). This tank should never be filled past its full line (when the engine is cold) or it will overflow as the engine warms up. Naturally, if the car is still warm following servicing, you'll have to wait for the following morning to know exactly what its level is. But if you ever find this tank filled to the brim when you pick up your car, return to the shop immediately so that *you* don't have to clean up the mess.

❑ The same holds true for the power steering reservoir, (*§*3F, p.106), the brake master cylinder, (*§*3D, p. 105), and the clutch master cylinder, if applicable (*§*3E, p.106).

❑ If your service included complete level checks, you'll need to weed through all of Chapter 6.

❑ Level checks typically includes lubrication – door and hood latches, hinges, locks. You can see if they've been lubricated by looking for a shiny, wet, slightly oily substance on, around, and drooling down beneath them. Most shops won't do door lock cylinders without being asked because the overflow dribbles down the door. Unlike brake fluid, it won't discolor paint, but it does attract road dirt like a magnet.

❑ How about the air filter? (*§*2A, p.188 or *§*2D, p.219) Is it new? A factory replacement? If not, is the filter material thick and cottony, or is it made of coarse paper?

❑ On higher-mileage cars (60K and up) following major or minor servicing check the throttle plate for sticking. (*§*2A, p.211)

❑ If your battery has just been topped off, you will usually find water droplets scattered randomly over its upper face. If you find a few drops, wipe them up with a *disposable* rag – it's dilute sulfuric acid. If you find a puddle, have the shop clean it up. In either case, **be sure to recheck the battery in several days to ensure that it has remained dry.** If its surface is wet again, the battery was overfilled. Read *§*3C, p.102, then take it back!

❑ If you were billed for repair of an oil leak, did they clean up the area so you won't get sold the same repair the next time in?

❑ If you were billed for a cooling system flush, inspect the expansion tank. Is the bottom inch of it full of crud? (*§*3, p.151).

5

❑ After you've driven three to five miles, check the automatic transmission fluid level. (**§**1A, p. 357)

❑ Check the lights. Do all the side markers, tail lamps and tag lamps work?

❑ Are the windshield washer nozzles aimed at the glass in the approximate center of the wipers' travel? Do the wipers clean the glass in one or two swipes?

For the more devious among you:

❑ Before going in, lower the air in one tire, enough so that it is obviously different from the rest. Was it refilled? See if anyone mentions that you should recheck that tire in a month.

❑ If you're going in for a brake check, put a dab of mud across the threads of one of your wheel lugs, one per wheel, or a bit of mud into a wheel lock. You will then know if you received a two- or four-wheel brake inspection. Compare the number of clicks on your parking brake before and after.

9. Dealing With Comebacks

If you must return your car to the shop because something just isn't right with the work that was done, first take the time to cool down. A problem car can cause frustration on both sides of the service desk; if you can rise above your urge to lash out, both you and your car will be better off. The following three truths always apply:

- Getting the problem resolved is far more important than venting your anger.
- Everyone makes mistakes every day.
- The human being that you will be speaking with has feelings much like your own. Insults will play with him about as well as they would with you.

Equally as important, be prepared to communicate exactly what you know about the symptoms of your problem. It's been my experience that as often as not, cars come back a second time because of the owner's inability to describe the problem adequately the first time around. If technical speaking is a challenge for you, refer to Tables 3-1 & 3-2 for help (pp. 56-63).

When you're ready to pick up the phone, remember to avoid using the always provocative "Ever since you worked on the car . . ." in favor of the more gentle "I'm still having some problems with the car and need to bring it back." Be friendly, patient, and understanding; but always ask to speak with the service manager. Although it's possible that your mechanic was a butcher, it is equally possible that the information you provided got lost in translation between you and the shop floor. Because the service writer is your medium, you will want to politely sidestep him in favor of his superior.

As with the staff, so too with the boss! Avoid recrimination and sarcasm, and stick to the facts. Describe the work that was done and the symptoms of your current problem. Don't lose sight of the fact that the size of your bill gives you no justification for hostility. Remember, getting your car fixed is far more important than venting your frustration!

On the day of your appointment, arrive with a concise, written description of your car's symptoms – one copy for them, one copy for you. Make sure that the description of your problem as written on the repair order is just as detailed and accurate as your own. If it's not, you may be wasting another day. But more to the point, if this problem turns ugly, you will not have adequate documentation to support your case. This can be especially true for cars still under warranty[1] And be sure to speak with the service manager. He needs to know that your car is there, and he might want you to refresh his memory. If yours is a cold-starting problem, the car has to be dropped off the day before. If it's a driveability problem or an intermittent noise in motion, you should volunteer your willingness to wait until he or his shop foreman can go for a drive with you, but don't insist on it. At this stage you should still assume that he knows how to do his job. In any case, whenever possible let him know that you are willing to leave the car with him until he is satisfied that it's been fixed.

If your second visit does not resolve the problem, continue the even-handed approach, frustrating though that might be. Ask the service manager to set up an appointment to meet with both him and his shop foreman personally. If you are working with a dealership, insist that this meeting be scheduled for a day when the factory rep[2] will be there – especially if it's still under warranty.

That meeting should be cordial and very specific: Describe the problem, when it occurs, how it occurs, etc. Be sure to bring along your written description of the problem – with enough copies for each of them – and again, insist that their repair order be specific and accurate.

If it's a driveability problem, this time insist that the shop foreman and/or factory rep road-test the car with you. Admitting to some frustration would be perfectly understandable at this point – as long as you refrain from an aggressive venting of anger. Understand and acknowledge their frustration as well.

5

---

1 As an example, let's say that your transmission slips when upshifting between second and third gear. When the car had 7,500 miles on it, you weren't really tuned in to what was happening; something just felt strange some of the time. Your repair order read, "Check trans." The mechanic's written response was, "Trans checks OK." By 22,500 miles, you know exactly when the problem occurs, you verbally describe it to an interested and seemingly helpful service writer, and he responds by writing "Check trans for shifting." The mechanic's written response comes back: "Trans is shifting fine." In all probability, the service writer didn't exclude your information intentionally. Maybe he was just never properly trained. In any case, you could run through your warranty like that, then have the thing shred itself into teeny little pieces 6,000 miles later. Ouch!

If you are specific in your description of the problem and insist upon precision in their writing, then your argument that this problem predated the end of the warranty period becomes much stronger. That second repair order should have read "Transmission slipping during the 2–3 upshift, intermittent. Inspect."

2 Factory representatives typically visit each dealership monthly. They are usually perceived as handling new car problems exclusively, but many would welcome the chance to get involved with something more substantial than their common fare of squeaks and rattles, paint discolorations, etc. Even if he is not technically trained to roll up his sleeves for you, his awareness of your problem brings the service manager's feet a bit closer to the fire.

Whenever possible, talk to the factory rep privately. It may be that the dealership handling your car does not have the personnel to deal with the problem. The rep may want the car inspected by a particular technician in some other dealership. In any case, be sure to get the rep's office phone number should further follow-up be required.

A large number of second and third revisits could be eliminated if the car were not returned to the mechanic who has been struggling with the problem since it first came through the door. But because of the nature of most commission shops, no one else will have the slightest interest in wasting their precious time curing little Johnnie's screw-up.[3] Be that as it may, if this is your third time around, you have every right to demand a more qualified mechanic. Leave it to the service manager to pick his man; at this point, he has the responsibility to fix your car. And be sure to reiterate that you will leave the car in his hands until he is personally satisfied that the problem has been completely resolved. In most cases, that means a rental car. In this book, as in baseball, it's three strikes and you're out! If your car is still not right, you should have no further interest in getting anything from this shop except your money back. Virtually every community has some sort of consumer affairs office; ask them to (1) recommend a shop that could give you a second opinion and (2) assist you in mediating the dispute. You'll want a written statement from that second shop, detailing their view of the problem as well as their suggestions for repair. If mediation seems to be leading nowhere and vengeance seems all the more sweet, let it be known that you will be calling your local TV station. Nothing like the threat of a bit of bad press to bring a recalcitrant shop around! If all else fails, you can always hire a lawyer, but attorneys don't come cheap. You might find that by the time he gets done, you will have just spent enough to pay for a new car!

5

Not all comebacks are the mechanic's fault. Two thoughts:

❑ Perhaps as many as 25% of all comebacks are due to the customer being more in tune with his car following repairs than he would ordinarily be, noticing things about it that he formerly hadn't. As an example, a man stormed into our shop last week, outraged by the fuel pump oil leak that he claimed occurred immediately following our oil change. After showing him the adjacent coolant hose that had become so distended with oil over the course of the past year or so that it was about to burst, and the repair order dated six months earlier that suggested its replacement, he finally cooled down. But the bad taste of his unrelenting hostility still hangs in the air. I'm not sure that I'll ever forget him.

❑ A manufacturing flaw shared by many of the same models could well be a significant part of the problem. In these cases, the manufacturer is painfully aware of the problem – the dealership network has been screaming for a solution. These situations are usually handled by intensive, though quiet, research, followed by an equally discreet bulletin to the dealers detailing repair procedures and an authorization for "good will" repairs. These so-called "secret warranties" are cataloged by the National Highway Traffic Safety Administration and are available to the public on their website (www.nhtsa.gov). Another resource is the Center for Auto Safety, a nonprofit watchdog group (www.autosafety.org). Naturally, you'll have to go to the dealer for these repairs.

If this is becoming a warranty claim, a carefully written letter to the auto manufacturer can be a very persuasive instrument. Send it by certified mail, return receipt requested, addressed to the president of the company. Your letter should detail the history of the problem and include photocopies of every relevant repair order.

---

3   "For one thing, he's the one who got paid for it; for another, who's going to pay me?" Inherent in the structure of a large commission shop is the natural process of reducing grown men to something less, more intent on whining and scrapping over their turf rather than working for the common good.

So what wisdom can I possibly impart to those of you who have exhausted your options and are still wedded to a car that won't fix? Simply this: You'll either need to dump the thing (A bit excessive, don't you think?) or find a better shop. But this time you'll have to choose more carefully; reread Chapters 1 through 5 before you even begin the search. Good luck!

5

Notes

Chapter 6. Maintenance Routines

6

Notes

If working on a car is not something you do every day, please read Section 2, p. 113 (For the Novice Interested in Tinkering), and Section Three, p. 114 (Safety Considerations) before you go anywhere near your car. Thank You.

6

Maintenance Routines

1. Once Every Month

If you develop the following routine one morning each month, you will extend your car's life by at least several years and save hundreds, possibly thousands of dollars. Once you do it a few times, you should be able to complete this inspection in less than five minutes.

- Before starting the car, with the engine stone-cold, inspect the coolant level in the surge (coolant expansion) tank, §1B, p.95.
- Before driving, check your tire pressures (§1D, p.96) and washer solvent (§1C, p.96).
- Run a few errands to warm your engine to operating temperature, then check the oil level on flat ground with the engine off, §1E, p.97.

- If you have an automatic transmission, check the transmission fluid level as well (engine fully warmed, running at idle, gearshift set in Park), §1F, p.98.

1A. Motivation

❑ **Oil Level**

I am well aware that checking the oil level can quickly become monotonous, particularly on (most) newer engines, because that dipstick always seems to register full or nearly so during the first several years of the car's life . . . Need some motivation?

I can't begin to count the number of customers who have rolled in for an oil change around 40K with absolutely no oil on the dipstick. Here stands a once beautiful car, not even close to being paid off, that is destined to burn oil down through the years! Then that plaintive wail goes up, "But how could this happen? The oil light never came on!" The answer is that **there are very few cars on the road today that will inform the driver that the engine's oil level is low.** (Those that do are predominantly higher-end models.) The oil light indicates a loss of oil *pressure*, not a loss of oil. When that light does comes on, you'd better be looking for a place to pull over, because it's telling you that the oil pump can no longer scavenge enough oil to lubricate the motor. In all probability, your oil level has been low for weeks. (No wonder they call it an idiot light . . .)

As with the oil light, so too with the oil gauge. These register oil pressure, which only corresponds to oil level when it is already too low to adequately lubricate the piston rings.

Auto manufacturers could build oil level sensors into every engine. We do, after all, have bluetooth capabilities and heated mirrors. So why don't they? The cynical response is that after the car is sold, the manufacturer can't make a dime on it unless you buy their parts. Engines have lots of parts.

So don't get complacent! Develop good habits, then stick to them. Lapsing into neglect can spell BIG TROUBLE, particularly in the 30,000 to 40,000 mile range – the time in an engine's life when oil usually starts to disappear.

6

❑   **Coolant Level**

Overheating the engine is one of those catastrophic mistakes that is almost always completely avoidable. A monthly inspection of coolant level coupled with routine, periodic checks of the temperature gauge virtually eliminate the possibility of trouble. Unfortunately, all those miles of trouble-free driving stretching out behind you have the same effect as continuously finding the crankcase full of oil.

This might help: Once an engine overheats, it takes no more than five to ten minutes of additional driving before the head gasket blows. Minimum cost to repair? At least a thousand bucks. Continued driving can lead to many thousands more of additional engine damage, culminating in power loss as the engine begins to lock up.

> **If you ever see the temperature gauge climbing toward the red range, turn on the interior heat full blast, pull over, turn off the engine, then turn to §1, p.149. Do Not Remove the Radiator Cap!**

❑   **Tire Pressures**

Every car manufactured for the US market since 2007 has a tire pressure warning system. I'm sure they've prevented many accidents. I'm also certain that their presence offers an assuring sense of security to a great many of us:

> "No need to bother with checking tire pressures. Remember last fall? That warning light came on as soon as it turned cold. We'll know if they're low."

All of which assumes the system continues to work and that it's properly calibrated. It also means that you're willing to settle for tires that might be under-inflated by as much as 25%,the point at which your least-inflated tire will trigger the dashboard warning light (Fig. 6-1). Assuming that a tire's correct inflation is 32 psi (pounds per square inch), a 20% loss means that tire will be carrying less than 26 psi.

Even moderately under-inflated tires produce increased tire wear, decreased fuel economy and consequent higher pollution. Tires running 10 psi or more below their suggested inflation are extremely dangerous when cornering, particularly when swerving to avoid hitting a pothole. Severe under-inflation can trash a tire in short order. Under-inflated tires usually don't look like they need attention until they get down around 15 psi, which is damned near flat!

Fig. 6-1. Tire pressure warning light

Over-inflation is not addressed by tire pressure sensors. Excessively high pressures dramatically reduce traction–particularly noticeable on wet roads. Hot weather or high-speed driving can produce blowouts. And just like severe under-inflation, too much pressure can quickly ruin a tire.

Significant over- or under-inflation exerts undue stress on the plies (layers) of the tire and their bonding one to the other. This type of abuse creates ply separations just as easily as smacking a pothole. For a deeper discussion, see §1 through §4, pp. 317-325.

> If your steering ever starts to change noticeably from what you're used to, one of your tires could be going flat. Check your tire pressures right away. Don't wait till you've ruined the tire – or your car.

1B. Coolant Level Inspection

All fluids expand as they're heated – it's a basic law of physics. To accommodate that fact, every cooling system has an expansion tank to handle the "swelling". Cars without an expansion tank were called "Model Ts". They had huge radiators that still needed water on a very regular basis. (No, I wasn't around back then . . .)

Expansion tanks come in two flavors: siphoning systems and gravity-fed, Fig. 6-2A and B. The first is characterized by a radiator cap – under pressure whenever the engine is running – paired with an unpressurized expansion tank. The second always resides above the rest of the cooling system; it's pressurized whenever the engine is running; there is no radiator cap. Because siphoning systems have been around since your grand-dad was driving, it has earned the right to be called an expansion tank system; the gravity-fed sytems will henceforth be called surge tanks.

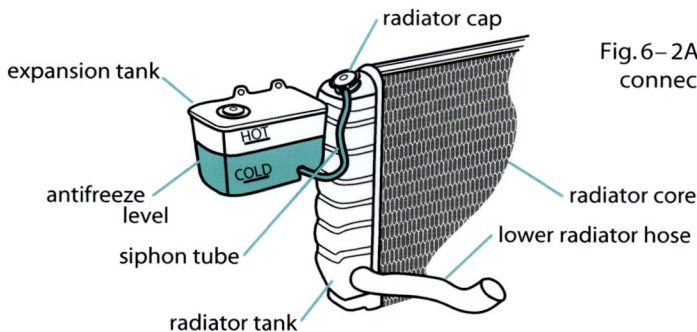

Fig. 6–2A. An expansion tank and its connection to the radiator. © PT

Fig. 6–2B. A surge tank

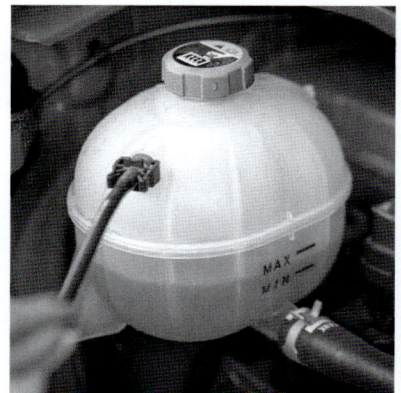

Routine inspections require nothing more than a glance at the coolant level within the tank. As long as it's sitting between the Hot–Cold (or Max–Min) lines, your work is done . . .

One reason for doing this inspection on a cold engine is that your cooling system develops pressure as the engine warms up. Consequently, the coolant level within the overflow tank rises with increasing engine temperature. Checking coolant level on a cold engine produces a consistent baseline from which to measure consumption. Since healthy engines don't consume coolant, a noticeable loss will be easy to spot.

While a consistent baseline is nice, the primary reason for doing this inspection on a cold engine is that if the tank is either empty or overflowing, then the problem needs to be addressed while the system is cold, §3, p.151. Removing the radiator cap on an engine that's run for even a minute can result in coolant loss; perhaps just a dribble, but more often than not, enough to make a mess of your driveway. And you might get burned! The same can't be said of surge tank caps–as long as its tank is not full (in which case **don't even think about it!**)–but just like a radiator, its contents are pressurized. You'll find a Do Not Open When Hot notice on every surge cap and on every radiator cap.

Coolant is slippery, it's sweet, and it's poisonous to pets. It does not clean up easily.

> **DO NOT** remove the radiator cap or the suge tank cap on an engine that
> **has run within the past two hours.** At best, you'll achieve the above result;
> **at worst, well, Just Don't Do It!**

Assuming your system needs less than a quart of coolant, just top it off with water–distilled if you have it. If the system has lost more than that, you'll need to find and repair the leak, §8, pp.156-159. For the short term, topping off with water is perfectly satisfactory–assuming that your weather is temperate. Freezing or desert temps necessitate a 50/50 mix of antifreeze and water.

1C. Washer Solvent Level

The washer bottle typically sits in one of the front corners of the engine compartment. Do not confuse the washer solvent bottle with the expansion tank of the cooling system– cleaning up antifreeze that's been squirted all over your windshield would be a nightmare!

Windshield washer solvent is preferable to plain water under all circumstances, but particularly in winter cold: Ice can destroy your washer motor. Washer solvent comes both premixed and in concentrate, the concentrate is significantly cheaper.

1D. Checking Tire Pressures

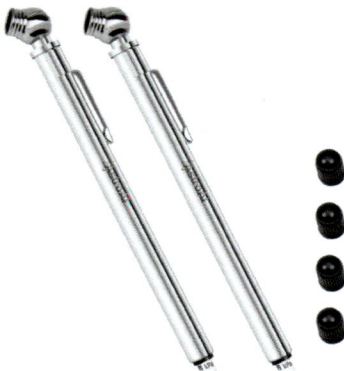

Fig. 6–3. Tire pressure gauges
Courtesy of Astro AI

You need a gauge for this, Fig. 6-3. Buy one for your glovebox; better yet, leave it on your console until you get in the habit of using it. An eight dollar gauge will last until you lose it or drive over it, and in most cases it will be more accurate than driving to the gas station. Why? Because tire pressures increase as you drive: Frictional forces warm the tires, the compressed air within them expands with increasing temperature, the pressures go up. Right now your tires are truly cold. Furthermore, many of the gauges found in gas stations are inaccurate: Most have been abused, few are routinely calibrated. The hand crank style is the worst offender.

The factory's tire inflation recommendations can usually be found on the driver's side door jamb. If not, check your owner's manual. Because of the difference in a car's relative weight front to rear, the inflations recommended for the front are usually different than those for the rear.

If you do need air, the most sensible approach for most people involves driving to the nearest gas station and setting the air pump at two to four psi above factory spec. Assuming a relatively short drive, adding those few extra "pounds" will compensate for the increased pressure produced by driving over there. But recheck them the following morning, just to be sure.

Low tire pressures in three or four tires usually indicate a significant drop in temperature or owner neglect. One low tire out of four usually indicates that you've picked up a nail that has punctured the tire. Tires, steering and alignment are covered in Chapter 13.

1E. Oil Level Inspection

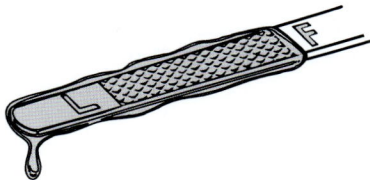

Fig. 6-4. Oil dipstick, business end
Toyota Motor Company

Check the **oil level** on flat ground with the engine fully warmed. With the **engine off**, pull the dipstick, wipe it off, reinsert, then pull it again. The level should be at or near the full line at the top of the low-full grid. If the level seems a bit low, it could be that you didn't give the oil sufficient time to drain to the crankcase. Many owner's manuals advise checking the oil level two to three minutes after shutoff; follow their recommendation.

Back in the day, the Low-Full grid shown in Fig. 6-4 represented one quart. That's no longer necessarily true. The gap between L & F on many late-model cars represents just a half-quart. So if you find your engine oil level down near the Low line, don't just add a full quart mindlessly. Start with a half-quart and see how far it takes you.

6

Fig. 6-5. Funnel

I'm a big fan of funnels, particularly the type shown in Fig. 6-5. They're cheap; using one will ensure that the oil goes *into* the engine rather than all over the top of it. Customize it by cutting off the small end with a utility knife so that the funnel fits your engine's oil filler neck tightly.

Don't ever add oil without having a rag handy . . . Drape it around the engine's oil inlet *before* you start to pour.

Assuming you're pouring from a quart bottle, hold it so that one of the wider faces (the side with the label on it) will be horizontal when you start to pour. Doing so will enable the oil to flow smoothly out of the neck. Pouring oil with those faces positioned vertically will produce a glug-glug sound as the oil pulses forward, then back. That might not be a big deal if you're using the large-mouth funnel shown;

it's a very big deal if you want to hit the hole without one! When removing the funnel, lift it onto the rag to avoid dribbling oil all over the engine, your bumper, etc. Install the oil filler cap, then double-check that it's tight.

Changing the oil, oil leaks, etc. are covered in Chapter Eight.

1F. Automatic Transmission Fluid Level Inspection

As with the cooling system, so too with automatic transmissions: Change is in the air. But unlike cooling systems, that change has left many of us with no access! Look around your engine compartment for a second (transmission) dipstick. If you find one, keep reading. If you don't, see §1A, p.357.

The automatic transmission requires pulling a dipstick as well, but with the engine running at idle, the transmission fully warmed (driven for three to five miles), and the gearshift in Park. Before the actual inspection, run the trans through every gear, at idle with your foot on the brake.

Clean Dexron and Mercon are ruby red in color, clear, and thin – much like a 5W–10 motor oil. The older it gets, the darker it becomes. (Transmission fluid contains cleaning agents that pull dirt and worn clutch pack material right into solution.) Fluid that has turned opaque brown, yellowish-green, or black should be replaced. See §1B, p.359 and §7, p.374 for information on automatic transmission servicing, etc.

Consult your owners' manual for more information.

2. Other Routines

❑ One day each week: Drive with the radio off, just listening to your car. You'll develop a keen sense of your car's normal sound patterns while providing yourself with ample opportunity to hear the warning sounds of impending repairs. Because many sounds evolve gradually over time and many just come and go, consult Tables 3-1 through 3-3, pp. 56-63 before you charge off to the fix-it shop. The better you know your car, the less it will cost you to maintain.

❑ Every five minutes: Glance down at your dash, particularly at the engine temperature and oil (pressure) gauges, to assure yourself that they're reading normally. Build this routine into your driving, regularly shifting your gaze from the road to the rear-view mirror, to the dash, and back to the road.

❑ When parking, be sure to come to a complete stop prior to changing gears (from forward to reverse and vice versa). Rolling forward while engaging reverse can cause severe damage to both automatic and manual transmissions.

❑ When filling up with gas, buy the good stuff at least once a month: Name-brand fuels have additive packages that are formulated to clean the pinholes through which gasoline is sprayed into the combustion chambers. Dirty fuel injectors produce a dribble of fuel rather than an umbrella-shaped mist, robbing you of both performance and mileage. For more on this, see §5D, p.234.

6

❑ Change your oil routinely in accordance with your owners' manual specifications. And read *§1, pp.31-33 and §3, p.34.* Insist on an OE or OEM oil filter, as well as synthetic oil if that's what your manufacturer calls for. Check the oil level immediately following every oil change, then routinely once every 1,000 to 1,500 miles without fail.

❑ Wipers need helping hands. Dirt gets trapped along the wipers' rubber edges and wears them down. Glass cleaner soaked into a paper towel works wonders. As for bugs, a plastic dish squeegee or single-edged utility blade applied to the glass at a 30-degree angle work a whole lot better than 100 strokes with your wiper blades.

❑ In summer, if you're parked in the hot sun, keep a pair of windows–one driver's side, one passenger's side–open just a bit to provide some ventilation. Temperatures can easily reach 130 degrees inside your car–tough on plastics, windshields, and leather seats.

❑ Winter Driving

- If you have been driving in snow, sleet, or freezing rain, be sure to turn off your wipers *before* you turn off the ignition. If you don't, the wiper motor will do everything it can to start moving as soon as the ignition key is turned to the Accessory or On position. Even if you remember the following morning to turn off the wiper switch before turning the ignition key back on, the wipers will still attempt to cycle themselves back to their parked position at the base of the windshield. If your wiper blades have long since been frozen to the glass, they could either bend, be distorted, or pieces of their rubber inserts could be torn right off the blades. If the blades won't move and the motor's drive crank refuses to remain motionless, it will pop right out of its union with the wiper link, Fig. 6-6. If you're lucky, simply reconnecting the two will suffice (probably an hour's labor); if you're not, you'll have to replace the broken link (probably 2½ hours work, plus parts).

6

Fig. 6-6. Honda CRV wiper link, front and rear views
        Courtesy of AH Parts Dismantlers

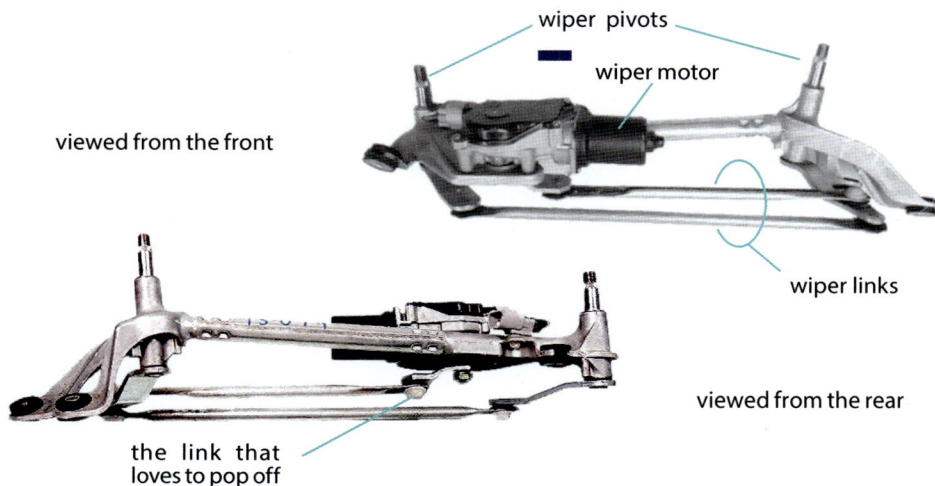

wiper pivots

wiper motor

viewed from the front

wiper links

viewed from the rear

the link that
loves to pop off

- Speaking of wipers, using them to clear snow from the windshield at the start of your driving day tears up their rubber inserts' crisp edges. Use a scraper instead–one with a brass blade works far better than those plastic ones. Use your gloved hand for the area surrounding the wiper pivots. In a pinch, the edge of a credit card works fairly well as a scraper. A can of de-icer simplifies all of this.

- Don't lock your doors in sleet and freezing rain: You might not get back in till spring! Nobody in their right mind will be out stealing cars in this weather anyway. If you do end up with a frozen lock, a can of de-icer works a lot better than a hair dryer–all those extension cords! Whatever you do, don't pour hot water down the side of your door. It might seem like a stroke of genius, but all it will do is create more ice.

3. Other Levels – Every Six Months to One Year

The following list itemizes all of the remaining fluid levels in your car. **The first two should be checked prior to trips of any significance;** otherwise, just once every six months. The remaining six need attention once a year. Naturally, as the car ages (80K-100K miles +/-) the frequency of these inspections should increase.

- Coolant level in the surge tank or radiator (cold!)
- Spare tire pressure
- Battery electrolyte
- Brake master cylinder
- Clutch master cylinder (where applicable)
- Power steering reservoir
- Manual transmission
- Differential

It's very rare for any system to just fail without warning. Routine attention to your car's levels should provide ample warning when a component has started to leak.

> **To safely open a radiator on a car that has been recently driven, the engine must be allowed to cool for at least two hours, longer on a hot day.**

3A1. Coolant Within the Surge Tank

The coolant level within the surge tank should be checked every time you open the hood – once a month. Check the antifreeze protection level twice per year – spring and fall. **Do not open the surge tank unless the engine is stone cold.**

3A2. Coolant Within the Radiator

The coolant level within the radiator and its protection level (percentage of antifreeze) should be inspected twice yearly, more frequently in older cars.

Significant pressure builds within the cooling system as you drive. The impatient among you will be surprised by blistering hot coolant spurting out as the radiator cap is first turned. Those lucky enough to be holding a rag – and with some shred of composure remaining – will re-close the cap before they get burned.

Because of the dangers involved, most shops don't inspect the level of coolant in the radiator (or its degree of antifreeze protection) unless the car has been brought in for major servicing or cooling system problems. Naturally, if the coolant expansion tank is low, topping it off is all that's needed – in most cases. But if it's empty, even a couple of quarts added to the overflow tank might never reach the radiator, because expansion tanks are typically connected to the radiator by a siphon tube rather than by gravity feed, Fig. 6-2A, p.95. If your cooling system is not full to the brim, an air pocket at the top of the radiator can block any siphoning from occurring, effectively isolating whatever coolant is in the overflow from the system that needs it. As a result, a fair number of drivers are presently well on their way to roasting an engine, believing all the while that they covered their bases last week with that "full service" oil change.

Antifreeze protection level presents a similar issue. Without removing the radiator cap, the only measurement possible is the strength of the mixture within the overflow bottle. The differences in dilution between cooling system and overflow can be dramatic.

❑ **Inspecting Coolant Level**

- Start by waiting for a stone-cold engine, four to eight hours past its last outing.

- To remove most radiator caps: Press the cap downward with the palm of your hand, turn counterclockwise approximately 1/4 turn to its first (safety catch) position, then push downward again and rotate counterclockwise another 1/4 turn to permit lifting the cap straight off.

   Some aftermarket radiator caps for older cars have a lever on top that must be raised first to relieve pressure, prior to rotation.

- The radiator should be full to the brim – right up to the lower flange, Fig. 6 – 7.

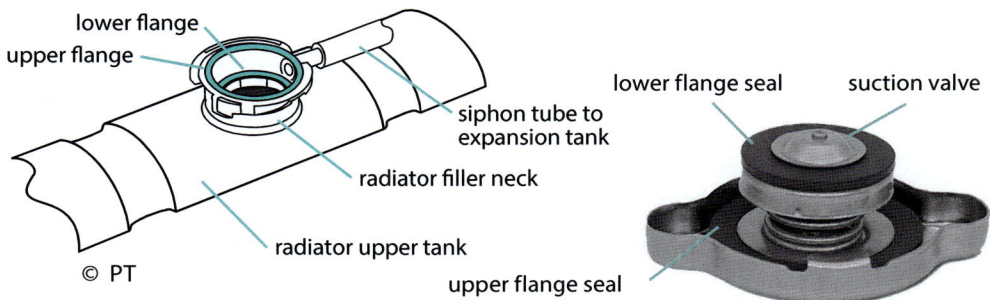

Fig. 6-7. Radiator upper tank with radiator cap flanges.

- If you have an antifreeze tester, take a sample. A 50/50 mix gives a reading of -34°F, ideal for winter and summer driving under normal conditions.

- Reinstall your radiator cap; it must be on straight and turned fully clockwise.

- Check that the coolant expansion tank is filled to its full line, located about 2/3 of the way up its side.

Check for leaks over the next several days by glancing at the pavement as you pull out. If you suspect a slow leak, park over a clean strip of concrete whenever you can, then check the pavement prior to driving away.

Cooling systems are covered in more detail in Chapter 9.

3B. Spare Tire Pressure

As for location and access, you're on your own. Most of you will find a space-saver "donut" – don't get me started . . . They typically require inflation of 60 psi, roughly twice that of a normal tire.

3C. Battery Electrolyte Level

The following discussion relates to traditional automotive batteries; it does not encompass hybrids or battery-powered vehicles.

Electrolyte is the liquid inside every car battery. It's composed of sulfuric acid and water. Electrolyte burns holes in clothing and paint; it's tough on cuts and excruciating to the eyes.

With the exception of maintenance-free batteries, all others require "watering," typically once per year. Failure to do so significantly shortens their life span. But overfilling is perhaps even more critical because of the damage that it can cause via corrosion, Fig. 11-8, p.227 and §A3a, p.226.

Maintenance-free batteries are labelled as such, and most can be distinguished by their flat, sealed upper surface, Fig. 6 – 8. All the rest have access covers, typically two elevated rectangles, occasionally six threaded plugs instead.

Now here's the fun part: There are a whole bunch of batteries out there that look just like that one in the center of Fig. 6-8 that are labelled maintenance free, and yet their rectangular covers pop right off. I never did know what to do with those, other than to service them if they were low on electrolyte (Fig. 6-10, p.104). Then again, there are those whose rectangular covers don't budge whether you're swearing or not. They're labelled "maintenance free" and I believe 'em.

If you're off to buy a battery, choose the one on the left.

maintenance-free   low-maintenance?   low-maintenance

full line

or
maintenance-free?

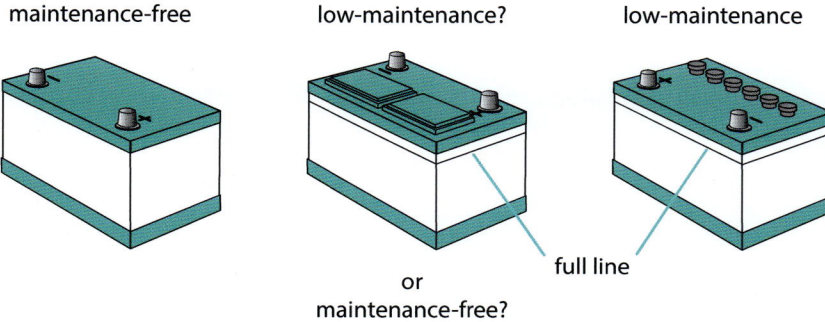

Fig. 6 – 8. Typical battery configurations.© PT

❑ Inspection

The electrolyte level within true maintenance-free batteries cannot be checked. The following discussion applies only to traditional batteries with removable upper covers or plugs – typically labelled as low-maintenance.

> **Never use the battery as a tool tray: If a wrench, screwdriver, or other metallic object accidentally bridges the gap between the battery's positive terminal and any adjoining metal surface of the car, it will create a short circuit, producing an intensely hot spark** with sufficient energy to melt (weld) the offending steel right into place. The most likely result is a severe burn to your hand as you fumble to remove it. Other possibilities include fire, a meltdown of the car's wiring harness, or explosion of the battery.
>
> **Never light a match or use a flame of any sort near a battery:** Flying shards of plastic bathed in sulfuric acid can really dampen a guy's enthusiasm for this kind of work.
>
> **Wear eye protection, just to cover your bases.**

All that being said, batteries are, in general, perfectly safe to work around. [1]

The typical battery has two rectangular plastic covers on its upper surface. To access the six cells housed beneath, these covers must be pried loose with a screwdriver. To prevent contamination of the battery's innards, first clean off any grunge that may have accumulated on the battery's upper surface. An aerosol can of battery terminal cleaner chased with water works best, but plain old baking soda and water will do in a pinch. Brake cleaner or electrical parts cleaner also work nicely, but tend to dissolve any painted markings (name brand, cautionary statements, etc.). Use a small wire brush to break up the crud.

---

1   In thirty-five years of working on cars, I've only seen one battery explode. It was defective (the car had been towed in because of a stone-dead battery that, as it turns out, had an internal short), it was hooked up to a battery charger pumping current at its maximum rate, and it took all of about twenty seconds before it went off. Fortunately, the mechanic had just turned away. It was still ugly.

positive terminal cover

negative terminal

battery strap outlined in green

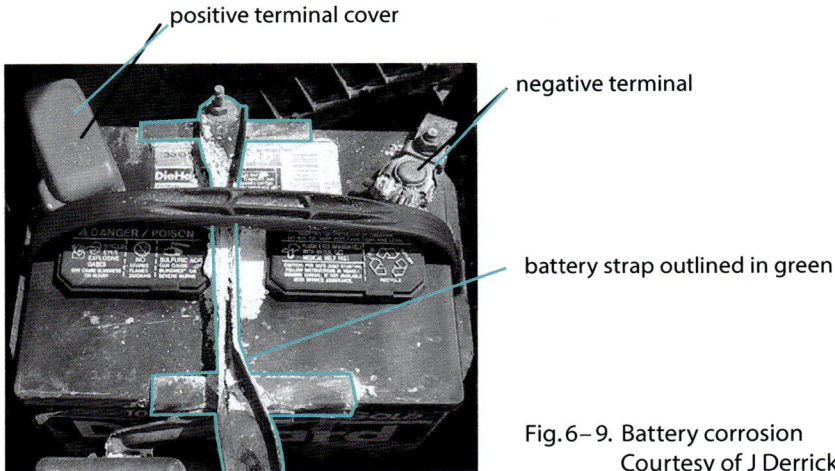

Fig. 6–9. Battery corrosion
Courtesy of J Derrick

On cars that carry a replacement battery, battery cap removal is often blocked by the holddown strap. Because removal of the strap will often litter the top surface of the battery with corrosive dust, remove it *before* your cleanup.

If your battery strap looks anything like what you see in Fig. 6-9 it will have to be replaced. Ditto for the vertical rod(s) that secure(s) it to the battery tray. The negative terminal shown is a replaceable part; it too should be replaced after carefully cleaning the terminal post. Any/all white dust must be removed. It's always best to attend to this before you open up the battery cells. All of this is discussed in more detail in *§*A3, p. 224. Fortunately, most of you won't have to deal with any part of this . . .

Once the battery caps are removed, you will see six chambers (cells) with a hollow plastic collar projecting downward into each cell. Each cell must be inspected individually – their levels are independent. **The electrolyte level should never be higher than the lower edge of each collar, Fig. 6-10 (A&B).**

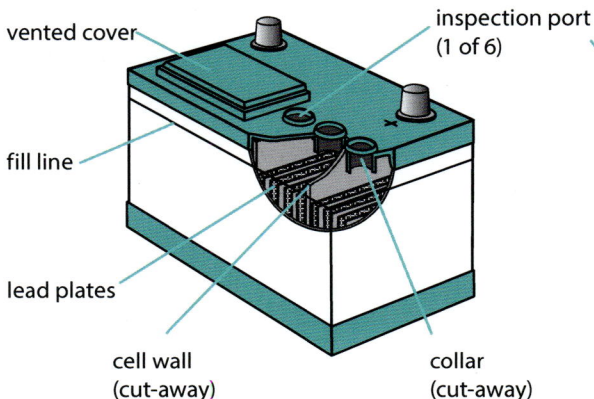

vented cover

inspection port
(1 of 6)

fill line

lead plates

cell wall
(cut-away)

collar
(cut-away)

Fig. 6–10A. Battery, cut-away view © PT

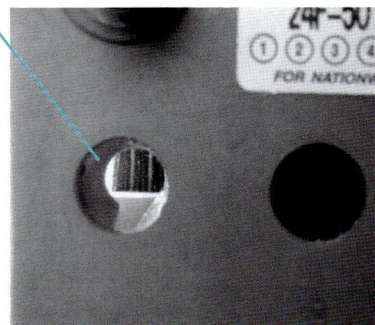

Fig. 6–10B. Inspection port

> **Overfilling a battery** – raising its electrolyte level above the full markers – **produces repeated boilovers of sulfuric acid until a proper level is reached. The resulting corrosion is far more costly to deal with than simply buying another battery,** because the corrosion that sets in will eat away at every piece of metal it contacts, blistering off any paint that might stand in its way.

If you can see the upper edges of a series of thin plates stacked vertically beneath any one of these openings, then the electrolyte level in that cell has been too low for some time. Sulfation is the likely result, reducing the battery's ability to accept and hold a charge.

In a perfect world, you would top off your battery with distilled water, but as a practical matter, 95% of the shops on this planet use tap water.

Following filling:

- The battery's upper surface needs to be dried with a disposable rag – standing water becomes dilute sulfuric acid in the presence of residual corrosion (on the battery strap, for example).

- Take care not to brush any crud into the cell's openings.

- Reinstall the cell covers, then the battery strap if necessary.

- Finally, check the positive and negative terminal connections for tightness by attempting to rotate them around their post – they should not budge. For more detail, see §A3, p. 224.

Recheck your work! **If the top of your battery goes from dry to wet over the course of a week or two,** then you probably overfilled at least one cell. A narrow-necked turkey baster can be used to suck out some of the electrolyte. Don't let it dribble onto the car!

Continue to check for signs of boilover for several weeks because this can also be caused by an alternator that is running amuck.

> Brake fluid can discolor your car's exterior paint if not wiped up quickly, followed by soap and water.

3D. Brake Master Cylinder

The brake master is located under the hood, directly in front of the brake pedal on the engine side of the firewall, Fig. 6-11, p. 106. Check the specifications page in your owners' manual for its correct fluid type – some use synthetic brake fluid, some do not.

Overfilling the brake master cylinder results in a gradual softening of its reservoir cap, Fig. 6-12, contaminating the fluid. Over time, this accumulates as sediment in the bottom of the reservoir and can unnecessarily shorten the cylinder's lifespan. These reservoirs typically have two lines that marks their maximum and minimum allowable fill levels. Overfilling can be dealt with by soaking up the excess in a shop towel.

For more on the brake master, see §18, p. 307.

6

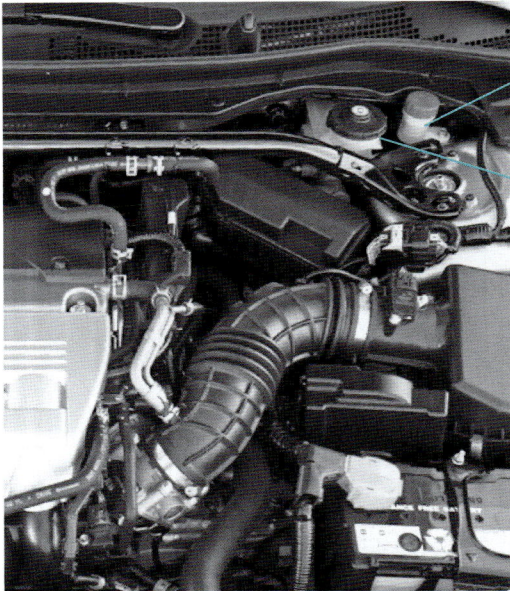

clutch master

brake master

Fig.6–11. Late-model Acura engine compartment, driver's side

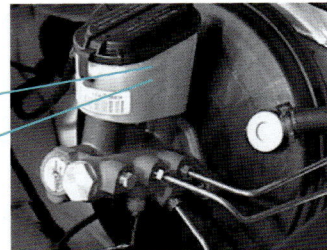

max

min

Fig.6–12. Brake master cylinder

3E. Clutch Master Cylinder (Manual Transmissions Only)

The clutch master is located just to the right of the brake master (standing at the front bumper as shown in Fig. 6-11). Your clutch hydraulics use the same fluid as your brakes do. All that was said in §3D above applies here as well.

See §2G3, p.369 for more on clutch hydraulics.

3F. Power Steering Reservoir

The power steering pump takes its power from one of the drive belts. The high-pressure line (metal tubing and flex hose) leads from the pump to the power steering rack. Fluid in the rack circulates back to the reservoir through a return hose (metal tubing and flex hose). A low pressure line from the reservoir to the pump completes the circle. Consult your owner's manual for the type of fluid it uses, and location if need be.

As with the cooling system, so too with the power steering: As the fluid gets hot, it expands. Unlike the cooling system, there is no separate expansion tank; the power steering reservoir has its own low and a full lines. But the same caution regarding overfilling applies: This tank should never be filled past its full line on a cold engine or it will overflow as the engine warms up.

See Chapter 13 for more information on the front end and steering.

3G. Manual Transmission and Differential Fluid Levels

Inspecting these levels is relatively simple if you own a lift, not so easy on the street – the car must be elevated and level. An inspection plug is removed from the side of each unit and, depending upon make and model, either gear oil or automatic transmission fluid is added until it starts to dribble out. On front-wheel drive cars these inspection plugs are not always easy to access; start by removing the splash pan.

Back in the day, most of these plugs had standard (albeit large) hex-heads – a six-point socket or wrench would suffice. These days, most of these plugs require specialty tools, identical to either allen-wrench or torx heads, but gi-normous.

I suspect most of you will head for the shop for these inspections. Some things you need to know:

- Back in the dark ages when most cars were rear-wheel drive, it was virtually a locked bet that the transmission and differential both ran gear oil. It didn't much matter who was doing the looking because there was very little that could produce a screw-up. These days many use Dexron, Mercon, or other specialized fluids. **These fluids are not interchangeable.**

- To further complicate matters, virtually all of the front-wheel drive transmissions share their exterior casing with the differential. But within those cases, some share their fluid while others are compartmentalized – isolated by seals – permitting the transmission to run Dexron, the differential to run heavier-bodied gear oil.

- Finally, the type of fluid used in a particular case is not visible without draining it.

> If any fluids are added to either the transmission or differential, make sure that your bill states the type of fluid used. Compare that information with your owner's manual – these components are extremely expensive!

How do you suppose Larry Luber is going to know what to use when he's topping off your trans or diff? He won't be using a boroscope, I can assure you of that! These are the times when an owner's manual lying open on the front seat can be worth more than you ever imagined – but you need to do your homework first.

- Mark the specifications page with one paper clip; mark whatever service interval you're on with another.

- Leave the book opened on the front seat when you drop the car off for servicing.

The former can provide your mechanic with quick access to the specs he needs – what type(s) of fluid are required and how much is needed for a complete refill; the latter can provide both of you with information regarding how frequently this work should be done – typically not often.

As for overfilling, that can only occur when sufficient time is not allowed for any excess to drain off prior to replacement of the inspection plug. The most probable result is leakage past a driveshaft seal.[2] But since these units don't employ a pump to create pressure, their leaks are usually more passive than those you might find on an engine or automatic transmission.

Check for leaks over the next several days by glancing at the pavement as you pull out. If you suspect a slow leak, park over a clean strip of concrete whenever you can, or take it back for a recheck. To pin down the presence and location of a leak, see §5A, p. 133.

4. Lubrication

With the advent of sealed ball joints and other steering couplers, under-chassis lubrication has become a piece of history. With the exception of the occasional undercarriage creak over bumps, lubrication is now just about door hinges, latches and locks, an occasional sticking hood latch or creaking trunk hinge–and the mind-numbing buzz of ill-fitting dash panels.

Most shops use graphite in door lock cylinders, an aerosol spray lubricant on door and hood latches, hinges, etc. Judicious application of an aerosol lubricant can often take away a dash rattle. A slip of cardboard from a matchbook cover will often turn the tide on larger gaps.

The creaks and groans endemic to suspensions usually result from aging – rubber bushings drying out, wearing down. An accurately placed shot of lubricant offers at least a temporary fix, occasionally a permanent cure. A good front-end man or specialist in your particular make of car will probably know just where to shoot.

Where present, wheel stops engage a surface, typically shaped like a flattened C, that is tack-welded to either the front or rear of the lower control arm, right in the vicinity of the ball joint. The wheel stop contacts that surface only when the steering wheel is turned hard right or hard left. One aspect of lubricating the front end is to apply a wad of grease across the center of those mating surfaces.

Most wheel stops have a plastic cap (cover). These wear away over time, then fall off. If one is missing, the steering will creak whenever the wheel is turned to the end of its travel, whether its mating surface has been greased or not. Replacement caps are available at the dealer.

---

2 A driveshaft seal – any seal for that matter – can develop a leak simply as a result of wear. Overfilling does not necessarily have to be part of the picture.

6

❑ Lubricants

We use five primary lubricants, depending on the job:

- CRC 5-56, WD-40 and PB Blaster are all good for door, hood, and trunk hinges; throttle body pivots, heater control valves, wiper pivots, door and ignition locks–any place you need a lubricant that can work its way through a crack into an area that squeaks or binds. Buy an aerosol can with a pin-point sprayer tube.

  Any of these products can be used to attack nuts and bolts that are frozen into place, but in cases involving rust, use PB Blaster.

- The teflon aerosols are a cleaner and dryer alternative for dash panels, seat tracks, etc. Buy one with a pinpoint sprayer tube and hold a rag below the dash or behind the track while spraying to protect carpeting, glass and paint.

- Anti-Seize is a silvery goo, excellent for more exposed areas and for heavier duty needs: the hood latch, exhaust manifold studs, etc. It's very slippery and won't wash away. Don't get it on your clothes and don't put it any place that's visible–it's hell to get off!

- In addition to the obvious, lithium-based wheel bearing grease is the goo of choice for lubing engine, transmission and all other oil seals (● 9, p. 380, Fig. 14-3, p. 363 and Fig. 14-18, p. 381).We use it in grease guns to lubricate just about anything.

- Use a molybdenum high-temperature lubricant on brake caliper guides and slides. Make sure that the package specifically states that the grease is compatible with brake caliper parts (rubber boots and seals). If it's not, the caliper boots will soften over time, then swell out of shape.

Shop rags or your sleeve are essential companions to any of the above.

6

6

Chapter 7. Tools and Such

7

Notes

Tools and Such

1. Stuff Every Car Should Have on Board

- A roll of paper towels or two rags–a nice clean one inside the passenger compartment for wiping down the inside of the windshield on cold, wet mornings, and a nasty old one to stuff in some cranny inside the engine compartment.
- Tire pressure gauge
- Flashlight
- One quart of engine oil of the correct viscosity and type
- Premixed washer solvent
- In cold climates:
  - Ice scraper, preferably with a brass blade
  - Lock de-icer (Keep it in your house, not locked in your car!)
- In hot climates:
  - Window cleaner or bug solvent
  - Bug scraper: preferably an ice scraper with a brass blade or a Scotch pad, though they do get messy . . .
- Whenever you're in for service or repair: A file folder containing all your service records, preferably in chronological order. An accordion-style folder with foldover top is by far the best, preferably with either velcro or string to keep it shut.

2. For the Novice Interested in Tinkering

Depending upon how handy you are, you could either save yourself a considerable amount of money or you could ruin your car. Start off small: level checks (Chapter 6), some diagnostic road tests (*§*2, p. 259) and (*§*2 & *§*3, pp. 317-322), possibly a battery service (*§*A3, p. 224). Buy as few tools as possible until you feel comfortable with your abilities.

If you own a car that's still under warranty, don't even think about changing your own oil until you've watched a professional do it several times **and discussed the procedures for record-keeping with the dealership's management.** Make absolutely sure that you won't be voiding your warranty.

> Read the next section – Safety Considerations – before you begin *any* service work. Reread it whenever you feel the urge to get dirty again. Repeat until competent.

7

3. Safety Considerations

A healthy respect for things that can hurt you can help to maintain both caution and vigilant attention. Attempting repairs that exceed your level of competence can produce dangerous conditions both while performing the procedure and further down the road. If you feel under-qualified, under-equipped, or just plain confused after studying a procedure and examining the particulars on your car, by all means seek out the security of a good shop. All those bold-lettered cautions scattered throughout this book are there for good reason:

> ### Safety Must Come First!

- ❑ Long hair, neckties, and loose fitting sleeves are most definitely not appropriate around an engine compartment, especially with the engine running. Likewise, watches, rings, and bracelets should be removed.

- ❑ Never run an engine in an enclosed space. Adequate ventilation is essential to your health and well-being. Exhaust fumes–carbon monoxide–can kill you.

- ❑ Friction materials such as brake dust and clutch disc dust are terrible for your lungs. Never use compressed air to blow it out. Avoid breathing in its vicinity, particularly if the dust is floating around. Use an aerosol brake cleaner – preferably a non-chlorinated brand – to wash the area, standing back at arm's length. Using a respiration mask is always sound practice.

  Put some newspapers down beneath your work, then bag and dispose of them before the brake cleaner evaporates. As it dries, that wet dust reverts back to, well, dust.

- ❑ Wear safety goggles whenever you crawl under a car, particularly in the vicinity of the exhaust system: Just one piece of rust in your eye can make you crazy. Rust is a lot tougher than your eyeball; blinking just once can scratch your eye, and often results in a trip to the emergency room.

- ❑ Wear safety goggles whenever you are using aerosol sprays of any kind, or compressed air, and when working with batteries.

- ❑ The proper tool for the proper job! Slip-ring pliers are no substitute for the properly sized wrench; cinder blocks can be a deadly substitute for jack stands. The list goes on and on . . .

- ❑ An LED shop light or flashlight is a must. A head lamp can be indispensable: It frees up your hands while aiming the light exactly where you're looking. Exposed, un-shielded bulbs can be very dangerous – explosive, in fact.

- ❑ Put the ignition key in your pocket before you start poking your head (and hands) around in the engine compartment. The same goes for oil changes and climbing beneath the car.

- ❑ Don't reuse stripped or mangled nuts, bolts, cotter keys, or circlips. Never reuse a self-locking nut – they're designed to be used just once.

- ❑ They say to never leave oily rags hanging around – they can supposedly ignite. OSHA[1] requires a lidded metal container for dirty rags.

---

1 Occupational Safety and Health Administration

- ❏ Don't start any job unless you have all tools, equipment, and parts on hand.

- ❏ Read all instructions at least twice, with one reading adjacent to the car.

- ❏ Do Not Attempt Shortcuts! They'll give you no end of trouble.

- ❏ For information on lifting a car, see *§ 7A, p.139* and *§7D, p.140*.

- ❏ If you start to get rattled, take a break or walk away.

- ❏ Because of the danger associated with an airbag unexpectedly deploying, don't get involved in any steering column repairs, dash panel removal, or even door panel removal without knowing exactly what you're doing, preferably with the factory's service manual lying by your side. Don't go poking around under the dash with a test light, either, unless you know exactly what you're looking for.

Virtually every nut, bolt, and screw on this planet unscrews counterclockwise (↺) and tightens clockwise (↻). Certainly a simple statement, but it assumes that the observer is looking directly at the piece being removed. If your view is limited by the bolt's positioning, permitting access only from the opposite side, then take time to visualize what counterclockwise means to the bolt (screw, nut). For most of us, it's not that easy. Some might benefit from a mirror; others will only get more confused.

And finally, several lawyers I know have told me to declare something like the following:

> I can offer no assurance to you that the information contained in this book will somehow protect you from any ill-fated disaster that might arise through its use, either to you or your car.

I believe they mean more by that than what's obvious to me – that I would agonize over the knowledge that someone got hurt after finding inspiration from my book. So please, pay attention to what you're doing, be prepared for the unexpected, and always remember that some things you read here are not applicable to your car and in some cases may be flat-out wrong. For those discrepancies I am truly sorry.

4. A Tool Kit for the More Adventurous

This stuff isn't cheap; its purchase should not be entered into lightly. But if you like working on cars, you'll like it a whole lot more if you are properly equipped. Depending on where you buy them, the following items will cost you $600 to $1,800. If you choose wisely, they will last a lifetime. Start small. Buy some of the basics, then expand as needed.

4A. The Basics

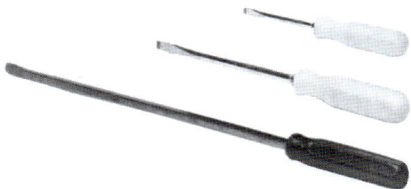

- • one standard-sized flat-bladed screwdriver

- • one big flat-bladed screwdriver (12"– 15")

- • 24 inch screwdriver (pry bar)

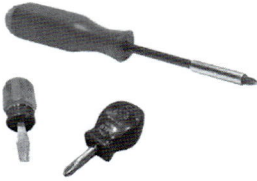

- Magnetic phillips-head screwdriver (most come with replaceable tips; flat-bladed heads and torx- heads are usually included)

- Stubby phillips-head screwdriver; stubby flat-bladed screwdriver

- Slip-ring pliers, needle-nose pliers, wire cutters, and vise grips

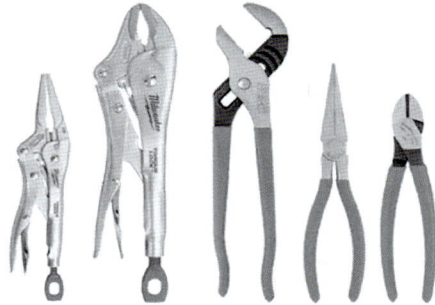

- Complete set of combination wrenches (open on one end, six-or twelve-point box on the other) in either metric or SAE sizes, depending on your car's particular needs.[2]

- Complete set of six-point sockets (3/8th-inch drive), short and medium extensions, a swivel, and an appropriately-sized spark plug socket. If the sky's the limit, a set of 1/4-inch drive sockets as well. A nice ratchet to go with the above; two if you spring for the 1/4-inch drive sockets.

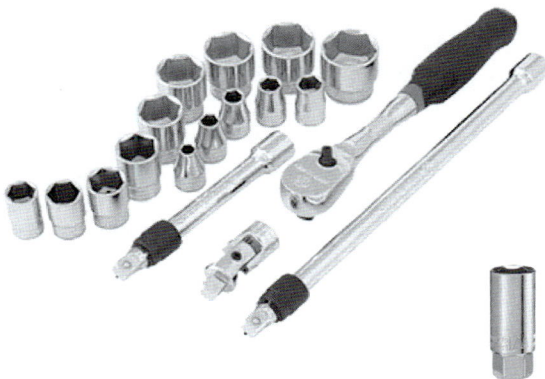

7

Courtesy of Craftsman Tools

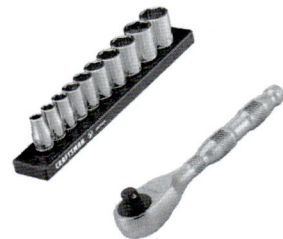

---

2 I personally prefer six-point box-end wrenches. They are much less likely to strip the shoulders off rusty or frozen bolt heads, which means fewer busted knuckles. Their only downside is that nuts and bolts located in tight spots with a limited range of access can be impossible to remove without another tool, typically a twelve-point box-end wrench.

- Shop light – the LED lamps are far superior

- Swivel-mounted mirror on a telescoping handle

- Magnetized wand with telescoping handle

- A four-pound hammer
Courtesy of Crescent Tools
and rubber-headed hammer

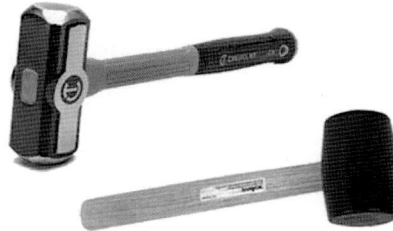

- Breaker bar, 1/2 inch drive, 18-24 inch length, with the appropriately-sized socket for your wheel lugs.

Courtesy of Snap-On Tools

- Utility knife and gasket scraper

7

- Testlight for electrical diagnosis

- Battery terminal cleaner, small wire brush

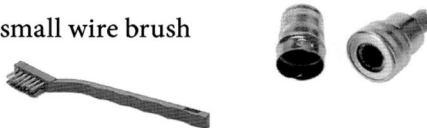

- Safety goggles

4B. Electronic Tools

- OBD 2 code scanner
  and multimeter

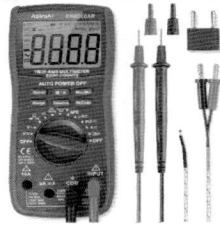

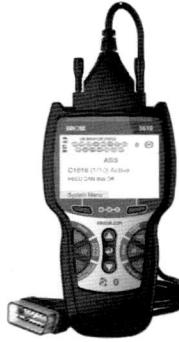

Innova 5630 scanner

- AstroAI 6000 multimeter with temperature probe

- Access to service and repair information for your car. Back in the day, you could actually buy a factory service manual specific to your car, straight from the dealership's parts department. A real book that could absorb your greasy fingerprints and occasional oil spills! These days, you need to buy a subscription to an online source, providing you with the opportunity to trash your laptop . . . (Print out whatever you need for the job at hand, then save those sheets in a folder for your next campaign.)

  The following websites are among the best known: alldatadiy.com, chilton.cengage.com, themotorbookstore.com, and eautorepair.net. Other resources can be found using the following search criteria: online factory diagnostic and repair information; automotive labor time guide free; car manuals online. YouTube is another, invaluable source of repair information. Some of the national parts jobbers – AutoZone comes to mind – have good information. But remember, the information you can find for free might not be totally accurate. If you need a wiring diagram for your 2012 Camry, or the pin configuration for every ECU and relay on that diagram, or the specs for an engine temperature sensor, you'll be much better off with a subscription to one of the subscription services underlined in green above.

4C. Tools for Raising the Car

- Floor jack– A 2½ ton jack is sufficient for most cars and light trucks (You'll only be lifting half the car at a time). The higher the jack will lift, the easier it is to work both on brakes and beneath the car. Low profile jacks are occasionally necessary for low-slung sports cars.

- *Four* professional-grade jack stands–Professional stands are significantly taller, which means that the car can be elevated somewhat higher. Each stand has four legs, which makes them far more stable.

24"

4D. Tools for Changing Oil

- Jack and four stands.

> Crawling under your car without a carefully placed pair of jack stands supporting the car's weight can lead to disaster. Before you crawl under, compare what you intend to save against the cost of permanent injury or death.
>
> Do not go under the back half of the car without a *second* pair of jack stands supporting the rear.
>
> See *§7A, p.139 and §7D, p.140* for explicit instructions on lifting and supporting a car.

- Ratchet with the appropriately-sized six-point socket to R&R (remove and reinstall) the drain plug. In cases where the plug has been overtightened, you might need a six-point box-end wrench and a four-pound hammer.

- The drain pan shown in Fig. 7-1 is terrific–it will catch your oil, store it, and provide clean transport to your local gas station. The bigger the catch-basin, the happier you'll be.

- Buy a long-neck funnel. Cut off its small end with a utility knife so that it fits your engine's oil filler neck tightly. That way, the oil will go *into* your engine rather than all over the top of it.

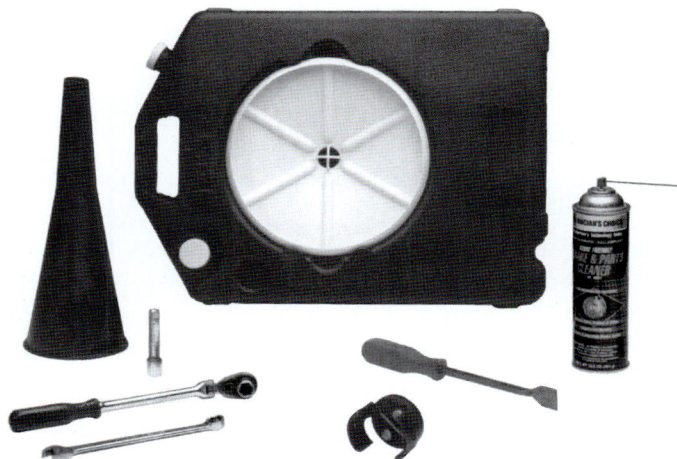

7

Fig. 7–1. Tools for changing oil

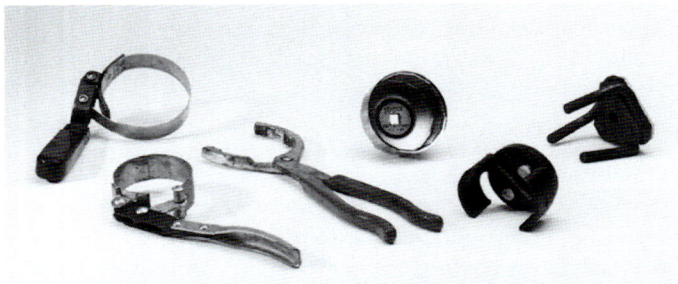

Fig. 7–2. Oil filter wrenches

- Professional oil filter wrench–Fig. 7-2 illustrates some of the many variations available. I prefer the one shown in Fig. 7-1, particularly in tight spaces. Made by Lisle, part number 63600 (for small filters) or 63250 (for large).

- Gasket scraper to remove the drain-plug gasket.

- Rags and some aerosol brake cleaner to clean up the inevitable mess.

> See §7, pp. 139-144 for explicit instructions on how to change your oil.

4E. Supplies

- Aerosol spray lubricant: Common around-home lubricants such as CRC 5-56 and WD-40 are fine for most purposes, but where rust is involved, use PB Blaster.

- Spray cleaner (Spray Nine is great, even on greasy floor-mats)

- Wheel bearing grease (lithium-based)

- Brake caliper lube

- Varsol (1 gallon)

- Hand cleaner: Skip the waterless brands. Those that require water are much better, particularly the ones that contain pumice.

- Brake cleaner: Aerosol cans with pinpoint sprayer tube. The non-chlorinated ones are supposedly kinder to the respiratory and nervous systems. Get several cans.

- Throttle body/air intake cleaner (aerosol with pinpoint sprayer tube)

- Battery terminal cleaner and battery terminal protector

- Duct tape: Particularly useful to stop the bleeding while you fumble around for the appropriately sized Band-Aid.

- Electrical tape: Can be used to pinch-hit for the above but not quite as effectively . . .

- Latex gloves are now in vogue, great for keeping your nails clean, but beastly hot in summer. Heavy-duty disposables are available from both Snap-On and Mac Tool dealers

- Rags – pounds of them.

- Come to think of it, a box of Band-Aids does make some sense!

4F. Fluids

Consult your owners' manual for the appropriate types, viscosities etc. for each of the first five items.

- Engine oil (§1, pp.125-128)
- Automatic transmission fluid (§1A through §1C, pp.357-360)
- Brake fluid (Footnote 2, p. 269)
- Antifreeze (§5, p. 153)
- Washer solvent

7

Notes

Chapter 8. Engines and Oil

8

Notes

Engines and Oil

1. Oils and Filters

Go to any parts store and you find four types of motor oil: conventional, full synthetic, blended synthetic, and higher mileage. Synthetic motor oils are commonly recommended in late-model cars, with a replacement interval of once every 10,000 miles under "normal" driving conditions. Synthetics start out as conventional oil but they are treated to a string of chemical alterations that make them, well, synthetic.

With older cars, semi-synthetics or conventional oils will do just fine; remember that our cars got along just fine for well over a century on conventional oils. Naturally, the oil change interval should be shortened accordingly, but because of constantly improving specifications in everything from lubrication to heat resistance to detergent quality, an interval of 7,500 miles is commonly recommended.

Oils made for higher mileage cars contain additives that will actually swell aging oil seals and recondition their rubber. They're available in both conventional and synthetic formulas.

Consult your owners' manual for the manufacturer's recommendations on types of oil that can be used and the frequency with which it should be changed. But be advised: many begin by recommending that you use their particular oil, available only through their dealership network . . . Keep reading–they all get around to providing a table of "acceptable" substitutes. Purchased through a parts jobber or megastore, you'll pay half as much or better.

> **Always replace the oil filter with an OEM or OE oil filter.**
> **Check your level routinely, don't run your engine low on oil!**

I suggest you use nothing but name brand oils marked with the API (American Petroleum Institute) starburst emblem and "donut", Fig. 8-1. SP is the highest quality presently available for gasoline engines. This issue of grading becomes truly critical on direct injection engines because of their penchant for pre-ignition (p. 198 top center), which, if you're careless, can shorten the life of your engine considerably.

Fig. 8–1. API emblems

1A. Read the Label!

> The Petroleum Quality Institute of America website (pqia.org) cautions
> consumers that there are any number of stores selling outdated oils with
> certification ratings that don't meet current standards.

I suspect that practice is limited to certain smaller independent parts jobbers and generic
gas stations; I've never noticed that behavior in any of the national or regional parts
jobbers. But let's be clear:

- Many of these obsolete oils may be perfectly appropriate for older cars; the
  specs in your owners' manual are as valid today as when they were printed.
  But these oils should most certainly be priced significantly lower than SP oils.

- The SP oils offer the latest technological refinements as of 2022, but they too
  will be superseded in coming years.

- Regarding the frequency of oil changes, most (all?) manufacturers make a
  distinction between "normal" and "severe" driving conditions. Read what
  they say! Or read what I say (§ 1, p. 31). Or do both . . .

You know where I said that current specifications will change? Apparently, they already have!
General Motors has their own spec; actually two now: Dexos1 and Dexos1Gen2. Fortunately,
the Asian and European markets are satisfied by the ILSAC formulary[1], which–at least of
this writing–conforms to the API standard. Or is it the other way around?

Let's try this – As of this writing, the following four brands have offerings that meet all
of the above criteria: Castrol, Kendall, Mobil 1, and Sunoco. There are probably more; I
just don't know about them. Read your owners' manual, read the labels, do your best . . .

1B. What's with the Numbers?

Viscosity describes how oil flows at a particular temperature. Imagine some maple syrup: the
warmer it is, the better it pours; the colder it is, the thicker it becomes. You don't want oil to act like
that; if anything, you want it to do just the opposite. The multi-viscosity oils (0W-20, 15W-40)
behave like a thinner, low-viscosity oil in cold temperatures and like a thicker, high-viscosity
oil when hot. The W stands for winter and describes flow characteristics at 0° F. The lower
that number, the faster the oil will flow in extremely cold temperatures, but the thinness of
the oil film it produces can lead to a loss of sufficient lubrication when stressed by extreme
heat. The higher the second number, the greater its resistance to flow. Consequently, the
oil is more resistant to thinning out at high temperatures, providing greater protection
in extremely hot engines.

Virtually all of the wear to an engine occurs during the first few minutes of driving.
Because the lighter weight oils (5W-30, 10W-30) flow more readily when cold, they are
the clear choice in colder climates. But heat can be tough on the thinner oils; those living
in the summer South might do better with a heavier-bodied oil (10W-40). Your owner's
manual contains recommendations regarding the appropriate weight (viscosity) for your
climate. Follow their advice. You'll find a range of weights to choose from; select from

---

1 ILSAC stands for the International Lubricant Specification Advisory Committee

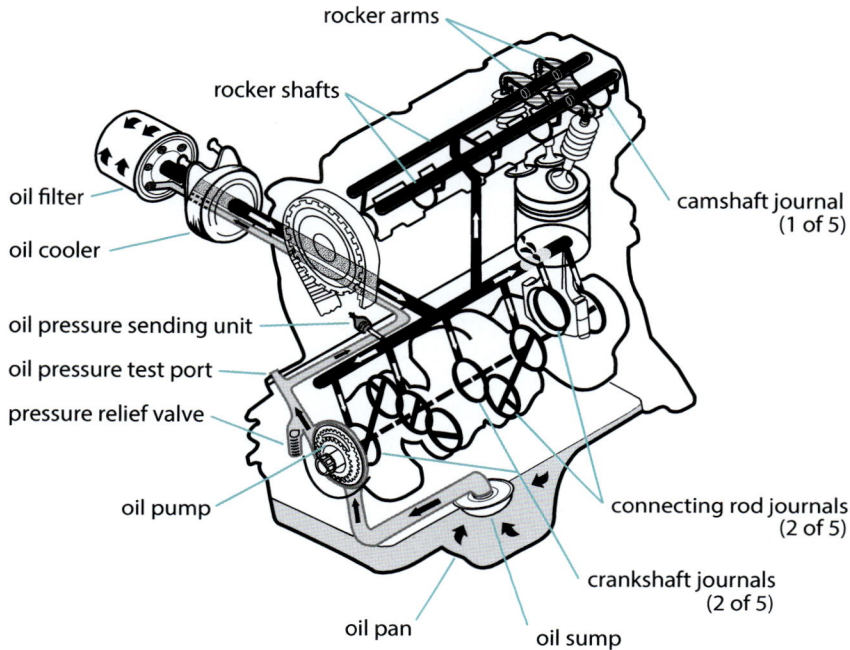

Fig. 8–2. Oil feed on a typical four-cylinder, overhead cam engine. For the sake of visual clarity, the illustration greatly exaggerates the distance between the oil filter and the main oil passageways: In real life, they're positioned adjacent to one another. The circular bands of oil separate the crankshaft and camshaft from their bearings (main and connecting rod). The narrowness of these bands (a few thousandths of an inch) gives some meaning to the term "tight tolerance." Courtesy of Mazda Motors of America, Inc.

the lighter oils suggested during the first 50,000 to 75,000 miles (while your engine is still running tight tolerances[2]), then switch upward to a heavier oil when needed to reduce oil consumption.

So why not buy a 5W-50 oil and have the best of both worlds? Because all multi-viscosity oils require additives to keep them from separating when stressed by high temperature. The greater the viscosity range, the more of these additives are required. Unfortunately, some of these additives lay down carbon deposits as they break down with heat. Over time, these deposits can reduce oil flow through the narrower oil galleys (passageways that feed oil throughout the engine, Fig. 8–2). Whenever possible, stick with a narrower band: (0W-20, 15W-40, 5W-30, 10W-30.)

What about oil additives? Every source that I've read advises against their use, recommending instead that high quality motor oils be used alone. These benefit from having been engineered from start to finish as a single product; why run the risk of an adverse chemical reaction between two conflicting additive packages that might lead to the type of deposit formation mentioned above? In my opinion, there are two exceptions to this rule:

---

2 Engines are built with incredibly thin clearances (tight tolerance) between their moving parts. As engines age, these clearances widen due to wear.

- The old heap that burns excessive amounts of oil can often benefit in the short run from certain supplements. See *§*5B-*§*5D, pp. 135-137.

- Those with noisy lifters (collapsed or worn) can benefit from a can of Marvel's Mystery Oil or a synthetic lubricant like Slick 50.

Note that both of the above problems arise from neglect–excessively long periods between oil changes or inadequate lubrication produced by running the engine low on oil. Words for the wise . . .

I'm a strong advocate of factory (OE) or OEM oil filters, simply because they have the engineering skills of an entire factory behind them. Superior filtration coupled with adequate rates of flow, a one-way valve built into them to assure that the filter doesn't drain while the car sits between start-ups: These are qualities that many generic filters don't possess. However, there are filters on the market today that actually exceed OEM specifications. I have no idea whether they're worth the additional cost.

Don't let anyone – including the factories – talk you into an oil change without oil filter replacement. That's like heading to Beijing for a breath of fresh air!

2. At the Shop

There is a lot to be said for a business that can take your car on short notice, change your oil, check basic levels and get you back out the door in a half hour or so. But if they're not using good quality oils and filters, then your car will suffer for the sake of your convenience.

The American Petroleum Institute routinely tests oils sold in the United States. The results of their latest study are unsettling:

> "API purchased and tested more than 1,000 motor oils dispensed from bulk tanks over the last five years and nearly 20 percent of the bulk oil samples tested failed to meet API standards. API compared the test results against thousands of licensed oil formulations to determine the identity of the oils and to verify that the oils met the performance level claimed.

> "API recommends that consumers visit MotorOilMatters.org to find a Motor Oil Matters (MOM) certified oil change location, look for the MOM symbol at approved locations, read important information on oil quality, and download the MOM oil change checklist to take to their next oil change."[3]

Of the five dealerships I worked in, only one used a first-tier motor oil. The other four all used the same brand – widely known within the industry as the cheapest of the major producers. That may have changed somewhat now that so many manufacturers are specifying synthetic oils, but I tend to doubt it.

Wherever you choose to go for service, ask what brand of oil they intend to use, its API grade and its weight *before* they drain your oil. Then take a moment to double check that the type of oil they'll be using is among those recommended in your owner's manual. If you're not satisfied, bring your own! But make sure to save that receipt and staple it to your repair order.

3   API.org – Data Finds 1 in 5 Bulk Motor Oils Fail to Meet Performance Standards; 10/18/2013

If you're having your oil changed at a gas station, speedy-quick outfit or general purpose garage, be sure to open your owner's manual to the specifications page (use a paper clip to keep it there), and lay it on the passenger's front seat. If you brought your own filter, leave it there as well, along with a spare OE oil drain plug gasket. Many higher-end models use a brass washer that seldom if ever requires replacement, but if your car does not, be forewarned: many of the franchise oil change shops stock a plastic replacement gasket that loves to distort when over-tightened, leaving a slow oil leak past the drain plug whether the engine is running or not . . . I like the fiber gaskets (not cork) that start their life about 1/8 inch thick, allowing some depth to crush through as they're tightened. Pick up four or five of these at your dealer or auto-parts store while you're there buying filters, then leave them in your car as insurance.

You should request (insist) that your receipt specify the brand of oil used, its API grade, and its weight. And before you toss that receipt into your file folder, take a moment to double-check that the type of oil they used is among those recommended in your owner's manual.

I have it on good authority that even though one of the major nationals proudly advertises that it uses a name brand oil, many of its stores actually pump generic oils unless the owner specifically asks for the name brand . . . Consider yourself duly warned.

Regardless of where you have your oil changed, regardless of what it cost, you should check your oil level before you leave the premises. If that seems insulting, then drive around the corner first, BUT DO IT! You would be amazed at how many oil changes result in grossly overfilled or underfilled engines.

3. Checking the Oil Level (Following an Oil Change)

❑ Check your oil level on flat ground, preferably over top of a clean piece of pavement. With the engine off, pull the dipstick, wipe it off, reinsert, then pull it again. The level should be at or near the full line at the top of the low-full grid, Fig. 8 – 3. If the level seems a bit low, it could be that the engine needs to warm up a bit – oil expands with temperature. It could also be that you didn't give the oil sufficient time to drain to the crankcase – many owners' manuals advise checking two to three minutes after shut-off – follow their recommendation. If the oil level is above the full line on your dipstick by as much as or more than the height of the low-full grid, take it back – the engine is overfilled by at least one quart. Significant overfilling builds excessive oil pressure that can work to blow out an oil seal, either gradually or all at once.

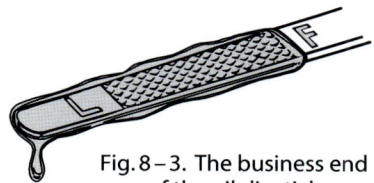

Fig. 8 – 3. The business end of the oil dipstick.
Toyota Motor Company

❑ If your level seems appropriate, let the engine idle for several minutes to establish that no oil is leaking to the ground. If the pavement stays dry, your work is almost done: Simply check that the oil filler cap is on tightly before closing the hood. A loose cap can vibrate right on out of there, resulting in an incredible mess. Oil gets thrown everywhere, rapidly depleting your engine of its life blood. Most owners won't become aware of the problem until they smell the oil burning off their exhaust manifold or see its smoke. The truly oblivious must wait for their oil light to come on.

❑   Some of you, possessed of an inquiring mind, will probably kneel down to inspect the undercarriage for drops of oil. Don't be surprised if you find some – most mechanics don't take the time to wipe up the oil that invariably runs down the engine block as the oil filter is removed. If you don't see oil actually dripping to the pavement, then what you're seeing is residual from the oil change. If your temperament demands it and your clothing allows it, lying on your back will usually provide sufficient access for wiping it up. Then if you find fresh drops forming, you can deal with the problem while still in the neighborhood.

Some of you will pull that dipstick and reel in horror when you find that the oil is still black. Is it possible that they didn't change it? Sure, but it's not probable. Oil turns black quickly in a dirty engine. If you've been negligent down through the years – changing the oil irregularly and infrequently, running it low – you'll find that by 40,000 to 50,000 miles the oil will turn dark within a matter of blocks, pitch black within a few days. If there is some reason to question their integrity, inspect the oil filter. Aside from the oily fingerprints, it should be shiny and clean.

❑   Recheck for leaks over the next several days by glancing at the pavement as you pull out. If you suspect a slow leak, park over top of a clean strip of concrete whenever you can, or take it back for a recheck.

❑   An oil change can turn ugly . . .

- If you suddenly develop an oil leak following an oil change – even if it's just a small puddle beneath the car – consider it to be major and have the car towed to prevent engine damage. Hand the shop your bill.

- If Butch cross-threads your oil drain plug, he'll certainly ruin the plug, possibly the oil pan's matching threads. Beware the rubber plug![4] It should be considered a temporary solution at best, enabling you to limp the car over to a real shop for repair.

Now consider the flip side: The drain plug could have been stripped on arrival, "butched" during your last oil change. Overtightening can distort the drain plug's threads, making it hard to remove and even harder to reinstall. It might make sense to offer your mechanic the benefit of the doubt – assuming that he hasn't suggested the rubber plug remedy – leaving the car in his care while he orders a factory replacement drain plug and gasket. That gives him the opportunity to rectify the problem, and you the chance to sidestep what might become an expensive bill (oil pan replacement).

4. Oil Pressure Warning Light

It never ceases to amaze me that people use the oil *pressure* warning light (or gauge) as if it were an oil *level* warning light. It is categorically not intended as the lazy man's cue to check the oil. The oil pressure light indicates that the engine is in very real danger of locking up! Most of these lights are designed to come on when the oil level is two or more quarts below capacity. In most cases, that's about half what it should be.

---

4   Yes, they actually do sell rubber plugs of varying sizes to jam into the oil pan. These rank right up there with cooling system stop-leak as "solutions" to be scrupulously avoided.

Alright, I confess: I wrote the above paragraph years ago when that statement was absolutely true. Imagine my horror when the oil light came on on my 2014 Subaru last year in the middle of a trip . . . I had checked the oil before I left home . . . yeah, it was a little low, but . . . Turns out that on my car, at least, the oil light kicks on when the engine is a half-quart low. Will miracles never cease! My advice? Err on the side of caution.

4A. If Your Oil Pressure Warning Light Comes On While Driving

❑ Shut off the engine as soon as possible, certainly within the first block.

❑ Check your oil level on level ground.

- If you find no oil on the stick, do not restart the car until you've added some. Add one quart, wait 30 to 60 seconds for it to reach the crankcase (the engine's lower end), then recheck the level. If there's still no oil on the stick, add another half quart, repeat the waiting game, then recheck. Once you start to see oil on the stick, add no more than 1/4 quart at a time. Do not overfill the thing by pouring in haste!

- Once the oil level is near the full line, peek under the car for signs of hemorrhagic leakage. If puddling is present, have the car towed. If the pavement is dry, restart the engine and confirm that the oil light goes out. Once assured of that, recheck beneath the car for leakage. (Engines seldom leak unless they're running.) Read *§*5, pp. 132-138 for clues on where all that oil went.

- If the dipstick indicates that the engine has plenty of oil, first recheck your level – perhaps you forgot to wipe off the dipstick before reinsertion. If the level is good but the oil light comes back on, then either the oil sending unit is defective, the oil filter is blocked, or the oil pump is failing. Of these three, the first is by far the most likely.

If the warning light comes on only at idle, it's a fair bet that little to no damage will result from driving to the shop – as long as the engine is full of oil. Rev the engine slightly at traffic lights, just enough to keep the oil light from coming on, after you put the transmission in Neutral – not Park! Severe damage can result from slamming an automatic transmission into Park while the car is still moving.

Perish the thought, but if your engine is now cranking over more slowly than it used to, it could mean excessively high resistance within the engine itself – the earliest phases of engine lockup.

4B. Possible Causes

❑ Your mechanic will probably begin his assessment by inspecting the oil sending unit. If it's leaking, have the oil sending unit replaced with a new OEM part, then verify that the light no longer comes on except when the engine is first started. In its earlier stages of deterioration, leakage from an oil sender will be slight – more of an ooze than an actual leak. The oil pressure warning light will blink on momentarily, seemingly at random, almost always at idle. As the leakage becomes more pronounced, the light comes on more frequently and for longer periods. Unfortunately, a weak or leaking oil pump produces the very same effect.

8

❑ An aftermarket oil filter is the next most likely candidate – most generic filters skimp on the amount and quality of their filtration material, resulting in premature blockage and a consequent decrease in oil flow. Keeping the idle speed slightly elevated increases the engine's oil pressure sufficiently to force the filter's relief valve open, which at least permits oil to flow – even if it is unfiltered. Replace the offending filter with an OEM replacement.

❑ The Oil Pump: Many oil pumps bolt directly to the timing cover at the front of the engine. Inspection for leakage is therefore a simple matter. If the pump is leaking, it will have to be removed for inspection and resealing or replacement.

Pumps that are not mounted externally cannot be inspected visually. If replacement of the oil sender and oil filter have had no effect on the oil light, then an oil pressure test is called for. An oil pressure gauge can be installed in place of the oil sender, permitting the mechanic to take readings at various RPMs. This procedure provides real answers to an expensive question. Normal readings follow:

At Idle(850 RPM)	At 3,000 RPM	At 5,000 RPM
15 – 30 psi	30 – 60 psi	50 – 60 psi

Readings below 10 psi at idle do not speak well for the oil pump – or for the engine. Insufficient pressure translates to inadequate lubrication, particularly for components in the engine's upper end – the camshaft, rocker arms, etc. Have the oil pump replaced, and ask that all accessible oil journals – particularly the passageway between pump and oil filter – be treated to a generous dose of compressed air to ensure that they are clear.

> **The consequences of low oil pressure can be calamitous both to your engine and your pocketbook. Attend to this problem now!**

4C. High Oil Pressure

Excessively high oil pressures indicate an engine with too much oil or reduced flow through one or more oil journals. Blockage is produced by sludge – the consequence of consistently running on excessively dirty oil. High oil pressures lead to blown oil seals, an expensive proposition.

5. Oil Consumption

What's a normal rate of loss? Every brand-new engine burns a little oil as it's breaking in, as much as a quart every 1,500 miles. This consumption should move toward negligible by 4,000 to 6,000 miles,[5] with the engine running clean until it gets into the 75,000 to 100,000 mile range. By then, engines start to consume a bit of oil as a result of normal wear, maybe a quart every 1,000 to 3,000 miles. All of that assumes that you've been routinely checking your oil so that the engine won't suffer from a lack of it.

---

5 I've read anecdotal accounts of several automakers continuing to struggle with oil consumption at low mileage.

> Engines fed a staple diet of synthetic oil from birth demonstrate less wear than traditional oils. All things being equal (which they never are), oil consumption should therefore be lower and develop later.

To determine the rate of oil loss, put a small memo pad in your car so that every time you need to add oil, you can jot down the mileage, the date, and the amount added. Check the level every time you fill up with gas – at least until you know how often you need to check – preferably when you first turn off the car (warm engine). Most of you will find that a check every third or fourth fill-up is sufficient as long as you aren't commuting long distances at high speed. Unless your owner's manual says to, do not check your oil first thing in the morning on a cold engine; gravity works all night to produce the illusion of more oil than there actually is.

5A. Oil Leaks

Roughly 90% of the oil leaks we find are unknown to our customers when they drive in. Their almost universal response to the news is that the leak must be just starting or not sufficiently serious to worry about, because as yet they have not seen puddles on the ground. There is a fundamental misconception in that conclusion: Leaks that actually do leave puddles under a parked car are either so massive that they threaten imminent destruction of the engine, or they're passive leaks resulting from a botched oil change that has caused oil to drip whether the engine is running or not.

A defective oil drain plug gasket is the most likely cause of passive puddling; when used over and over they distort or crack. Even if Rapid Ray installs a new one, all vestiges of the original must be removed or a path is formed for leakage.

Perhaps the most common oil gusher – as contrasted with the passive dripper – is the oil filter that's been installed over top of the preexisting filter's oil seal. These seals occasionally have the bad taste to stick to the block when the filter is removed. Needless to say, Ray needs to watch what he's doing.

Another gusher can be created by a loose oil filler cap that vibrates off while you drive. This oil seldom makes it to the ground. The driver will become aware of the smell of oil followed by smoke as it reaches the exhaust manifold. I've never seen oil catch fire under these circumstances, but what I have seen is one helluva lot of smoke!

With the exception of leaks past the oil drain plug gasket, all other engine oil leaks ooze only while the car is running. The oil coming out moves downward and rearward, diffusing over, adhering to, or filling surrounding components. These unknown leaks can be quite serious, not only because of the oil that's lost, but also because the parts to which that oil adheres can fail as a direct consequence. Primary examples include coolant hoses whose sidewalls soften when soaked in oil, then balloon and burst; and electrical components – starter motors and alternators, for example – which can short out when loaded with oil. Wiring harness plugs, other types of hoses (fuel line, vacuum tubing), the rubber boots on steering racks and driveshafts, fan belts, and clutch parts are other components that can suffer from being exposed to oil. You don't need to be losing buckets for this to happen; you just need to be unlucky enough to have oil ending up in the wrong spot, and sufficiently oblivious to the problem that enough time can pass for damage to develop.

8

Certain oil leaks are not worth the cost of repair, particularly if they're small and collateral damage is not an issue. One example is a head gasket oil leak which over the course of a year might bleed a quart down the side of the block. The block certainly won't be hurt by it, and assuming that there are no coolant hoses or a starter motor in the way, a savings of $1,000 or more might be realized as long as the owner is conscientious enough to watch the oil level. Another example might be a front or rear crankshaft seal in its early stages of deterioration. A bad seal can produce an ugly leak in its advanced stages, but that might take two years from the point of first detection to the point of unmanageability.

> Front and rear crankshaft seals, driveshaft seals and the like tend to leak much more rapidly in cold weather, particularly during the first few minutes of driving.

If you're around when a leak is discovered, ask to see it. A BIGTIME oil leak will soak an area, leaving drops of oil hanging from one or more low-hanging projections. A light to moderate leak produces nothing more than some wetness or discoloration – a trail that's somewhat darker and shinier than the surfaces surrounding it. Ask the mechanic to point out any parts in the vicinity that might be damaged should the leak be allowed to continue. Only those parts that are below and behind the origin of the leak are candidates – remember that oil always leaks downward and blows rearward; very few leaks actually spit oil up or out in a spray. Notable exceptions include:

- Leaks that drip onto fan belts, where their motion throws the oil outward
- The gasket(s) that seal the engine's oil pump (on models that mount these pumps external to the block)

If we have any question regarding the source of a leak, we recommend that the surrounding area be cleaned with a power sprayer, blown dry with compressed air, then returned in 250 to 500 miles for a recheck. Naturally, an active dripper won't require more than a few miles of driving, but most leaks take some time to resurface. If you put off the return inspection, you will again obscure the source.

A specialist in a particular line of cars can probably identify 80 – 90% of the oil leaks entering his bay. Furthermore, he'll know the consequences of not fixing the problem. But he will not know how much oil the car is consuming. He can tell that a particular engine is leaking a lot of oil or that it's been leaking for a long time; but it's impossible to distinguish between the two unless it's a gusher. Only the owner can tell him that, based on an accurate record of consumption. If you prefer to wait on repairs to see just how bad (or inconsequential) a particular leak might be, put a memo pad in your car to help you keep track.

A light to moderate leak will not cause internal engine damage as long as you keep up with oil consumption. Unfortunately, most of us aren't that responsible – particularly in winter or when we're rushed. As incentive, read §1A, pp. 93-94 and §5B through §6, pp. 135-138.

8

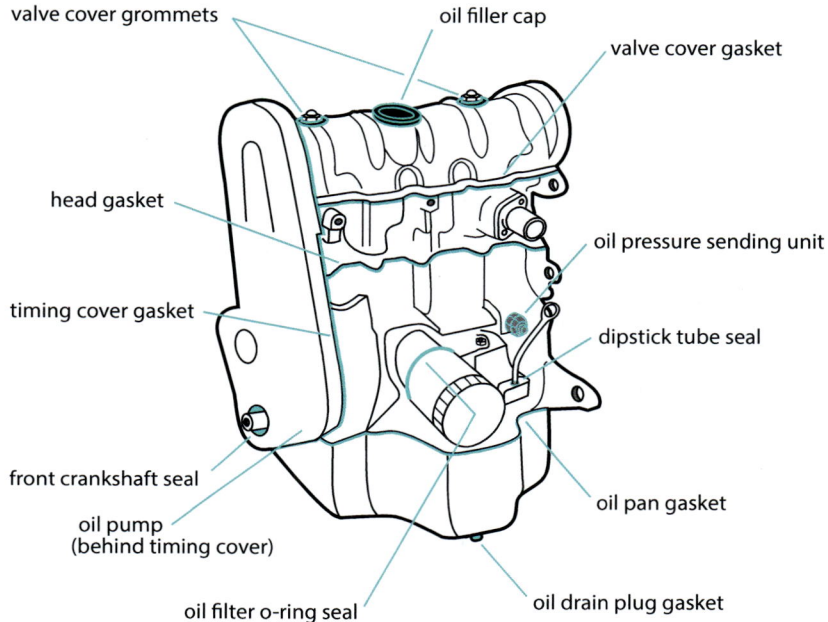

Fig. 8 – 4. Principal sources of oil leakage.

If necessary, you can document the existence of an oil leak and identify its location by laying a large piece of cardboard beneath the engine/transmission area as soon as you park. Mark both the front of the cardboard and the placement of your wheels. If you have off-street parking, tape the board down with duct tape.

In this age of bottled oil and resealable caps, it makes no sense to wait till you're a full quart low before adding oil. By staying ahead of the problem, you not only give the engine its full complement of oil, but also forestall the possibility of running it low again. If you need a third of a quart, add it!

5B. Engines That Burn Oil

Engines that have been run low on oil have a tendency to burn oil from that point onward. This is particularly true on neglected engines – those run consistently and continuously low on oil. This has become a real problem with the decline of full service gas stations. People just don't check under-hood levels with anything near the frequency they should, especially in light of the increased complexity (and cost!) of the modern engine. The only lubrication that piston rings (Fig. 8 – 5) receive comes from oil that's sloshed around in the crankcase;[6] the less oil there is, the more friction they must suffer. This translates to excessive heat, reduced temper (springiness), and rapid wear. The consequence? The

---

6   There's a hollow cavity inside the block (the lower portion of your engine) that houses the crankshaft and connecting rods. As these components spin, they dip into a "pond" of oil, then splash it everywhere. The lower your oil level, the less there is to throw around.

piston rings' extremely tight fit, required to prevent oil from entering the combustion chambers, is worn away. So part of the oil that's left literally burns itself away, exiting your tailpipe as smoke.

The rate at which an engine burns oil varies with the type of driving you do, so don't assume that just because your car consumes a quart every thousand miles when driving around town that you can go to the beach and back without checking the dipstick. A high-speed trip of an hour or more burns oil at a much faster rate than short trips around town due to the continuously high temperatures and increased oil pressure of high RPM driving. **If your car burns oil, check the oil level whenever you stop for gas on the highway.**

Once again, for emphasis: In this age of bottled oil and resealable caps, it makes no sense to wait till you're a full quart low before adding oil. By staying ahead of the problem, you not only give the engine its full complement of oil, but also forestall the possibility of running the damned thing low again. If you need a third of a quart, add it!

5C. Smoke From The Tailpipe And What It Signifies

Every tailpipe emits a cloud of vapor during the first few minutes of driving, particularly on cool to cold mornings. Engines, like people, need a little extra push to get themselves going first thing in the morning: A richer fuel mixture – more fuel and less air relative to what's required once the engine has warmed up – provides the requisite "kick" to overcome those cold combustion chamber walls, § 2, p. 211. But that creates an emissions problem because the catalytic converter (§3D, p. 196) needs a few minutes to heat itself into efficiency. As a result, even on a warm morning, tailpipe emissions are significantly higher for the first several minutes than they are after the temperature gauge starts to rise. Part of the "cloud" you see is just that: The catalytic converter transforms exhaust fumes into water vapor as one of its products.

An excessively rich mixture turns the exhaust black, a fairly common sight on older, carbureted cars. On fuel-injected engines, black smoke indicates that fuel regulation is *seriously* out of whack. Both carbureted and fuel-injected engines should produce a perfectly clear exhaust stream within three to five minutes following start-up.

Engines that burn oil can always be identified by letting them idle for five to ten minutes (transmission in Neutral or Park), then blipping the throttle – quickly revving the engine to 3,000 – 4,000 RPM, then just as quickly releasing the accelerator pedal – while watching for a plume of grayish smoke to exit the tailpipe. Really bad cases of oil burning show up as grayish smoke in the rear-view mirror under hard acceleration up hills.

The very same test, running under high RPMs up a hill, produces black smoke on engines that are running on a fuel mix that's excessively rich. That black hue defines unburned fuel, just as it did in the cold start discussion above.

Turbocharged engines that suffer turbocharger failure emit thick clouds of grey oil smoke under hard acceleration.

8

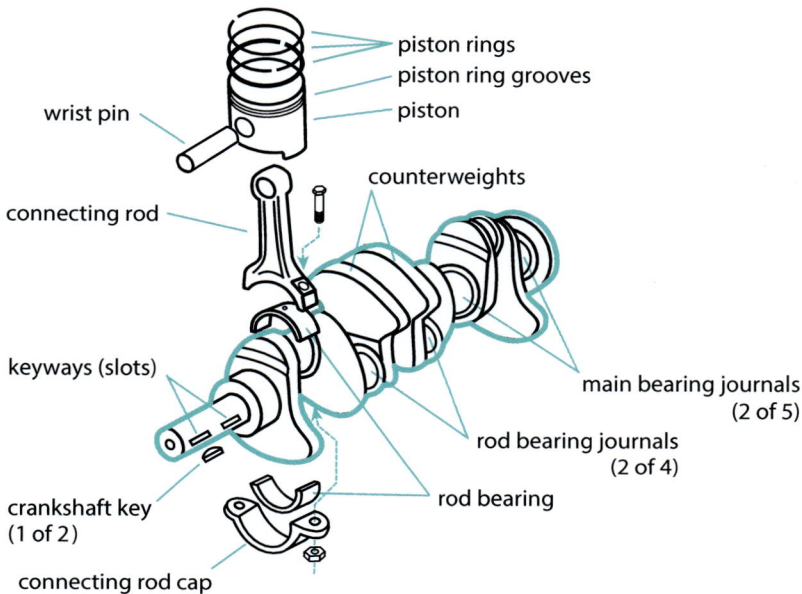

Fig. 8 – 5. The crankshaft of a 4-cylinder engine with piston #1 and associated parts. The crankshaft's main journals are inline along its central axis; the connecting rod journals are all offset.

If a car sits for an hour or so, then produces a plume of smoke when first started, it suggests valve seal wear. Oil that remains in the upper end of the engine after it's turned off always seeks a lower level. A worn valve seal permits the passage of oil down the valve stem and into its combustion chamber, where it lays on top of the piston until it's burned at startup. Smoke produced by a defective valve seal is bluish-gray in color, and is evident every time the car is fired off following a one to two hour lull.

There's one other color that exhaust can display: White smoke is indicative of a blown head gasket, produced by coolant as it "burns" within one or more cylinders. Swallow hard, then see § 13, pp. 163-167.

5D. So How Do You Deal with an Oil Burner?

- Try an oil change with a 5W-50 or 10W-50 motor oil. Don't waste your money on synthetics.

- Having never used any of the following products, I'm not about to attest to their value, but if your only option is engine replacement, you might as well give one of them a try. Add a can to the crankcase, either at your next oil change or now if it won't overfill the engine.

    Alemite CD-2 Oil Burning Formula
    Engine Restorer & Lubricant by Restore, Inc, or
    STP Smoke Treatment

None of these approaches will have any impact on oil leakage, nor can they permanently help the big smokers. If your consumption is worse than a quart every 600 miles and oil leakage is not a contributing cause, then start considering a new engine, a new car, or cab fare.

6. Engine Replacement Options

Every factory offers brand new engines or the primary components thereof–short blocks and cylinder heads–straight out of a box. Clearly, a brand new engine beats every other option, but you'll pay dearly for it. Three ways to save (some) money:

- Price the engine (or its components) online whenever possible, and always from multiple dealerships. The prices dealerships quote online are often lower than the prices they charge at their parts counter. If necessary, pay for it yourself to avoid installer markups.

- Use OEM parts for cooling system components and other exterior parts that might be required,[7] but always use OE parts for any remaining *interior* engine components.[8]

- Employ the best *engine* mechanic you can find, preferably someone who works for a shop with whom you already have a relationship.

Having your own engine rebuilt can end up being far more expensive than simply replacing it. You'll be paying labor charges in two shops–to the mechanic who will remove and replace (R&R) the engine, and to the machinist who will rebuild it. There are *always* hidden costs that cannot be determined until the engine is laying in pieces on the machinist's bench. A lack of new engine availability would be the only reason I can think of for going this route voluntarily.

If you need to take this road, I'd suggest you seek out the best machine shop in your area and insist that factory replacement (OE) parts be used throughout the interior of the engine. OEM parts should be perfectly satisfactory for the more easily accessible exterior components. Machine shops typically do not do the R&R work; for that you'll need a very competent mechanic.

- ❏ The least expensive alternative is an engine swap – installing a used replacement from a junkyard. You'll be looking for a low-mileage engine, preferably with less than 60K on it. Most carry a 90 day guarantee. These run anywhere from $1,200 to $4,000 depending upon mileage, condition, their complexity and availability. The labor to put one in can range from 10 to 20 hours. Additional parts required[9] can cost $250 to $600.

- ❏ To my mind, a better option–more expensive than the junkyard, significantly less than a factory replacement–is an engine that's been remanufactured to OEM specs. Many carry a three-year, 36,000 mile warranty. Labor and additional parts charges

---

7 Hoses, radiator, clamps, air filter, etc.

8 Engines are typically supplied as complete units, but short block and cylinder head components will require an engine gasket kit, water pump, thermostat, timing belt, etc.

9 A new thermostat and oil pressure sensor, a new rear main seal, a front crankshaft seal (if easily accessed), possibly a new timing belt. Putting in a new clutch can make a great deal of sense on cars equipped with manual transmissions. New oil and filter, some tune-up gear, 10 to 15 hose clamps, a few coolant hoses, some coolant, and solvents.

will be somewhat higher on these "reman" engines than on junkyard units because there are significantly more pieces that need to be changed over,[10] and because you'll be inclined to authorize more coolant hoses, etc.

7. DIY – Changing Your Own Oil and Filter

The do-it-yourself oil change is not as easy as it might seem. It requires a significant investment in tools to facilitate the job and if you spend the money necessary to ensure safety, most of you won't save a dime for several years.

7A. Lifting the Car

Unless you own a truck or an SUV with high clearance, you'll need some way to elevate your car. Spare tire jacks are fine for changing a tire, but for this kind of work, they're dangerous – their bases are just too small. By the time you raise one corner high enough to accomplish anything, that jack has become a very tipsy affair. In addition, most tire jacks can't lift the car high enough to accommodate a jack-stand.

Ramps are a cheap alternative, but they create a tremendous slant, usually reducing the engine's ability to drain. And they're useless if you ever intend to remove a wheel. As if that weren't enough, ramps also like to slip forward as you try to drive onto them.

A hydraulic hand jack and *two* pair of four-legged jack-stands (§4C, p. 118) are by far the best way to go; but if the above approach is too rich for your blood, there is one other method I used frequently in my younger years; namely, driving either the right or left two wheels up onto a curb. You do this *gently*, at a point where a driveway or alley intersects the street – the gentle slope offered by the driveway provides a natural ramp – one that won't shift. One problem with this approach is that you've got to be skinny. Another liability is that once again, the car is not level: Position the car so that the drain plug is on the lower (street) side.

> Crawling under your car without carefully placed jack-stands supporting the car's weight can lead to disaster.
> Never go under the back two-thirds of the car without a second pair of jack-stands supporting the rear.
> Never use cinder blocks (they crumble).

7B. Tools for Changing Oil

You'll need every one of the following items unless, of course, you want to make a mess, hurt yourself, or engage in some swearing.

- A means to collect and transport your waste oil. The drain pan shown in Fig. 7-1, p.119 is terrific – it will catch your oil, store it, and provide clean transport to your local gas station. The bigger the catch-basin, the happier you'll be.

---

10 Motor mounts, the brackets for AC compressors and power steering pumps, intake and exhaust manifolds, oil pan, flywheel, etc.

■

- Six-point wrench or six-point socket and breaker bar, correctly sized to fit your oil drain plug. The twelve-point stuff loves to slip – busting knuckles and rounding off the head of the drain plug so that only vise grips can grab it . . . In cases where the plug has been overtightened, you might need a four-pound hammer.

- Gasket scraper to remove the old drain plug gasket. A single-edged razor blade will work in a pinch, but it's slower and more dangerous.

- Long-necked shop funnel – Cut off its small end with a utility knife so that it fits your engine's oil filler neck tightly. That way, the oil will go *into* your engine rather than all over the top of it.

- A pound of rags (possibly excessive, but not necessarily)

- Aerosol can of brake cleaner

- Flashlight or shop light

- A good quality oil filter wrench. There are a number of good alternatives to choose from, some of the most common are shown in Fig. 7-2, p.122.

7C. Parts

In addition to an OEM oil filter and the proper grade and weight of oil, be sure to have an OEM drain plug gasket on hand. If you've had leakage issues following previous oil changes, pick up a new oil drain plug as well.

7D. Procedure

> To protect both your safety and your pocketbook, please read each part of Section 7 in its entirety – including all other referenced sections – at least twice before beginning. Chapter 7 contains two related sections: For the Novice Interested in Tinkering, §2, p.113 and Safety Considerations, §3, p.114. I suggest you read them both.

❑ The engine should be warm enough that the oil will drain quickly, but not so hot that you could burn yourself (run it for 3 to 5 minutes)

❑ Start with a check of oil level, §1E, p.97.

❑ Next, verify that your oil level warning light works. It should come on with all your other warning lights when the ignition key is turned to the On position (engine off). If the oil light is not working, don't proceed with the oil change until you have that resolved. In most cases, you'll find that either the oil pressure sending unit is unplugged, leaking, or defective, or the dash light is burned out.

If you have an oil pressure gauge, you may not have a low pressure warning light. If so, observe where the needle sits with the engine turned off, and when it's idling.

8

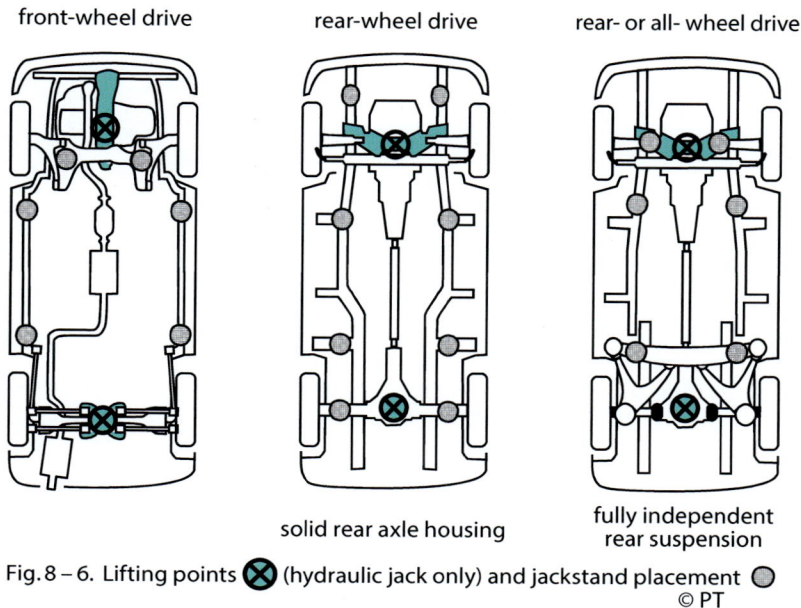

front-wheel drive   rear-wheel drive   rear- or all- wheel drive

solid rear axle housing   fully independent rear suspension

Fig. 8–6. Lifting points ⊗ (hydraulic jack only) and jackstand placement ⊙
© PT

❑   Put your keys in your pocket.

> If you intend to remove the wheels for any purpose, now is the time to loosen the wheel lugs, §10A, Step 2, p.283. Changing the oil does not require doing so.

❑   Lift the car on level ground.

- Engage the parking brake. Address slight inclines by blocking the rear wheels.

- Lift the front end at the center of the front crossmember ⊗ directly below the engine, Fig. 8–6. Position your jack-stands behind the front wheel openings, directly beneath the reinforced body seams. ⊙ These points should correspond with the illustrations presented in your owners' manual showing proper tire-jack placement when changing a flat tire .

- Lift the rear from beneath the differential (rear-wheel drive) or at the center of the rear crossmember (front- or all-wheel drive).

  Note: It may seem impossible to do what I just said because your jack won't slide under the rear differential. Do this: Slide your jack in from the side, ahead of the rear wheels, and to avoid shifting the front end as the jack lifts higher, roll the jack in toward the lift point as you pump the rear higher.

- Position the rear stands at the same points your owner's manual recommends for the jack when changing a flat tire.

❑   Remove the oil filler cap to promote drainage, and insert your funnel now to serve as a reminder not to start the engine until you've poured oil through it.

8

Fig. 8 – 7. Loosening the oil drain plug.

For the sake of visibility, my left hand is absurdly placed. Grip the wrench tightly with your whole left hand, high up by the oil drain plug. That will keep the wrench from slipping and put some distance between the hammer and your hand.

❑ Collect the following beneath the car: oil drain pan, rags, drain-plug wrench, hammer, gasket scraper, oil filter wrench, can of brake cleaner, and light.

❑ Remove the oil drain plug, unscrewing counterclockwise (while looking straight at it). A tight plug will respond nicely to a sharp smack applied to the free end of your wrench.

Fig. 8–7. (Use the hammer, protect your hand!)

Oil will start to dribble as soon as the plug is backed off a half-turn or so, it rushes forth as soon as the plug comes out.

If the oil is not too hot, it'll flow smoothly and predictably into the pan. But **hot oil splashes as it hits the pan, bouncing right back out** and splattering all over whatever lies in the way, usually the inside of your tire and wheel. If so, quickly make a screen out of rag #1 to arrest the mess.

> **Make sure that you aren't caught in the crossfire – you could get a nasty burn.**

❑ The arc that the oil follows as it spills from your engine changes with time. As you remove the oil filter, recheck every few minutes to ensure that your waste oil is actually dribbling into the pan.

❑ Some oil filters are accessed from above, some from below. All oil filters unscrew counterclockwise, just like the oil drain plug. Hopefully, the tool that you selected will be a good match for the job, because if it isn't, your composure will deteriorate rapidly.

Fig. 8 – 8. Oil filter seal

Oil filters dump a considerable amount of oil once the seal between filter and block is broken, so be sure to have the oil drain pan properly positioned.

Verify that the old filter's o-ring seal came off with the filter, then place the thing seal-side down in your oil drain pan – used filters drain dirty oil for some time.

❑ Inspect the oil filter's circular sealing surface on the engine block. Use a *clean* rag to wipe it down, then run your finger around it as a double-check against grit or the previous oil filter seal. Don't screw this up – **having two o-rings between the filter and the block will create a torrential oil leak**, either immediately or – worse still – three miles down the road.

❑ Clean the block below the oil filter with a rag and some brake cleaner, this to provide assurance that no oil is leaking once the engine is up and running.

❑ Dampen the oil filter's o-ring seal with clean motor oil, using your fingertip to spread the oil around its circumference. Install the filter clockwise. Spin it on till it just contacts the block, then hand-tighten another 1/2 to 3/4 turn till it's nice and snug. You don't want to twist that thing till your veins start to bulge – that will just make it harder to get off the next time.

❑ When you cracked the drain plug loose, did it then turn out easily by hand? Or did you need to continue using the wrench turn after turn? If you needed the wrench's leverage, it's important to determine whether it was the drain plug *gasket* that provided the resistance, or whether the threads of the *drain plug* itself were stretched as the result of it's being severely overtightened in the past.

The way to distinguish between these two is to remove the drain plug gasket – it's usually stuck to the oil pan, Fig. 8 – 9 – then try to install the drain plug by hand. If that proves difficult, you'll usually find that the threads on the plug are not uniformly spaced – the threads literally stretched during some earlier oil change. You'll need a new drain plug, preferably for this oil change, certainly for the next.

God forbid you find that your new drain plug won't thread in cleanly. If that's the case, the threads in the oil pan are stripped. You'll need a tap & dye set for repair. .

Drain plug gaskets come in three flavors: fiber, brass, and plastic. Fiber gaskets are by far the most common and the most likely to produce the binding described above: As you tighten a drain plug, you can watch a fiber gasket's outer edge start to bulge outward – that's when you should stop tightening. If overtightened, the inward bulge pushes fiber into the drain plug's threads, causing it to drag during removal.

Brass gaskets crush rather than bulge when they're tightened; they rarely create the problem described above.

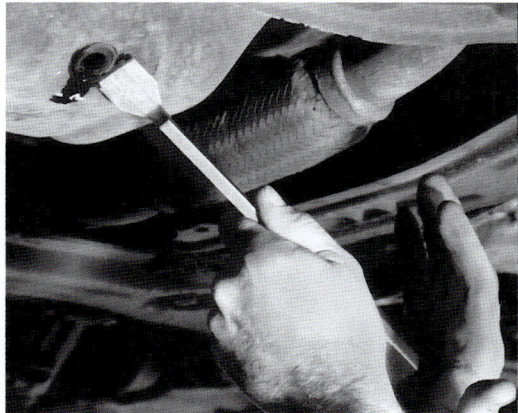

Fig. 8 – 9. Removing a fiber oil drain plug gasket

Plastic drain plug gaskets don't crush or bulge; they distort, usually into an oval shape that offers a highway for oil loss. **Replace the o-ring with an OE gasket.** It's highly unlikely that you will ever come across a plastic o-ring unless you've been frequenting speedy-quick lube joints. If you have no choice but to reuse plastic, tighten till snug, then add no more than another 1/4 to 1/2 turn.

❑ Inspect the oil pan opening. If the gasket or a portion of it is stuck to the pan, remove it by placing the blade of your gasket scraper against the outside edge of the gasket

8

as shown in Fig. 8 – 9. One sharp smack should drive off the gasket in one piece. Perform any cleanup that might be necessary – you will soon want to verify that the plug area is not leaking.

❑ Wipe the drain plug clean. Check that its head wasn't rounded off by the use of some improper tool in the past. Install the drain plug with a new gasket (typically not necessary with brass o-rings), tightening down 1/4 to 1/2 turn past snug, or until the fiber gasket starts to bulge outward.

❑ The amount of oil required by your engine is listed in the specifications section of your owner's manual. Start with 1/2 quart *less* than their recommendation, this to prevent overfilling. When removing the funnel, lift it onto a rag to avoid dribbling oil all over the engine, bumper, etc. Install the oil filler cap, then peek underneath the car to assure that no egregious errors have been committed thus far.

❑ Before starting the car, first verify that the oil pressure warning light still comes on when you turn the ignition key to the On position. (If not, you have accidentally unplugged the oil pressure sending unit. Look for it near the oil filter.)

When starting, watch the oil light until it goes out, usually within 3 to 7 seconds following startup. If it does not go out within the first 15 seconds, shut the engine down and recheck your work. Those of you with an oil pressure gauge will be watching for the needle to move upscale within the same period of time, usually to about 15 psi.

❑ Assuming that all is well inside the car, leave the engine running for a minute or two while you check underneath. Inspect the oil pan surrounding the drain plug and the area beneath the oil filter.

Keep clear of drive belts, electric fan blades, and other moving parts.

❑ Turn off the engine, and – assuming that your car is level – inspect your oil level. If it's no more than halfway down the low-full grid, wait till after your road test to top it off. (Oil expands with heat.) If lower, top off as needed. Check that your oil filler cap is on tightly.

❑ Finish with a brief road test (1 to 2 miles), final leak check, and inspection of oil level. Again, recheck that your oil filler cap is tight.

❑ Clean Up! Remember that waste oil is toxic. Dispose of it and your filter with consideration for others, in accordance with local laws.

8

Chapter 9: The Cooling System

9

Notes

9

The Cooling System

1. Dealing With an Overheating Engine

If you ever have the misfortune of seeing your temperature gauge climbing toward the top of the scale, first **turn your heater control to its highest setting and turn the heater fan on full blast.** If it's the middle of summer, deflect the heat away from you by moving the heater/vent control to Defrost. **Turn off the AC.**

If this stops the upward climb of the temperature gauge and keeps you below the red zone, you can continue driving to the nearest convenient spot. Watch the gauge! Listen for any unusual sounds.

If **the temperature gauge has already pegged in the red range and does not respond promptly to your attempts to dissipate heat** into the passenger compartment, **then pull over** *immediately* **and shut the engine down** while there's still something left to save. **Blowing a head gasket takes a matter of minutes.** Repair costs start around $1,000. Further driving can lead to all kinds of internal engine damage.

> If the temperature gauge is in the red zone, turn off the engine and walk away. Wait until the engine has cooled back down before you open the hood.
>
> Do not remove the radiator or surge tank cap under any circumstances. The fountain of boiling coolant waiting to erupt could disfigure you for life.
>
> If you see steam coming from under the hood, leave it closed.

If the engine is "just" running hotter than it should, you may want to investigate whether your electric cooling fan(s) are working (engine on). See Fig. 9-2, p. 149. If not, you've discovered why the car is overheating. If you have two fans, one of them is for your air conditioning. Turn on the AC long enough to see if one or both fans kick in. If they do, leave the AC on (with the heater control still set to Hot) and watch the temperature gauge to see if the engine's temperature starts to drop. **If the fans don't come on immediately, shut off the AC.**

2. Anatomy

The basic components of every cooling system are illustrated in Fig. 9-1 p. 148. Most modern designs are more complex than this–more hoses, for one thing–but unless you own an old air-cooled VW Beetle, you'll find every one of the parts illustrated there under your hood, plus at least one temperature sensor threaded into the thermostat housing. Common variations follow:

Back in the day, the water pump on every engine was prominently displayed on the front of the block, directly behind the radiator, coupled to a metal fan, driven by a fan belt. With the addition of air conditioning and power steering, the fan belt evolved into one of several drive belts, usually powering more than just one component. Over time, those multiple belts morphed into a single serpentine belt, Fig. 9-6, p. 159.

9

Electric fan assemblies attached to the radiator have replaced the mechanical fan. These are thermostatically controlled, designed to hasten warm-up and keep engines running about ten degrees hotter, thereby improving mileage and reducing emissions. That change meant that the water pump could be placed just about anywhere on the front of the engine; all it needed was access to a drive belt. Many pumps are now driven off the timing belt, hidden away behind the timing belt cover. Regardless of their position, every one of them is connected to the lower radiator hose (radiator outlet hose), either directly or through a connecting pipe.

With the advent of transverse-mounted engines (Fig. 9-2), the thermostat housing often migrated to the rear of engine. As with the water pump, so too with the thermostat: It is always connected to the upper radiator hose(radiator inlet hose), either directly or through metal piping. Regardless of its location, the engine side of the thermostat housing is always connected to the inlet side of the water pump, (Fig. 9-1). The reason? Because as the car is warming up from a cold start, that thermostat is closed. Since the engine's coolant can't circulate through the radiator, a bypass hose must be preseent to dissipate some of the pressure created by the water pump.

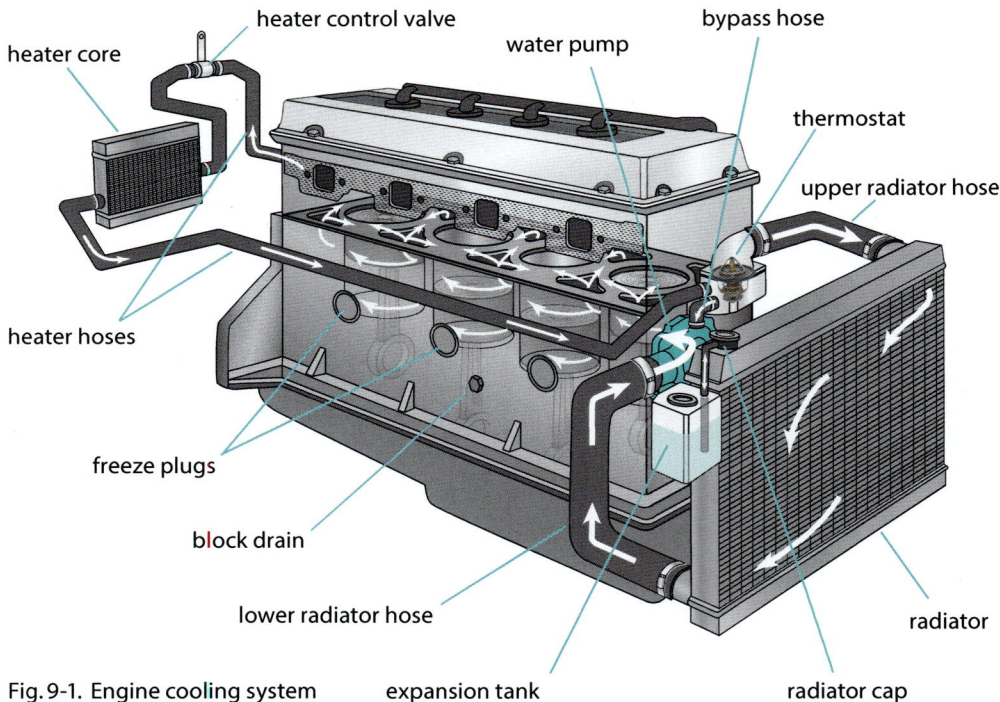

Fig. 9-1. Engine cooling system
Arrows indicate coolant flow
Radiator cooling fan not shown

It's virtually impossible to illustrate every component of the cooling system; doing so reduces clarity in a jumble of arrows and phrases piled one atop the other. So Fig. 9-1 lacks a radiator cooling fan, mention of the thermostat housing and its temperature sensors, and the inlet and outlet hoses that pipe coolant to the intake manifold. Fig. 9-2 skips the thermostat, its housing, the bypass piping and several other hoses. Which leads to:

❏ Fig. 9-1 illustrates an expansion tank, Fig. 9-2 shows a surge tank. The first is always adjacent to the radiator; the two connect through a siphoning tube emanating directly below the radiator cap. Surge tanks can be found just about anywhere as long as they're positioned above the engine's coolant level. Two tubes connect them to the system. Surge tanks are pressurized, expansion tanks are not. Systems designed with an expansion tank require a radiator cap. Surge tank systems typically employ a threaded cap at the top of the tank.

❏ In Fig. 9-2, notice that the radiator hoses enter and exit close to the radiator's midline. Depending upon circumstance, this book uses the terms *upper radiator/thermostat outlet/*and *radiator inlet* hose and *lower radiator/water pump inlet/radiator outlet hose* interchangeably. I know it's confusing; my apologies.

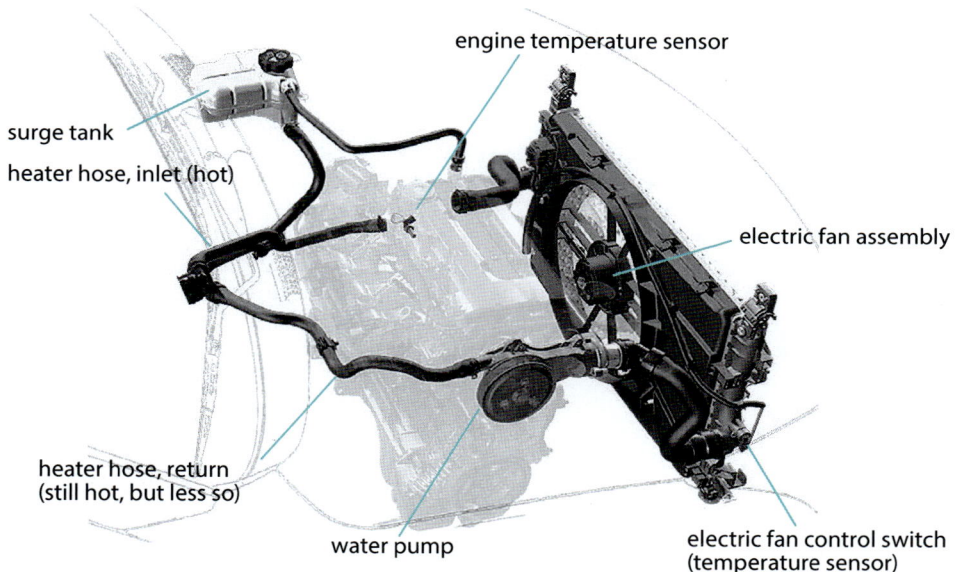

Fig. 9-2. Late-model cooling system, transverse-mounted engine
Courtesy of AC-Delco

❏ In Fig. 9-1, the bypass hose exits the base of the thermostat housing. Had space permitted, you'd see one or two thermoswitches (Fig. 9-3) as well, threaded into that same housing on the engine side of the thermostat. The engine temperature sensor is always present, feeding information both to the instrument panel and to the electronic control unit (ECU). Systems that employ electric fans (all modern cars) typically have a separate thermoswitch controlling the fan(s). These fans only run when the engine gets hot, then cycle on and off as needed. This switch can be placed either in the thermostat housing or in the radiator (Fig. 9-2).

Fig. 9-3. Temp sensor (L)
Fan switch (R)

9

Many radiators sport two electric fans, one dedicated to the cooling system, the other to the air conditioning. The first is controlled either by the engine temperature sensor or by the cooling fan switch depending upon circumstance. The AC fan motor kicks on whenever the AC compressor engages; as it does so, the engine fan kicks in as well. On some cars it also comes on if engine temperature demands it.

As coolant flows around the outside walls of an engine's combustion chambers, it picks up an incredible amount of heat. The bulk of this heat is transferred to the surrounding air once it reaches the finned area of the radiator. The thermostat governs both when this flow begins and the volume of fluid that's allowed to pass. The water pump does the work of moving coolant through this loop.

All materials expand as they're heated–it's a basic law of physics. As the temperature gauge climbs from cold into the normal range, coolant tries to expand outward any way it can. In short, pressure starts building within the cooling system as soon as the coolant begins to warm.

❏ Expansion tanks systems: Assuming no leaks in the system, the coolant's only way out is past the radiator cap–a spring-loaded device designed to maintain pressures within the cooling system of 7 to 16 psi (pounds per square inch), Fig. 9-4. Creating this internal pressure raises the boiling point of the coolant, making it more difficult for the engine to overheat. That doesn't mean that all the coolant stays put; it can't because it's volume is expanding. That excess forces its way into the expansion tank by pushing the cap's lower end off the lower flange. As the engine cools back down, the contraction of coolant within the engine and radiator creates vacuum pressure which pulls fluid back into the radiator through the suction valve.

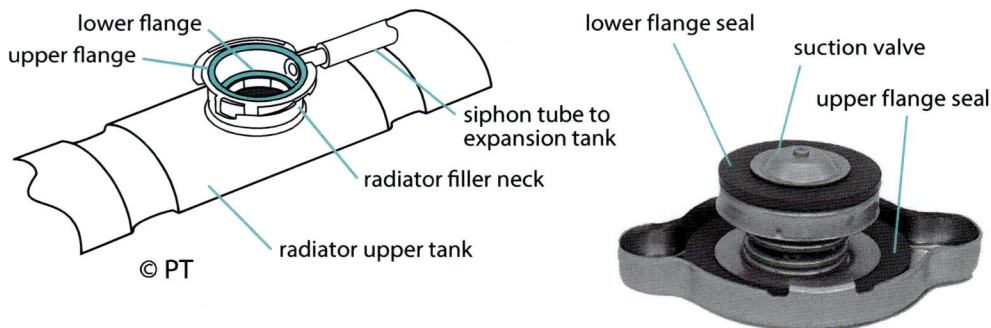

© PT

Fig. 9 - 4. Radiator cap, flanges and seals

❏ Surge tanks: Since these systems are totally sealed, the surge tank itself accomodates all expansion. Air at the top of the tank compresses as the coolant expands; the resulting build-up of air pressure has the same effect as the radiator cap's spring.

9

3. Inspecting Coolant Level and Condition

It's best to inspect coolant level first thing in the morning when the engine is stone-cold. This will ensure that:

- There is no pressure in the system
- The coolant occupies its smallest volume
- You will have a consistent baseline from which to measure future coolant consumption

Both surge and expansion tanks are marked with Low and Full lines.

Surge tanks require nothing more than a glance: On a stone-cold engine, the coolant level should be at or above the Low line, on a hot engine it should be no higher than the full line. Anything in between is good enough.

> Never overfill a surge tank! The volume of free space above the Full line is essential to permit coolant expansion.

Expansion tank systems are a bit more complicated. Levels cold and hot correspond to the specs set out above, but because these systems depend upon siphoning fluid back and forth between radiator and tank, the level within the radiator must be checked at least once every six months or so, and *always* if you find the expansion tank empty. If your cooling system is not full to the brim, the air at the top of the radiator can block any siphoning from occurring, thereby isolating whatever coolant is in the overflow from a system that is running dry. If you find that the finned area of the radiator is visible once the radiator cap is removed, suspect a coolant leak, §8A, p.157.

> DO NOT remove the radiator cap on an engine that has run within the past two hours. At best, opening the cap will cause coolant to dribble out all over the radiator–even if it's been running for just a minute. Radiator cap removal on an overheated engine can unleash a boiling geyser capable of scalding your hands, face, or chest.
>
> Unscrewing the cap on a surge tank is not as dangerous or messy–unless the engine is overheating–but it's still a bad idea because it disrupts the tank's internal pressure balance.

To remove most radiator caps, press the cap downward, turn counterclockwise approximately one quarter-turn to its first (safety-catch) position, then push downward again and rotate counterclockwise another quarter-turn to permit lifting the cap straight off. The radiator should be full to the brim, right up to the lower flange, Fig. 9–4.

Surge tank systems: You won't be peering into your radiator because you don't have access.

Topping off an expansion/surge tank can be as simple as adding some distilled water, §4, p.152. But if you encounter an empty tank, you may well be in the market for a gallon of antifreeze. If so, see §5, p.153.

9

Use a clear-sided antifreeze tester to inspect the coolant's condition. Pull a sample from the radiator (expansion tank systems) or surge tank. Testers use either a pivoting float (Fig. 9-5) or a series of balls to indicate protection level.

> Antifreeze is a poison, capable of killing pets . . . Don't be that genius who used his wife's turkey baster, presumably to save time. (true story)

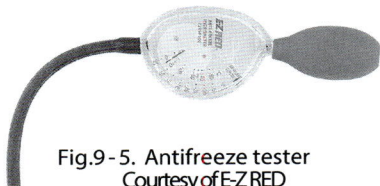

Fig.9-5. Antifreeze tester
Courtesy of E-Z RED

If you find that the solution is clear and its protection level is satisfactory, proceed to §4. If your coolant is murky (dark brown or rusty), if you find crud floating in it, or if the bottom end of the radiator cap is buried under a slimy gel, have the cooling system thoroughly flushed, §6B, p.155 and §15, p.170.

Fortunately, most of you will never have that problem; but expansion tanks are another matter–they tend to collect sediment. Their siphon tubes stop a few inches from the bottom of the bottle, so whatever crud that floated to the top of the radiator will ultimately take up residence there. That doesn't necessarily mean that the entire system needs to be flushed. Assuming that the radiator's fluid is clear, that its protection level is satisfactory, and you're not yet due for coolant replacement, then simply clean the expansion tank and replace its contents, §15, Step 8, p.171.

Surge tank systems don't offer a backwater for concentrating crud; the coolant you see in the tank is identical to whatever is flowing through the system.

4. Topping Off the Cooling System

Every car "consumes" a bit of coolant–not much during their early years–primarily due to evaporation. In a perfect world, we'd all have the specific antifreeze the manufacturer specifies (§5 below); we could blend the proper mix of antifreeze and water and top off our systems. Yea, right . . . Fortunately, every manufacturer provides a cushion: The 50:50 dilutions in most cars protect against freezing down to -34°. Even a 40:60 dilution (my wife's VW, for example) offers protection to -13°F. So adding a pint of water once or twice won't spell the end of the world.

Make that distilled water: If you ever put a pot of water on the stove and let it boil for awhile, the observant among you–well maybe just the ones who did the dishes–noticed a ring of deposits circling the pan. Using tap water in your cooling system will do exactly that. Over time, those deposits will interfere with heat transfer within the engine, and can ultimately plug a radiator. Every auto manufacturer, every antifreeze maker, recommends using distilled water.

9

Every manufacturer specifies the type of antifreeze to use in their cars. Adding a different brand might create problems–sludge formation is their biggest concern–because blending different additive packages alters the chemical properties of both.

So, on to the how-to –

Antifreeze is slippery and hard to clean up without brake cleaner and compressed air, so take the time and precautions necessary to avoid making a mess.

Ever noticed how a stream of liquid being poured from a plastic jug jumps back and forth as the bottle sings glug-glug-glug? Because the openings into the surge/expansion tank and radiator aren't about to move back and forth to accommodate this fact, using a plastic funnel is a splendid idea. Use one that's smaller than the inlet opening; a funnel that forms a tight fit can result in a "stack" of coolant with nowhere to go except all over the outside of your radiator.

Another trick is to reduce the distance the coolant will jump. Rather than using the bottle's handle when pouring, turn the jug sideways so that its flat side will become parallel to the ground. This permits air to flow more smoothly into the jug.

If the level of coolant was down significantly, it's likely that the system will contain air pockets with the potential to produce severe spot overheating. Set the passenger compartment's temperature control all the way to Hot in order to purge any pockets of air from that loop. Leave it there until the engine reaches operating temperature.

Certain cooling systems have a bleeder at the highest point of their system to help dispose of that air. Check your owner's manual. If present, remove the bleeder plug (or open the bleeder valve) prior to filling the system (engine off). Address the heater core circuit as described above.

Before you start the car, verify that the radiator cap/surge tank cap is securely tightened.

You'll need to recheck the level once the car has cooled down—at least three hours later—preferably the next morning before you start the car.

As with any coolant work, you'll need to perform a follow-up inspection, *§*7, p.156.

5. Antifreeze

The 20th century evolution of antifreeze produced three basic varieties: American, Japanese, and European. It's my understanding that this had nothing to do with national pride, but rather with the composition of their water–hard (high mineral content) or soft. Antifreeze is usually mixed 1:1 with water (50% antifreeze, 50% water), so what comes out of the ground matters a great deal to chemists. Mineral content affects what additives can be used to inhibit corrosion. Mixing various types of antifreeze can produce a blend of different additive packages. Whether doing that actually reduces its effectiveness is a question for others, but I've seen plenty of sludge in cooling systems; whether that was due to incompatible mixes or to owner neglect was never clear to me. Whenever possible, match what's in the system. While not foolproof, finding the same color of antifreeze is a good place to start.

9

Fortunately for us, contemporary cars come equipped with greatly improved antifreeze: These days, maintenance schedules call for replacement once every six or seven years rather than the three-year-36,000-mile rule of yesteryear. Having to mess with it less often should mean there's less opportunity to screw things up . . . Consult your owners' manual; **use the type of antifreeze your manufacturer recommends.**

Antifreeze used to be sold exclusively as a concentrate; now, certain brands market 50:50 dilutions of antifreeze and distilled water. You can't beat the convenience if they sell what your car requires, particularly if all you're doing is topping off a surge tank. But it's a pricey way to go if you are flushing the system.

The easiest way to produce a 1:1 dilution is to pour half a gallon of concentrate into a second, empty gallon jug, then fill both with distilled water. Each jug should be capped and shaken to mix its contents. You'll find that the jug gets warm as you do so – chemistry in action! Sadly, most shops use tap water for their dilutions.

So how important is that 1:1 dilution? Well, if you live in Maine or Arizona, it's far more critical than if you live in Virginia or along the mid-California coast. Antifreeze not only lowers the freezing point of water, it raises the boiling point as well. A 50/50 mix drops the freezing point from 32°F to -34°, it raises the boiling point from 212°F to 257°. Amazing. Because that level of protection is far greater than most of us will ever require, adding a pint of water every so often is just fine.

The only time you should pour straight antifreeze into a cooling system is when the car's antifreeze protection level has been tested and found to be very weak or nonexistent: +30°F down to 0°F.

Antifreeze is slippery and hard to clean up without brake cleaner and compressed air, so take the time and precautions necessary to avoid making a mess.

6. Replace the Coolant or Flush the System?

6A. Replace

Cooling systems, properly maintained, don't require flushing with chemicals. Modern day antifreeze formulations have all of the scale and corrosion protection you'll ever need. Consult your owners' manual for their recommendations regarding the type of antifreeze to use and how often it should be changed, §5, p.153. The timeline for most cars these days is twice what is was just twenty years ago–once every six years or so.

The cooling system has two drains: one at the bottom of the radiator and one in the block. Opening the radiator drain lets out most of the coolant above the water pump. Everything below that remains in place. That may not be a big deal if your coolant looks clear, but the odds are that six to ten years of sloshing around in there will render it less than pristine. The only way for you to know is to look at it, §3, p.151.

Because the block drain is so seldom touched, request that they coat its threads with Anti-Seize while they're in there. It shouldn't cost a penny more because presumably the block drain will be opened as part of the job. Its silvery hue is impossible to miss (assuming, of course, that it's visible from above–most aren't). Whether they bother with draining the block is another matter altogether, but at least by mentioning it, you put them on notice that *you* know they should.

9

Many shops don't stock premium brands of antifreeze, and many carry just one type. While that blend may be satisfactory for most cars, it begs the question whether or not it's suitable for yours. Most shops buy what their supplier is selling that month, and believe me, the First Law of most suppliers is profitability, not quality or end-market satisfaction! Ask what kind your shop uses. Better yet, bring your own–plus distilled water–sufficient to complete the job. Your owners' manual will specify the total volume of the system; buy accordingly.

You should also bring along a replacement o-ring for the radiator drain. This is a dealership item. Just because the car arrived without a leak does not imply that it won't leave with one: That o-ring has been crushed for how long now?

I strongly recommend replacing the thermostat in conjunction with this work, §10, p.161. There's no better time to do it: The system's drained, you'll be leak and level checking anyway. And unless you've already met with catastrophe, it's the original. Bring your own OE thermostat plus its gasket or o-ring seal. Lay these prominently on your passenger's seat, right next to your antifreeze, distilled water and radiator o-ring.

If you found that your the overflow bottle is full of crud (§3, p.151), let the shop know that, and specifically request that it be cleaned as part of their work. If the car has been running hotter than it used to, read §8B, p.158 before you go, and let them know about that as well.

When you drop off the car, verify that they have no intention of putting any additive packages into your cooling system (or in with your engine oil, for that matter). Many shops, dealers included, used to hustle additives: For every can that was added, the mechanic made a buck. I have no idea whether that's still true, but I have no reason to doubt it.

As with any coolant work, you'll need to perform a follow-up inspection, §7, p.156.

6B. Flush

Flushing the cooling system is a time-consuming process. But if the coolant is murky (dark brown or rusty), if you find crud floating in it, or if the bottom end of the radiator cap is buried under a slimy gel, you definitely should not wait. Cars running warmer than they used to may need to be flushed, but not necessarily. See §8B, p.158.

On systems that are full of slime, the radiator cap will often be overwhelmed by it. It will need to be cleaned or replaced. The expansion tank will have to be cleaned. The block drain will have to come out–twice. For more information on what's involved, and the differences between draining and flushing, see §14 and §15, pp.167-171.

As with any coolant work, you'll need to perform a follow-up inspection, §7, p.156.

9

7 Follow-Up Inspection of Cooling System Work

If the engine is stone cold, check the coolant level in the expansion tank or surge tank. That level should lie close to the Cold line, marked on the side of that tank. Those of you with a radiator cap should refrain from opening it: Wait till tomorrow morning when you *know* it will be stone cold.

As the engine warms, closely observe any and all areas that were repaired/disturbed for signs of leakage. Keep your hands and other body parts clear of all moving parts. And don't be dangling neckties or other loose clothing over the engine–drive belts and fan blades have no regard for carelessness. Stay clear of the exhaust manifold as well: It gets blisteringlyHOT almost immediately.

Verify that your radiator cooling fan cycles on and off once the engine reaches normal operating temperature. Watch for it to kick in and kick out: You won't have to wait long in the summertime, but in winter it might never cycle on.

Watch the temperature gauge frequently as the engine warms. Continue to observe the gauge routinely over the course of the next several days to ensure that its needle stays well within the normal range.

Recheck your coolant level the following day, again on a stone-cold engine.

- If you find that the system is full, you need only recheck it once more in a week to verify that no leakage is taking place.
- If you find that the coolant level has dropped again, top it off and recheck once more the following day. If no further loss is noticed, the drop in level can probably be attributed either to air pockets within the cooling system that had to work their way out, or to hasty filling.

Once you've accomplished a steady state for 24 hours–with both measurements taken on a stone-cold engine–you can wait for a week to recheck. To be safe, check again in a month.

If you have not already done so, check your antifreeze protection level as well.

8. Coolant Loss

The clearest indicator of coolant loss is an empty surge/expansion tank or low radiator level. A refill and subsequent recheck will clarify whether the level was low simply out of neglect or because it's going somewhere. To prove unequivocally that you're losing coolant, inspect it's level routinely for a month–cold engine only–to enable meaningful comparison from one week to the next.

Significant coolant loss leads to air bubbles coursing through the system. The chemistry involved in antifreeze + air + extreme heat produces a highly corrosive atmosphere.

There are two principal causes for coolant consumption: leakage or overheating.

9

8A. Leakage

Don't waste your time looking for a leak unless you've seen a consistent trend of coolant loss. A one-time occurrence can usually be chalked up to owner or mechanic neglect.

Bigger leaks leave a trail of antifreeze that can be traced upward and forward to its source. Its origin usually has a drop of coolant dangling there. If you find wetness but no apparent source, dry the area with brake cleaner, then watch while the engine runs, preferably from beneath the car.

Smaller leaks leave less obvious trails, often taking the form of a grey-green to white path down the side of the block. They are seldom wet unless the engine is good and hot. It takes a month or more for these trails to develop.

The most common "look" is just a crusty buildup of antifreeze salts (green, brown, or black), akin to a small collection of barnacles, deposited around the source of the leak. These are most commonly found at the end of a hose (loose or weakened hose clamp) or around the perimeter of a thermostat housing gasket (tightening won't suffice).

Another approach to leak detection – the cooling system pressure tester – is a hand pump designed to duplicate the pressures of a hot engine. Although this tool certainly has its place in locating more perplexing leaks, I personally don't like using them. They can apply upwards of 20 to 30 pounds (psi) of pressure to systems designed to withstand half that. If a leak doesn't reveal itself promptly, the natural inclination is to keep on pumping. How many leaks actually start out that way?

Spots that often leak include:

- The end of any hose due to seepage past its hose clamp: Simply tightening the clamp is usually all that's needed unless there's a considerable buildup of deposits. If so, see §9A, p.160. Hoses that are swollen or cracked should always be replaced, §9B, p.160 and §17, p.174.
- water pump seal, §11, p.161
- Metal piping is often used in place of hose to eliminate chafing along the block. These pipes typically connect the water pump to a heater hose. An o-ring, kept secure by bolting the pipe to the block, seals the pipe to the water pump. Repair requires removal of the pipe to replace the o-ring. The mating surfaces are cleaned first, the o-ring is installed with silicone grease or silicone glue.
- a split seam in the radiator, usually resulting from overheating, §12, p.162.

Man-made causes:

- The radiator drain plug, usually as the direct result of a drain and refill or cooling system flush during which the o-ring wasn't replaced.
- An ill-fitting aftermarket flush tee, installed by chopping a heater hose in half, usually near the firewall. Aftermarket shops love to install these things as part of their cooling system "service". I understand the appeal; once

9

installed, a garden hose can hook right up to it. But like so many other aftermarket add-ons, the fit varies from car to car. Tightening hose clamps overtop of a loose-fitting tee might hold for a week, but always give way once no one is watching. Possibilities on high-mileage cars include all of the above plus:

- A split in the side wall of a coolant hose
- Leakage from the radiator fins due to rot

In my opinion, **cooling system stop leak is not a good solution for coolant leaks of any sort.** These products are designed to plug any small flow of liquid and they do a fairly good job of that. Unfortunately, they can't distinguish between a leak to the outside and the normal flow of coolant through the fins of a radiator or heater core. Systems already compromised by mineral deposits are easy pickings: Every one of those already narrowed passageways down through the radiator core is targeted, leading to blockage and consequent overheating. As if that weren't enough, trying to flush your mistake away seldom clears the passageways. **If you have a leak, get it fixed; don't let anybody sell you a Band-Aid.**

8B. Engine Overheating

There's no such thing as a "little" problem with overheating. The consequences of ignoring the issue can be disastrous: Engines usually signal their distress either with steam rising from beneath the hood or with the gurgling sound of boiling coolant passing through the heater core, hidden up behind the dash.

Likely causes:

- low coolant level: leakage or neglect, §3, p.151, §7, p.156 and §8A, p.157
- loose, worn or slipping water pump drive belt[1]
- stuck thermostat: either partially or completely closed
- electric radiator cooling fan(s) that aren't turning on
- weak fan clutch, mechanical fans only, §19, p.177
- blockage in the radiator, §12, p.162

External (visible) water pumps: Drive belt tensions are checked with the engine off. (Duh!) Locate the belt's longest accessible span. At its center, use thumb and index finger to push and pull on it within the plane through which it travels, Fig. 9–6. Expect about a 1/4 inch of play; more than that should be addressed by retensioning the belt. Really loose belts produce a screeching sound when revving the engine, particularly while it's still cold. If the belt is soaked with oil or coolant, at minimum it will have to be replaced.

The best way to test for a stuck thermostat is to measure the temperature on either side of it once the engine has reached operating temperature. Healthy thermostats produce roughly equivalent readings on both sides; sticking thermostats produce a significant – perhaps frightening – differential between the engine and radiator sides. Multimeters equipped with a temperature sensor can be purchased for as little as $35.

1 Those driven by the timing belt can only slip if the belt is disintegrating; the engine always quits.

Electric radiator cooling fans only run when their temperature sensor tells them to-whenever the engine is *hot*. (Cool engines don't run as efficiently; they consume more fuel and pollute more.) The fan should kick in consistently when your temperature gauge reaches a certain point, somewhere between 2/3 and 3/4 of the way up the scale. Possible reasons for its failure include:

- bad coolant temperature sensor
- defective fan motor relay, *§*2D&*§*2E, pp.392-396 and *§*8 & *§*9, pp. 418-421
- defective fan motor, *§*8 & *§*9, pp. 418-421
- blown fuse or tripped circuit breaker, *§*11, p. 422
- the wiring in between

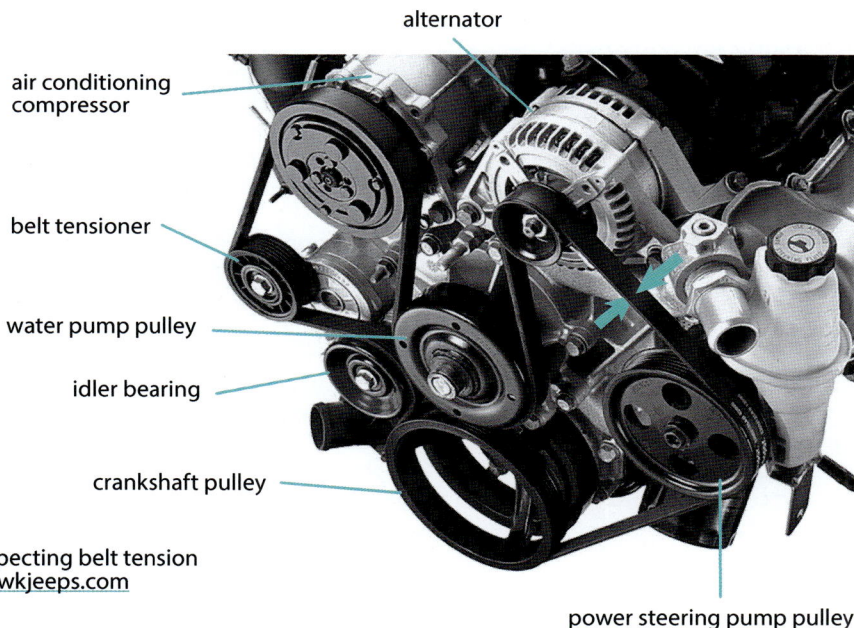

alternator

air conditioning compressor

belt tensioner

water pump pulley

idler bearing

crankshaft pulley

Fig. 9-6. Inspecting belt tension
Courtesy of wkjeeps.com

power steering pump pulley

The simplest way to test a temperature sensor is to just replace it. Obviously that's not always cost effective. However, testing a sensor requires that you remove it first, wait for it to get cold, put it in a pan on your kitchen stove, then heat it while you measure its resistance (in ohms) at various temperatures. In addition to the multimeter mentioned above, you'll need access to the specs for your particular sensor. Regardless of how you do it, you'll need to drain enough coolant to get its level below the sensor.

The specs for most temperature sensors are available online, but you can't really know if they're accurate. Joining an auto repair website (*§*4B, p.118) will provide you with those specs, plus pin configurations for every relay, wiring diagrams and troubleshooting flow charts.

9

Cars with air conditioning typically have a second fan that runs constantly whenever the AC compressor is running. In most cases, when it switches on the engine cooling fan kicks on as well.

9. Coolant Hoses

9A. Inspection

Most coolant hoses last well past 100,000 miles. The exceptions have usually been soaked in oil or gasoline due to a leak in the vicinity.

The most likely source of leakage runs under a hose clamp, then out the end of the hose. Leakage may be too strong a word; in most cases, seepage is probably the better term. These produce a buildup of crusty deposits just past the end of the hose. The trail leads back into the hose, past the clamp. If the deposits are dry, just tighten the clamp, top off the system, then recheck levels in a week or so. If the deposits are wet, or if there's a drip dangling there, then at least that one end of the hose will have to be removed for inspection and repair. Hairline cracks or a hose previously crushed by overtightening its clamp requires hose and clamp replacement. Scrape off the deposits from the hose flange/pipe with a single-edged razor blade), then sand the area smooth with emery cloth (100 grit). If the hose looks fine, just replace the clamp. But if the hose is more than seven years old, replace it.

Overtightening can easily distort a thin-walled copper pipe (heater hose inlet and outlet), creating a pathway for leakage. Start by tightening 1/2 to 3/4 turn. You can always tighten a clamp further, if needed, but making a crimped piece of brass round again is another matter altogether.

Any hose that you find soaked in oil or gas should be cleaned with brake cleaner, then monitored routinely for swelling. If all the hoses are swollen, it's because someone sold you a cooling system flush, then didn't adequately remove the chemicals he used. Get the system flushed properly. You may need to replace every swollen hose (ouch!) if the problem doesn't rectify itself. Replace the thermostat as well.

Now drive the car for 15 to 20 miles. Inspect your hoses with the engine still running. Any hose that appears swollen should be replaced. This ballooning outward in response to cooling system pressure can be localized in just one portion of a hose or it can involve the entire piece. If localized, it is usually due to an oil or gas leak that has soaked into the hose wall and weakened it.

If you find that the upper radiator hose collapses in on itself as the engine cools down, the cause is a defective radiator cap: The suction valve (Fig. 9-4, p.150) is either stuck or contaminated.

As with any coolant work, you'll need to perform a follow-up inspection, §7, p.156.

9B. Replacement

Virtually every hose on your car is preformed: Each has a distinct shape with precise angles molded into the rubber. Due to their great variety, many hoses are unavailable on a same-day basis. A spot of patience will ultimately save money and future stress: If you hold

9

a piece of straight heater hose in your hands and put even a mild bend into it, you'll notice that a kink forms at its most severe angle. That kink will reduce coolant flow. Although that might be OK in an emergency, it's a very bad solution for the long haul. See *§*17, p.174 for replacement procedures.

10. Thermostat Replacement

The thermostat is a prime suspect in all cases of running hotter than normal, running cooler than normal, and overheating.

Replace the thermostat in conjunction with either a drain and refill or a flush of the cooling system, strictly as preventive maintenance.

Always use an original equipment thermostat. *Always* use the type of antifreeze specified in your owners' manual. If this work is being performed in an independent shop, bring your own thermostat plus its gasket or o-ring seal. Lay these prominently on the passenger's seat right next to your name-brand antifreeze and two gallons of distilled water.

Your temperature gauge readings should match what they were before the flush or overheating problem surfaced. If the engine is still running warmer than it used to, you still have a problem. See *§*8B, p.158.

As with any coolant work, you'll need to perform a follow-up inspection, *§*7, p.156.

11. Water Pump

There are only four ways to verify that a water pump is bad:

1. By the noise it produces–a low-pitched, grumbling or growling sound, as if rocks were being chewed up in a grinder. See *§*18, p.175 for testing procedures.

2. Newer cars with an accessible water pump, Fig. 9-6, p.159: **Engine Off!** Loosen the drive belt sufficiently to eliminate all tension on the water pump pulley. Grab it with one or both hands and try to rock it back and forth. Any slop at all is more than you want; significant bearing wear will be obvious.

3. By a trail of antifreeze exiting from the weep hole located directly below the pump shaft: That can be tough to see unless you're looking up from beneath the car; impossible if the water pump is driven by the timing belt (because it's housed within the timing cover). Transverse-mounted front-wheel drive configurations (Fig. 9-2, p.149) can make it tougher still. But for those of you with more traditional layouts, a swivel-mounted mirror on a telescoping handle (p. 117) can make all the difference.

    Note that a dried trail of antifreeze salts indicates that the pump has been weeping. That's still one step away from an actual leak. Water pumps usually don't start actively leaking coolant until they're shot, by which point the grinding sounds it's been making are already impossible to ignore. How much coolant is the engine actually consuming? That question might take a month or so to answer.

4. Older cars with a mechanical fan: **Engine Off!** Grasp opposing blades of the radiator fan, Alternately, push one toward the engine while you pull on the other to determine

9

how much slop there is to the water pump bearing. Significant bearing wear will be obvious: the water pump pulley will rock back and forth. As with Step 2 above, you may need to loosen its drive belt.

When should the water pump be replaced? Right away if it's actively leaking or grinding – a bad pump has much in common with a loaded gun. Replacement should also be considered if you are about to embark on a major trip when any of the above symptoms are present.

You might also consider replacement if other repairs need doing that offer greater access to the water pump. This approach makes the most sense on front-wheel drive cars equipped with transverse-mounted engines – particularly on those with higher mileage. You could conceivably save a considerable amount in subsequent labor charges by not having to go into the same area twice. We often recommend this "shotgun" approach on cars with over 120,000 miles on them if they need work on the engine's front (passenger's side) face. For example, timing belt replacement is a scheduled maintenance item. The water pump pulley is often driven by it. Because the new belt will put more tension on that pulley, and by extension, on the pump bearing, you might need to have the water pump replaced within the year. Get the crankshaft and camshaft seals replaced at the same time.

Some shops like to sell water pump lubricant. It's unnecessary, it's oily, and some offerings result in slime coursing through the system. And it's useless-water pump bearings are sealed. If you ever see it on your bill, consider a thorough flushing of your cooling system.

As with any coolant work, you'll need to perform a follow-up inspection, §7, p.156.

12. Radiator

Radiators can leak as a result of age, overheating, or collision damage. Most leak past a blown seam between their central core and upper or lower tanks (Fig. 9-1, p.148) or side tanks (Fig. 9-2, p.149). Radiators can also fail by becoming clogged due to neglect, or by the presence of additives (leak stoppers or water pump "lubricants").

Once upon a time, all radiators could be removed, repaired, and reinstalled for a reasonable cost – except for those that were smashed. Their tanks were made of metal, their seams were soldered to the core. These days they're built with plastic tanks joined to the core using fifty or more small tabs that crimp over top of a long, spaghetti-like o-ring.

A badly plugged radiator actually has cool(ish) spots where coolant no longer circulates; the rest of the radiator feels like it's on fire. Easy to diagnose if you have access, but most of you don't: the air conditioning condenser[2] sits directly in front of the radiator, the electric cooling fans are parked directly behind.

It's possible to *infer* that the radiator is plugged by looking at the coolant: it'll be thick, probably brownish or rust-colored, but most of you won't be certain until after the radiator has been replaced. If the system was properly serviced in accordance with the manufacturer's recommended service interval, odds are that your overheating problem resides somewhere else.

2   The AC condenser looks just like a radiator and performs the same function, except that they're thinner and they connect via aluminum piping rather than hoses.

There are three possible ways to deal with a plugged radiator: an extensive cooling system flush, replacement, or recoring. Clearly, a preliminary flush is tempting because of its potential for savings, but it's usually a waste of time. OEM radiators typically cost much less than OE replacements.

Recoring is an alternative to replacement on old-style radiators with soldered seams: The finned core of the original is replaced with a new one. This approach can offer considerable savings on older cars for which parts are scarce.

As with any coolant work, you'll need to perform a follow-up inspection, *§*7, p.156.

13. Blown Head Gasket

The head gasket is designed to contain and seal cylinder compression, engine oil, and engine coolant. Those of you familiar with pressure cookers know that heat applied to an enclosed space creates tremendous pressure. In a car's engine, there are woefully few places for this pressure to find release. A small amount can escape to the expansion/surge tank, but its rate is limited by the radiator/surge tank cap. Another option is for a hose to burst, but for obvious reasons hoses are designed with the opposite idea in mind. In cases of severe overheating, coolant forces its way past whatever the weakest area of the head gasket happens to be.

In cases of a "minor" blow following a fairly brief dance with disaster, it may seem that you've dodged a bullet: the thermostat (or whatever) is replaced, all seems normal except that there's a tendency for the engine to consume coolant. While that may be manageable in the short run, it won't remain that way for long.

In more severe cases, you'll often hear the gurgling sound of boiling coolant as it forces its way into your heater core. Major, minor, whatever –

> Do Not Remove the Radiator or Surge Tank Cap until the engine has completely cooled–a minimum of three hours . . . For that matter, Don't Even Open the Hood! Head gasket failure will cause coolant to blow out of any opening in the cooling system–just like a geyser. Until you're certain what you're dealing with, keep your face and all other body parts clear of the engine compartment: That coolant can be hot enough to *melt* your skin.

Head gaskets can fail in a number of ways, Fig. 9-7, p.164:

1. Between one or more of the combustion chambers and the cooling jacket, which results in exhaust gases bubbling throughout the cooling system, visible in the expansion/surge tank.

2. Between the cooling jacket and an oil feed journal. In these cases, the oil begins to look like a chocolate milkshake–not a pretty sight. The inside of the oil filler cap is the best window into the engine's interior: A foamy, congealing mess the color of milk chocolate is adequate proof. A severe blow actually increases the volume of "oil" registering on the dipstick.

9

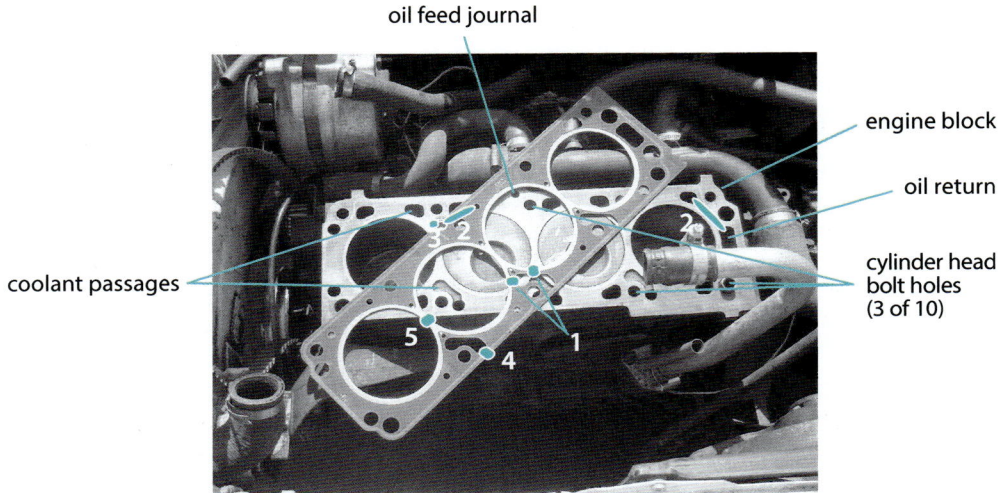

Fig. 9-7. Block and head gasket
Courtesy of Lewis Collard © 2006

3. Between the oil feed journal and the outside of the block

4. Between the cooling jacket and the outside of the block

5. Between adjacent cylinders

Version 1: Diagnosing this problem can be as simple as observing magnificent plumes of white smoke exiting through the tailpipe. Its sweet-smelling, noxious fumes are impossible to miss. You're witnessing "burning" antifreeze; it indicates a major blow between the cooling jacket and one or more cylinders: Fig. 9-7 (1). But that's end-stage failure. Earlier signs include:

- The engine overheated.

- It's been running hotter than normal.

- Consistent loss of coolant

- A rough-running engine, particularly evident at idle: The cylinder that's been violated has lost compression and can't fire properly.

- Bubbles of exhaust gas are rising to the surface in the expansion/surge tank.

Diagnosis at this stage starts with a stone-cold engine. Typically, you'll find that the expansion/surge tank is empty. Top it off with water. Expansion tank systems only: Top off the radiator as well, §4, p.152. **Close up the system.**

Set the interior temperature control to Hot, then start the engine. As it warms, observe the behavior of the coolant within the expansion/surge tank. Any car that is developing bubbles in there has a blown head gasket (as long as you're certain that the system is full). To confirm, remove all of the spark plugs. Any blown cylinder will have a squeaky-clean

> **Shut the engine down before it overheats. Do Not Remove the Radiator/Surge Tank Cap Under Any Circumstances!**

plug. All normal plugs will have some deposits or, at the very least, a grey or brown discoloration. Plugs from blown cylinders have pristine, white ceramic insulators; the electrodes will be "clean as a whistle". Occasionally its insulator will be cracked.

> **Do Not Crank the Engine with the Radiator Cap Off!**
> You could be treated to a fountain of coolant three to four feet high – a magnificent demonstration of the power of engine compression!

If you have a compression tester, you can further enhance your findings by using it, but I certainly wouldn't rush right out to get one. You already have all the data points you need.

Version 2: Here's an easy one – Remove the oil filler cap (engine off) to inspect for a foamy, chocolate-colored mess. What would prompt you to do that? Any one of the first four bullet points above, plus:

- Power loss: Engines that have been driven to the point of power loss are suffering from severely compromised engine oil, so diluted that friction has taken over. These engines are usually hard to restart and hard to keep running. Basically, the engine is locking up. Replacing the head gasket at this point is a waste of money; you need a new engine or a new car.

Versions 3 & 4: These two problems (oil or coolant leakage down the side of the block) are not necessarily due to overheating, and numerous cases require nothing more than consistent level checks. In many cases, you wouldn't even know about them were it not for the occasional drop of fluid in your driveway or your mechanic spotting the leakage during an oil change. The question then becomes how much of a nuisance this is going to be, and whether the problem deteriorates over time. Having the cylinder head retorqued is certainly a possibility, but the valve train configurations on most modern engines makes that hard to do (expensive). At best, doing so will only slow the leak, not eliminate it.

Version 5 : You can't develop a blow between cylinders without overheating the engine. Symptoms include a rough idle, misfiring, or power loss. A cracked head is a definite possibility.

Does every car that's run into the hot range blow its head gasket? Certainly not. How long does it take to blow? That's hard to say; five minutes? ten? A lot depends on the car, its condition, and the luck of the owner.

If the damned thing *is* blown, the cylinder head must come off for repair. While it's off, certain procedures should be performed. These include:

❑ Planing the cylinder head: Its gasket surface should be machined (cut) until it's perfectly flat. Aluminum heads warp when overheated. (So do aluminum blocks; fortunately, most are made of steel.) Unless the head is planed, it's likely that the new head gasket will blow again–usually after the warranty has expired. In order to plane the head, both manifolds (intake and exhaust) must be removed.

9

❑ Pressure testing the head: Severely cooked heads can crack, often in places where the crack can't be seen, resulting in symptoms ranging from (1) to (5) above. The ugly part of all this is that a mechanic and machinist can work together to produce what seems to them both to be a masterful repair, only to find out once it's running again that the engine is doing the exact same thing! It doesn't happen often, so for many, skipping this step might be worth the savings in labor. However, when it does occur, it's excruciating. Who pays? Obviously, the owner has to buy the new cylinder head, but what about all that labor – twice in, twice out – plus that second gasket set, the oil, the antifreeze, . . .

❑ Inspecting for other damage related to overheating: How do the exhaust valves look? Are they burned?

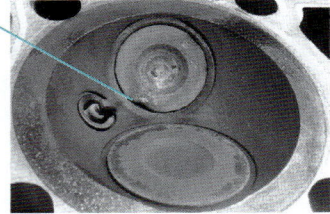

Fig. 9-8. Burned exhaust valve

What about the valve guides? Were they galled in the process? The only way to know for certain is to pull out the valves. One burned valve or a damaged valve guide necessitates a valve job.

Removing the valves for inspection should ensure a new set of valve seals – good insurance against future oil consumption, particularly considering the temperatures those seals were forced to endure.

❑ Engines with direct injection – where the fuel injectors spray directly into the combustion chamber – add another possible expense: Injector seals at a minimum, possibly the injectors themselves.

❑ Inspecting for age-related issues: The timing belt, water pump, coolant hoses, drive belts, whatever else that becomes accessible, should all be considered while the engine is in pieces. A new thermostat is a must; perhaps a new radiator as well.

Parts that should be replaced for no additional labor charge include coolant hoses, the thermostat, oil and filter, certain drive belts, and the spark plugs.

I've worked for two dealerships that would quote this job as if it simply involved cylinder head removal and replacement (R&R), then once the engine was lying in pieces on the bench, the service writer would call the customer to jam him with the news that the head was warped, that it needed to be planed, etc. The bill would go up by 2.5 hours labor; naturally, there would be additional parts. A second call might follow . . . You need to know *up front* what's included in their quote.

Find out who does their machine work. If the shop claims that they'll take care of it in house, keep looking: Machinists are highly skilled specialists using tools that are not present in 98% of repair shops. It's not a bad idea to speak with the machine shop directly, not only to verify that they have a strong working relationship with the shop, but also to get a feel for what's to come.

There's no part of this that's cheap, but it's a whole lot better to address everything that's necessary while the engine is still apart. What if you were to spend $1200 on a head gasket, only to find that the engine won't idle or that it burns a quart of oil every 200 miles?

9

The question of whether or not you are being screwed by double-billing is a delicate one. You don't want to imply a lack of trust by your questioning, but there is nothing wrong with being well informed. The best way to get at this is to work with the most experienced of their service writers – possibly the service manger or shop foreman-right from the start.

> Again, **if your engine is coming apart due to overheating, you need a good mechanic *and* a good machinist to inspect, repair, and advise.**

When you go to pick up the car, inspect your bill thoroughly *before* you pay for it. Be sure that you understand exactly what it says. If anything seems amiss, either through omission or addition, get it resolved *before* you pay the bill.

When you get the car back, you'll need to check the following:

- Oil level
- Coolant level and condition, including an inspection of the expansion/surge tank for crud – see *§*15, Step 8, p.171.
- That your radiator cooling fans are operating properly
- Recheck the oil and coolant levels the next morning, in a week, then again in a month, *§*7, p.156. If anything seems off, take it back.

14. DIY – Replacing Coolant

Cooling systems, properly maintained, don't require flushing with chemicals. Modern day antifreeze formulations have all of the scale and corrosion protection you'll ever need. Consult your owners' manual for their recommendations regarding the type of antifreeze to use (*§*5, p.153), its proper dilution and the overall volume of the system. Most of you will need one gallon of antifreeze and one or two gallons of distilled water. Obviously, if you're buying a 50-50 blend, you'll need more of the first and none of the second.

Before you go shopping, inspect your engine's coolant, *§*3, p.151. If it's murky (dark brown or rusty), if you find crud floating in it, or if the bottom end of the radiator cap is buried under a slimy gel, you'll need to flush the system, *§*6B, p.155.

You'll need every one of the following:

- antifreeze
- distilled water
- clean gallon jug
- radiator drain plug o-ring (OE or OEM)
- PB Blaster or comparable
- dab of Anti-Sieze
- large plastic drain pan
- a second, smaller pan
- automotive funnel, Fig. 6-5, p.97
- aerosol brake cleaner (2 cans)
- long-sleeved shirt

9

- work gloves
- plenty of rags
- garden hose and water supply
- tools to address the radiator drain, block drain, and air bleed valve (if present)

Use six-point sockets and/or wrenches whenever possible. You may need to remove the splash pan below the radiator to access its drain. If you are unsure about block drain or air-bleed location, now would be the time to buy that online subscription service (§4B, p.118).

I strongly recommend replacing the thermostat in conjunction with this work (§16, p.171). There's no better time to do it: The system's drained, you'll be leak and level checking anyway. And unless you've already met with catastrophe, it's the original. You'll need:

- thermostat, preferably original equipment (OE, not OEM)
- thermostat housing gasket or o-ring (OE or OEM)
- one or two new hose clamps
- possibly a new thermostat outlet (upper radiator) hose
- tools for hose clamp removal, § 17, p. 174
- single-edged razor blade(s)

Procedure

A few points about recycling: **Antifreeze is a sweet-tasting poison that can kill a pet. When dumped in a sewer, it ends up polluting ground water.** Store your used antifreeze in clean gallon jugs, then drop it off at a recycling center or cooperative garage. Most (all?) states require repair shops to recycle their used antifreeze. It's usually stored in 55-gallon drums that get pumped out routinely, hauled away by recycling trucks. Don't use that grimy drain pan you employ for oil changes! Contaminated coolant can cause recyclers to refuse the whole batch.

The engine should be cool.

1. Raise the car off the ground, support it on four jack stands, §7, p.139.

   Depending upon location and access to the radiator drain, the front splash pan may require removal.

2. To avoid splashing coolant all over yourself, keep the radiator/surge tank cap in place until after you've either (a) loosened the radiator drain valve, or (b) removed the radiator drain plug. Some coolant will immediately start to drain, so have your drain pan (and rags!) close by. Be advised: It's a whole lot easier to leak-check an engine that hasn't had coolant sloshed all over the place.

   Drain valves are nice because you can control their flow and you know exactly where it's going to go. Drain plugs offer very little upside: First, coolant will probably splash out of the pan when the radiator cap comes off–the hotter the coolant, the bigger the splash.

9

Second, their trajectory is less predictable: Unless the opening faces straight down, the arc of coolant will change over time, meaning that you'll probably need to reposition your drain pan once or twice.

Now remove the cap and let 'er rip. Once you're down to a dribble, move onto the block drain.

3. Reinstall the radiator cap to reduce the rate of coolant flow when the drain plug comes free.

The block drain can be ugly: bad location, often frozen in place, its bolt head just begging to be rounded off. Soak its threads in rust penetrant, then use six-point tools

If it simply won't budge, your options are to:

- use a pick to remove the rust around its point of entry, then let it soak some more
- smack it with a hammer, straight on against the bolt head, then let it soak some more
- heat it with a torch
- give up on it and flush the block with water

A water flush will dilute/replace whatever is in the block, but it produces two problems:

- You won't have the luxury of creating a precise blend of coolant. Certainly, you can fill most of the system that way, but at some point you'll need to add pure antifreeze. The question is how much.
- Your blend will contain some tap water. You may find comfort in the fact that perhaps 85% of our nation's mechanics have never used anything else . . .

If you are forced down this path, you will achieve somewhat better drainage by lowering the front of the car. Disconnecting one of the heater hoses may also help.

4. Before closing the block drain, lightly coat its threads with Anti-Seize. Replace the radiator drain plug's o-ring. Close both drains.

5. Clean out the overflow tank, §15, Step 8, p.171.

6. Fill the cooling system, §4, p.152. Remember to turn the passenger compartment's heater control to Hot.

7. Leak check before starting the engine, immediately following, throughout the warm-up process, and once it has reached operating temperature, §8A, p.157. **Keep an eye on that temperature gauge!** Verify that the electric cooling fans are kicking in and out. Once you're comfortable with your handiwork, take the car for a brief road test, then recheck your work.

8. Air bubbles are almost always trapped within the system during fill-up; reopen the radiator cap/surge cap/bleeder valve to top off the system once the engine has cooled back down–**a minimum of two hours.**

9

■

15. DIY – Flushing the Cooling System

Expansion tanks tend to collect sediment. Their siphon tubes stop a few inches from the bottom of the bottle, so whatever crud that floated to the top of the radiator will ultimately take up residence there. That doesn't mean that the entire system needs to be flushed. Remove the radiator cap to check what's beneath. Use a clear-sided antifreeze tester to pull some for inspection. If you find that the coolant looks OK, then simply swap out the coolant (§14, p.167) and clean the expansion tank using varsol and a brush, chased with an aerosol can of brake cleaner. You'll need to remove it first.

If your coolant is murky (dark brown or rusty), if you find crud floating in it, or if the bottom end of the radiator cap is buried under a slimy gel, you'll need to flush the system. Choose a name-brand, one-stage flush. Avoid the stronger two-stage flushes whenever possible. The first stage is very aggressive, the second stage is a neutralizer to arrest the process. Not performing an adequate neutralization step will soften (weaken) every one of your hoses over the next several months. Most can't be used on radiators with plastic tanks anyway (virtually all modern cars). Did I mention that it takes more than twice as long . . .

Whenever we do a cooling system flush, we replace the thermostat as well. There's no better time to do it and it may just prevent disaster: A thorough flush forces cool water across the thermostatic coils of a hot thermostat. Although this does not necessarily damage the thermostat, I've seen it happen often enough that I never want to see it again.

> Flushing a cooling system can be dangerous–not only to you, but to the engine as well. (If you start getting unlucky, things happen very fast.)
>
> The possibility of getting burned by hot coolant and even hotter engine parts is very real and must be carefully weighed before deciding to proceed.
>
> Keep your hands and other body parts clear of all moving parts.

Procedure

1. Turn the temperature control to Hot. Leave it there throughout this process.

2. Drain the system following the procedures of §14, pp.168-169, Steps 1-3.

3. Remove the thermostat, §16, p.171. Reassemble the thermostat housing without the thermostat using a new o-ring/gasket. Reconnect the hose.

4. Close both the radiator and block drains–just past finger-tight. (They'll be hot as jalapenos soon, make their removal as easy as possible.) Before closing the block drain, lightly coat its threads with Anti-Seize.

5. Add the flush, top off the system with tap water, §4, p.152. Follow the directions provided by the manufacturer, particularly regarding length of time, special considerations, waste disposal, etc.

9

> **The engine and its coolant are now *very* hot! Be extremely careful how you position your body to reduce the chance of a *nasty* burn.**

- Check that the drains are not leaking.
- Check the temperature gauge routinely to assure that it's staying in the normal range.
- Verify that the electric radiator cooling fan is cycling on and off once the engine reaches normal operating temperature.

6. Following the "chemistry" portion of your flush, turn off the engine. Roll down your sleeves and don your work gloves. Repeat Steps 2 and 3, pp.168-169, then drain the system.

7. Now comes the water flush. Take a tall, automotive funnel, cut off its lower end so that it makes a nice, tight fit with the radiator/surge tank opening. Use a garden hose to feed water through it until you see clear water pouring from both drains. Ten to twenty minutes of flowing water is immeasurably better than five to ten, because all chemicals must be removed.

    Because the block drain so often plugs with sediment, you might need to close the radiator drain to deliver additional water pressure to the block.

8. While you're waiting, clean the crud from the bottom of the expansion tank. Many of them just pull up and out. If necessary, use a spray lubricant to help it slide. Varsol and a bottle brush, chased with brake cleaner and water work best.

    Surge tanks generally clean themselves.

9. Install the thermostat, §16, below.

10. Lightly coat the threads of the block drain with Anti-Seize. Replace the radiator drain plug's o-ring. Close both drains. Fill the cooling system, §4, p.152. Remember to turn the passenger compartment's heater control to Hot.

12. Leak check before starting the engine, immediately following, throughout the warm-up process, and once it has reached operating temperature, §8A, p.157. **Keep an eye on that temperature gauge!** Verify that the electric cooling fans are kicking in and out. Once you're comfortable with your handiwork, take the car for a brief road test, then recheck your work.

13. Air bubbles are almost always trapped within the system during fill-up; reopen the radiator cap/surge cap/bleeder valve to top off the system once the engine has cooled back down–**a minimum of two hours**.

16. DIY – Thermostat Replacement

The engine should be cool. Most thermostats are reasonably accessible without elevating the car, but because the cooling system has to be drained down to (at least) that level, it will probably be easier to put the car up on jackstands, §7, p.139. If you plan on reusing your antifreeze, make sure your drain pan is pristine.

9

You'll need everything that follows:

- thermostat, preferably original equipment (OE, not OEM)
- thermostat housing gasket or o-ring (OE or OEM)
- radiator drain plug o-ring (OE or OEM)
- two new hose clamps
- tools to address the radiator drain, hose clamps, the nuts/bolts that secure the thermostat upper/outer housing, the air bleed valve (if present), and anything else that's in the way
- aerosol brake cleaner
- single-edged razor blade(s)
- PB Blaster or comparable
- large plastic drain pan
- automotive funnel
- plenty of rags
- other possibilities include . . .
  – antifreeze
  – distilled water
  – clean gallon jug
  – a new thermostat outlet hose

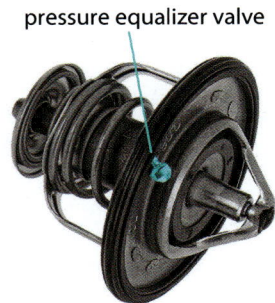

pressure equalizer valve

Fig. 9-9. Thermostat and o-ring

Procedure

1. Soak the nuts (or bolt heads) that secure the top/outer half of the thermostat housing with rust penetrant.

2. Drain the system following the procedures of §14, p.168-169, Steps 2 and 3.

3. It can be a lot easier to reassemble the housing if its positioning is not restricted by the outlet hose. Safer too–less likely to be misaligned, less likely to crack. Loosen the clamp that secures the hose and thermostat housing, enough to slide the clamp back three or four inches. Spray WD-40 or comparable around the seam, then insert a thin-bladed screwdriver under the lip of the hose, shoot in a bit more juice, then "walk" the driver around the housing's circumference till the hose slips free. See §9A, p.160 regarding removal of antifreeze deposits.

4. Remove the nuts/bolts securing the top/outer half of the thermostat housing, then pull the housing away.

   Thermostats typically have a specific orientation that you'll want to duplicate on reassembly. As the housing comes free, take note of the thermostat's position: Your surest reference is the pressure equalizer valve, Fig. 9-9.

5. Remove the thermostat. Many are encircled by a rubber o-ring that will pull out with it. The rest of you will be confronted with a paper gasket.

   Paper gaskets: Most are asymmetrical; before removing it, inspect the gasket's positioning relative to the base of the thermostat housing. Parts of the gasket will

9

stick to the housing, be attentive to the areas surrounding the studs/bolt holes. Use a brand-new single-edged razor blade to scrape them from both faces. Brake cleaner can be handy for softening the last remaining pieces.

6. Inspect the studs and nuts (or bolts). If any show significant corrosion, they should be cleaned (wire brush chased with brake cleaner) or replaced. Studs should be left in place for cleaning; if they need to be replaced, use a small set of vice-grips. Any components that you replace must be exactly the same length and thread pitch.

7. Wherever water has collected in a bolt hole, chase it with brake cleaner and compressed air. A cotton swab or soda straw can also be used. **Skipping this can lead to a cracked housing**.

8. Install the thermostat, placed in its proper orientation, with its:
    - O-ring: Slip it over the thermostat's base plate before insertion, Fig. 9-9.
    - Paper gasket: A thin coat of silicone gasket sealant can be used if it makes you feel better, but it's unnecessary if your clean-up work was sound.

9. Position the thermostat housing loosely overtop, engage all nuts/bolts. Alternate your tightening sequence from one nut to the next to ensure that both the thermostat and its housing settle down flat against their facing surfaces. Once snug, tighten another 1/4 turn (paper gasket) to 1/2 turn (rubber o-ring).

> Thermostat housings love to crack, particularly if they're overtightened or misaligned to begin with.

10. Resecure or replace the radiator hose using new clamps. Put them on the hose before you place it, loosely enough to permit easy positioning. When working in tight spots, take a moment to decide which way the clamp should face to simplify tightening. (Every clamp can be slid over the hose in two different ways). A thin film of CRC 5-56 or WD-40 wiped into the first inch of each hose end facilitates installation. Do the same to both flanges. Clamps should be placed in the center of their flange. Tighten until the rubber on either side has just started to bulge.

11. Fill the cooling system, §4, p.152. Remember to turn the passenger compartment's heater control to Hot. Remember too, that it's a whole lot easier to leak-check an engine that hasn't had coolant sloshed all over the place.

12. Leak check before starting the engine, immediately following, throughout the warm-up process, and once it has reached operating temperature, §8A, p.157. **Keep an eye on that temperature gauge!** Verify that the electric cooling fans are kicking in and out. Once you're comfortable with your handiwork, take the car for a brief road test, then recheck your work.

13. Air bubbles are almost always trapped within the system during fill-up; reopen the radiator cap/surge cap/bleeder valve to top off the system once the engine has cooled back down–**a minimum of two hours**.

9

17. DIY – Hose Replacement

You'll need all of the following:

- OE or OEM replacement hose
- new radiator drain plug o-ring or seal (OE or OEM)
- two new hose clamps
- clean, shallow, plastic drain pan
- automotive funnel
- utility knife
- single-edged razor blade(s)
- aerosol brake cleaner
- plenty of rags
- 1/4" drive ratchet and socket for hose clamp removal, or a wide-blade screwdriver
- tools to address the radiator drain, the air bleed valve (if present), and anything else that's in the way
- other possibilities include . . .
  – antifreeze
  – distilled water
  – clean gallon jug

Procedure

1. The engine should be cold. Open the radiator drain, then remove the radiator cap/surge cap. Keeping the cap in place will reduce coolant flow while you're backing out the drain plug. Be advised: It's a whole lot easier to leak-check an engine that hasn't had coolant sloshed all over the place.

2. As the system drains, loosen the two hose clamps and slide them back toward the center of the hose approximately three to four inches. Make a straight cut into both ends of the hose, back past each flange. It's safer to make two or three passes rather than trying to force your way through the hose with one. An added bonus: You'll be less likely to cut a groove into the piping! Use a screwdriver to pry the cut ends away from the pipe, then remove the hose. Never try to twist the hose off or remove it without first getting its clamps out of the way and cutting down to the pipe. Most of these pipes are made of aluminum or copper: They can crack, snap off, or distort in the face of even moderate force.

3. The end of each pipe should be inspected for deposits of crud that might create a path for leakage. Scrape off any deposits from the hose flange/pipe with a single-edged razor blade, then sand the area smooth with emery cloth (100 grit).

4. Clamps must go onto the hose before it's installed, loosely enough to permit easy positioning. When working in tight spots, take a moment to decide which way the clamp should face to simplify tightening. (Every clamp can be slid over a hose in two different ways.) A thin film of CRC 5-56 or WD-40 wiped into the first inch of each hose end facilitates installation. Do the same to both flanges. Clamps should be placed in the center of their flange. The clamps should be tightened until the rubber on either side just starts to bulge.

Overtightening can easily distort a thin-walled brass pipe, creating a pathway for leakage. Start by tightening 1/2 to 3/4 turn. You can always tighten a clamp further, if needed, but making a crimped piece of brass round again is another matter altogether.

5. Fill the cooling system, *§*4, p.152. Remember to turn the passenger compartment's heater control to Hot.

6. Leak check before starting the engine, immediately following, throughout the warm-up process, and once it has reached operating temperature, *§*8A, p.156. **Keep an eye on that temperature gauge!** Verify that the electric cooling fans are kicking in and out. Once you're comfortable with your handiwork, take the car for a brief road test, then recheck your work.

7. Air bubbles are almost always trapped within the system during fill-up; reopen the radiator cap/surge cap/bleeder valve to top off the system once the engine has cooled back down–**a minimum of two hours**.

As with any coolant work, you'll need to perform a follow-up inspection, *§*7, p.156.

18. DIY – Bearing Noise

Take a look at Fig. 9-10. Every one of those pulleys has at least one bearing associated with it. Water pumps, idlers and tensioners typically have just one bearing. The alternator has two; the AC compressor assembly has three. Directly behind all of that, hidden away behind the timing cover, is the timing belt tensioner and possibly another idler or two.

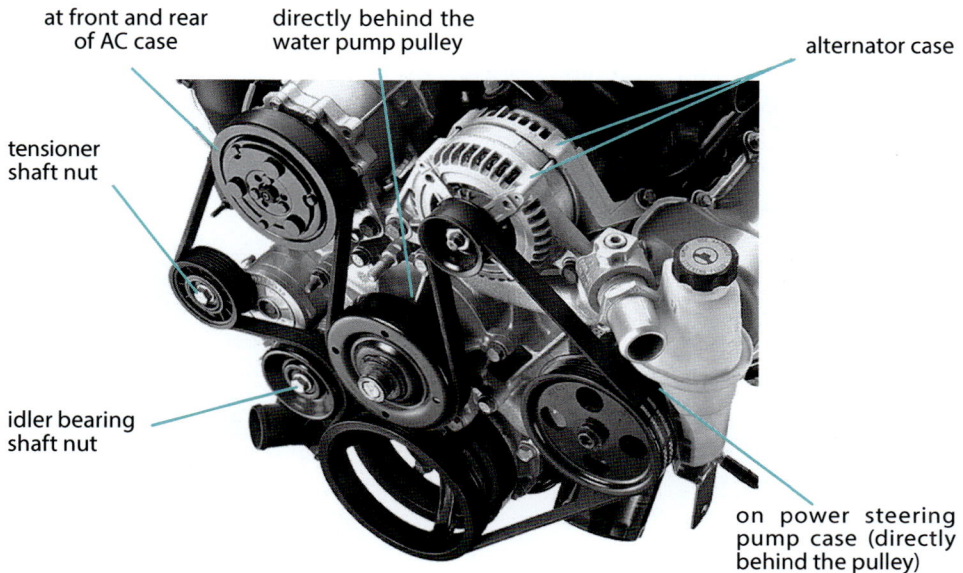

Fig. 9 – 10. Arrows indicate regions suitable for troubleshooting bearing noise
Courtesy of wkjeeps.com

Isolating one noisy bearing from that cluster is not necessarily all that hard, **but it can be dangerous–particularly for the careless**. You'll need a mechanic's stethoscope for this, Fig. 9–11. You'll be placing its tip as close as possible to each bearing while studiously avoiding everything else–pulleys, drive belt(s), fan blades, etc.

> **This can be extremely dangerous!** Before you begin, watch the engine run for a minute or two while you consider access. Do not proceed unless you are convinced that you've "got this".

The arrows shown in Fig. 9-10 indicate common listening points.

> Keep the stethoscope and all body parts clear of the engine, drive belts and pulleys! Failure to do so can damage anything in the vicinity – you, the radiator, a drive belt . . .

❑   Water Pump –

- Visible pumps: Pick any spot on the water pump's housing that's both accessible and safe. Figure 9-12 illustrates the bearing's position.

- Pumps housed behind the timing cover: Water pumps driven off the timing belt can't be seen directly, but all of them are joined to the heater core return hose in some fashion, typically through metal piping that's bolted along the side of the engine. Find the pipe, trace it to the pump, then access the unit from there. But be advised, there are all kinds of moving parts in the vicinity, any one of them could be producing the noise you're hearing.

❑   The alternator and AC compressor typically offer the easiest access. Their bearings are positioned at the front and rear of their cases; anywhere in the vicinity is fine. Read §2D, p.392 before dealing with the AC compressor.

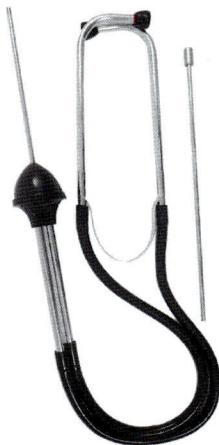

Fig. 9– 11. Mechanic's stethoscope
Courtesy of Lisle Automotive Tools

Fig. 9– 12. Position of water pump bearing
Courtesy of Toyota

❑ Listening to idler and tensioner bearings is a bit more dicey, in that their pulleys are positioned directly overtop of them. For these, you literally place the stethoscope tip directly onto the *stationary* bolt (or nut) that secures the bearing to its shaft, Fig. 9-13.

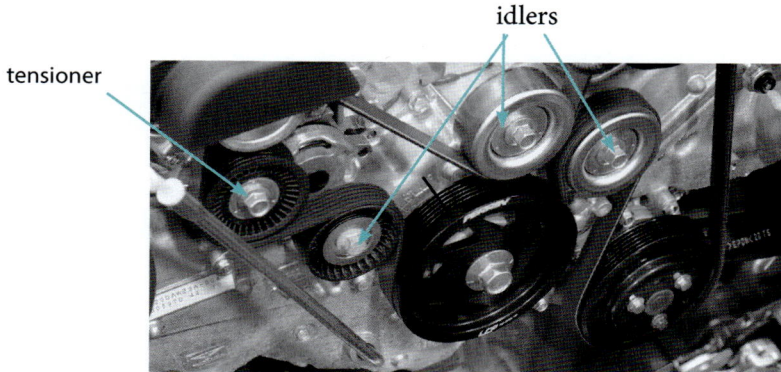

Fig. 9 – 13. One tensioner, three idlers
Courtesy of Jeff Perrin, tune86.com

If all of these seem quiet, then the sound is most likely originating inside the timing cover. Every timing belt has one tensioner and at least one idler, Fig. 2-1, and 2-3, pp.43 and 44).

19. DIY – Fan Clutch

These only exist on water pumps with mechanical fans. They bolt directly to the pulley that's driving the water pump, Fig. 9-14, p.178. Their purpose is to affect the speed with which the fan turns: While the engine is still cold, it spins lazily; as the engine gets hotter it locks up, spinning in sync with the water pump.A thermostatic coil inside the fan clutch controls the process by shifting a thick viscous fluid from the center of the clutch to its outer circumference, thereby increasing its angular momentum.

To test a fan cluch, start with a cold engine, an assistant, and a thick pair of gloves. Hold one of the fan blades, then have your assistant fire up the engine. Once it's running, let go of the blade and watch it's rate of acceleration. You should be able to brush your gloved hand lightly against the outermost edge of the fan blades to slow and gradually stop the fan (Cold Engine Only!). Let it go, drive the car till it's hot. Turn off the engine. Re-glove your hand, hold one of the blades while your friend starts the engine. You should feel strong resistance. When you let the blade go it should take off.

> **Do not attempt to slow down the fan once the engine is hot!**
> **Never attempt to grab a moving fan blade! Even plastic fans**
> **produce severe cuts.**

A worn fan clutch will demonstrate little difference between cold and hot. If you observe the same lazy acceleration you saw while the engine was cold, particularly if you didn't feel significantly stronger resistance while you were holding it stationary, then replace it.

9

drive belt tensioner

water pump pulley

water pump to
fan clutch studs,
nuts removed

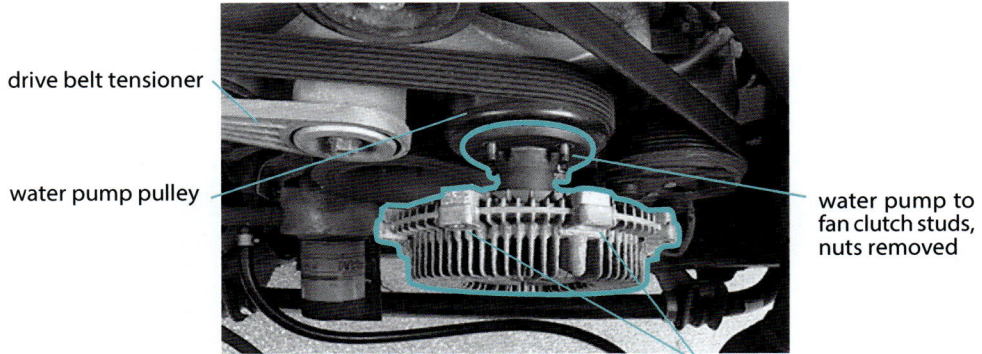

Fig. 9 – 14. Fan clutch assembly,
fan blade removed
Attribution requested

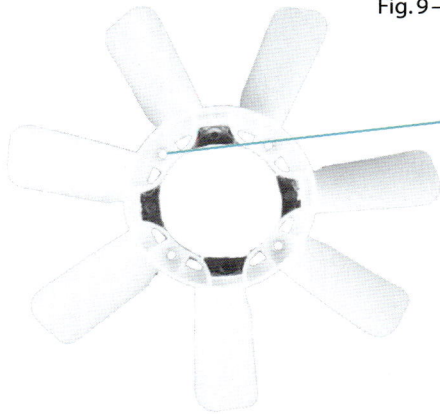

Central circle of fan blade assembly surrounds
the fan clutch, bolts secure it here and around
its circumference.

9

Chapter 10. Engine Performance

10

Notes

Engine Performance

1. Advancing Technology – The History of Change

It is often said that the sophistication of modern engines leaves the typical owner with no recourse but the professional mechanic. In several very real senses, that's just not true. Twenty years ago, virtually every engine required a bag of parts and a whole series of adjustments every year, just to keep them in tune (Table 9-1). These days, most engines need little more than a few maintenance procedures performed once every 30,000 miles: some parts get swapped, a few components get cleaned and lubed, a brief series of checks are performed and that engine is headed for its final road test. In fact, any adjustments you might want to make usually require that you first take it up with the electronic control unit – notorious for overriding the wishes of mere mortals!

❑ A set of points, housed within the distributor, required replacement with every tune-up, usually once every 12,000 miles. The points required precise adjustment or the engine simply wouldn't run. Points replacement would invariably alter engine timing – the point at which each spark plug fires relative to the position of its piston – requiring a corresponding adjustment to distributor position. Setting the timing is now relevant only when the timing belt or the crankshaft position sensor is replaced.

❑ All of the old carburetor adjustments – mixture setting, basic idle speed, choke fast idle, and throttle positioner – have become a thing of the past, supplanted by computer-controlled electronic fuel injection (EFI) systems.

❑ Starting in the mid-1990's, distributors were eliminated in favor of completely electronic engine control: Engines now employ a crankshaft position sensor coupled to the electronic control unit (ECU). The ECU responds by triggering a set of coils – one for each spark plug – which in turn "fire" the plugs, §1B, p.183.

❑ Before the advent of electronic fuel injection (EFI), all troubleshooting began with the know-how of the mechanic. Since the inception of the On-Board Diagnostic System (OBD-II) in 1996, most engine problems turn on a Check Engine light. The ECU stores both a problem code and many of the engine's parameters at the moment of failure. Some of these can be retrieved with an OBD-II reader, an OBD-II scanner can supply them all, §6, p.200.

❑ Valve adjustment has given way to valve systems that seldom require any attention other than clean oil, §1C, p.185.

❑ Advances in fuel technology, engine management, and metallurgy have resulted in spark plugs that only need to be replaced every 30- to 60-thousand miles. Recommended replacement intervals for air filters, fuel filters, and PCV valves have been altered – sometimes radically.

10

Since virtually no adjustments remain for the mechanic to address, just what exactly are you paying him for? That's my point! Many of you are overpaying – simply because you don't yet know what to ask for. The following pages and the lessons of Chapter 2 should help you create a list that details exactly what you need and when. By doing your homework, most of you will save a considerable amount of money, just by requesting the specifics of tuning when they're called for in your owner's manual, rather than blindly asking for a tune-up every year.

If your engine is running well – idling smoothly, accelerating smoothly, getting good mileage, passing its emissions tests – then in all probability, any "tuning" that might be required – replacing the air filter, spark plugs, fuel filter and PCV valve, cleaning the throttle body – could be handled by most anyone . . . including you! On the other hand, if your Check Engine Light comes on, or if the engine is not performing properly, you'll either need an OBD-II scanner to determine whether this is a problem you want to tackle, or you'll need a skilled mechanic. Don't tie his hands with a list of specific instructions unless you've checked the codes with a reader or scanner. If you get into it and realize that you're in over your head, be up front about it. Don't complicate the picture by cancelling the OBD-II code, *S6, p. 200.*

1A. Fuel Supply – From Carburetors to Fuel Injection

Engines cannot run without a fine mist of gasoline suspended in air. And the proportion of fuel to air in that mixture must be quite close to ideal (one part fuel to 14.7 parts oxygen) or the engine will pollute its surroundings, destroy the car's catalytic converter, or not run at all.

From the inception of the automobile through the 1980s, carburetors were used to create and deliver that mixture. But the need to reduce exhaust emissions ultimately doomed the carburetor, as their ever-increasing complexity overwhelmed the engine compartment with vacuum tubing, sensors, and switching valves, Fig. 10 –1. Fortunately, by the mid-1980s, fuel-injection technology had evolved sufficiently to fill the void.

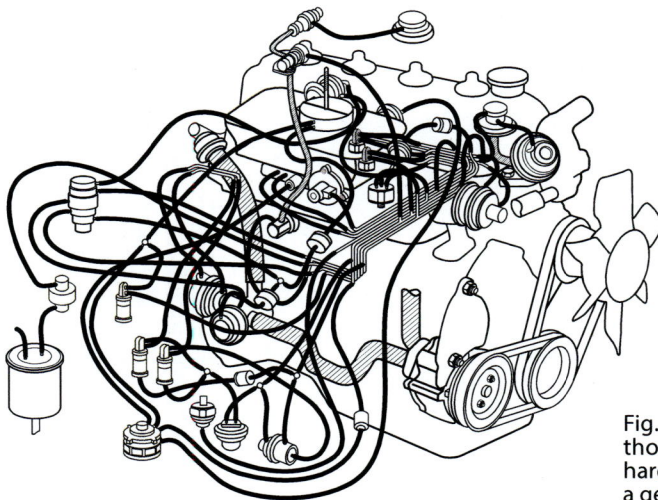

Fig. 10 –1. Vacuum tubing hell: If you thought that today's engines were harder to diagnose than those built a generation ago, think again! © PT

10

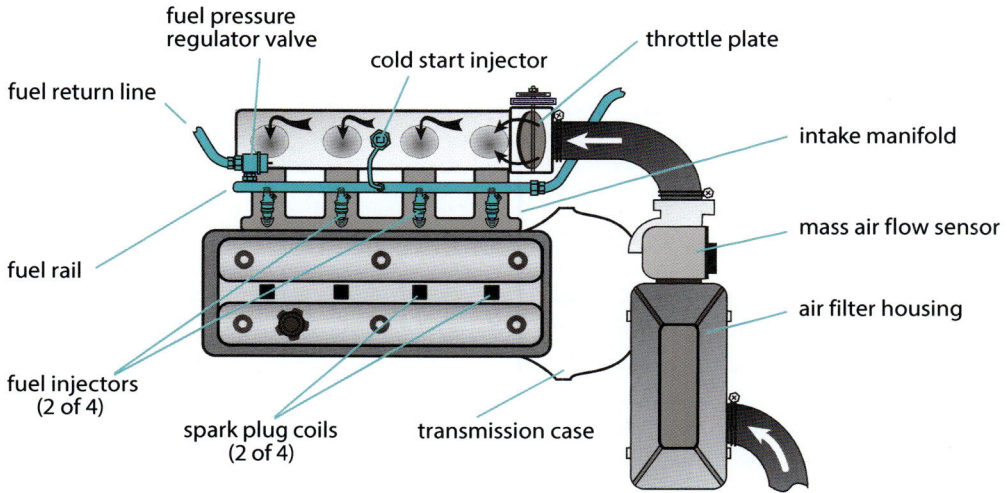

Fig. 10 – 2. Multi-port fuel injection on a four-cylinder engine. The large arrows indicate air flow. The fuel path is green. © PT

Virtually all engines built for the American market since the early 1990s have employed multi-port fuel-injection[1]. The multi-port system employs one injector per cylinder, Figure 10 – 2. As each piston moves upwards during its compression stroke, its fuel injector delivers an atomized spray of fuel. The ECU signals when it should open and for how long. Almost immediately, the ECU signals that same cylinder's spark plug to fire.

> Note that the terms *electronic control unit* and *electronic control module* refer to the same computer. For the sake of consistency, I've chosen to use the term ECU. I use the terms *fuel-injected*, *electronic fuel injection*, and *EFI* interchangeably; all refer to the multi-port systems.

1B. Ignition Systems – From Points to Computerized Engine Management

For an engine to do anything more than blow a hole in its muffler or simply refuse to run, that mixture of fuel and air must be ignited by a precisely timed spark. The spark must be delivered to the appropriate cylinder, and to every cylinder in its turn, about a thousand times a minute.

Remember the old KA-BOOM backfires of the 1970s? Probably not . . . Some sounded like a cannon going off! Most were caused by a set of points that had either slipped out of adjustment or had worn their way into oblivion.

---

1 Another form of EFI, throttle-body fuel injection, was produced for a number of years. It took all the problems of a carburetor – centralized fuel disbursement, atomization at some distance from where the fuel would actually be ignited, and significantly cooler release temperatures – and offered little in return beyond more precise metering.

10

Points were used to generate the baseline "timing" of the spark: Their opening and closing against a spinning cam – the central shaft of the distributor – caused the coil to fire, Fig. 17-18, p. 459. The distributor cap, rotor, and spark plug wires served to get that spark to the appropriate cylinder. The condenser – where present – was simply an adjunct to the system, a storage receptacle for energy while the points were open. The problem with points was that they wore out, requiring replacement every 12,000 miles or so. Even though the points themselves were cheap, replacing them so frequently was expensive.

The use of points and condenser declined precipitously toward the end of the 1970s, discarded in favor of an igniter. These solid-state designs dramatically improved on the points' mechanical approach to spark generation – no wear and tear, no scheduled maintenance. But they too had their downside. . .

Any of you ever get stuck with a bad igniter? The car just quit, simple as that. You might be lucky enough to be sitting at a light, but you could have the experience travelling at 70 mph in the outside lane! Fortunately, design engineers eventually broke free from the tradition of sticking all of their ignition components inside the distributor. Once the igniter migrated away from the engine's heat, those problems with cutting out dropped off dramatically.

During the 1990s, the emergence of ever-more-sophisticated sensors and computerized electronics led to a sharp reduction in the number of mechanical controllers required to adjust spark timing. We now enjoy engines that no longer even have a distributor! Retiring all of those moving parts meant that problems of wear, insufficient lubrication, and mechanical failure have become a thing of the past. Instead, we've been blessed (and only occasionally cursed) with fully electronic engine control – a full-fledged computer receiving input from numerous sensors and responding by commanding one of a set to fire, all the while adjusting for the specific conditions under which the engine is currently operating, Fig. 10 – 3.

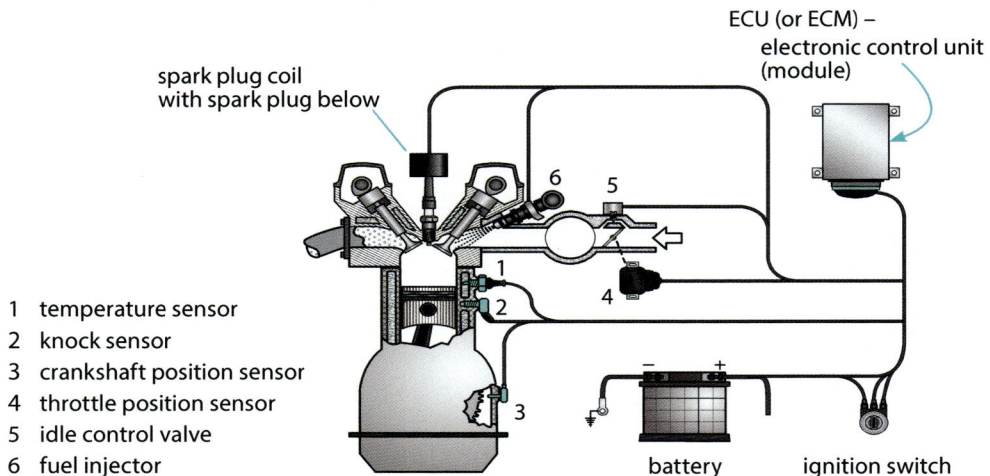

spark plug coil with spark plug below

ECU (or ECM) – electronic control unit (module)

1 temperature sensor
2 knock sensor
3 crankshaft position sensor
4 throttle position sensor
5 idle control valve
6 fuel injector

battery   ignition switch

Fig. 10-3. More of the sensors and activators involved with engine control. The mass air flow and oxygen sensors are not shown for the sake of simplicity.
Derived from an illustration by Robert Bosch, Inc.

10

1C. From Routine Valve Adjustments to No Maintenance Designs

Today's oils, metallurgy and fabrication precision have permitted the manufacture of extremely durable camshafts, shims and all other components in the valve train, allowing automotive engineers to move away from designs that require routine valve adjustments –as long as they are awash in clean oil . . . Maintenance tables for most of these engines call for no adjustment at all, while others suggest adjustment once every 60,000 miles.

Most engines in the 20th century employed a single camshaft operating both the intake and exhaust valves (Fig. 10-4). It wasn't until the mid-1990s that the transition from single-overhead camshaft (SOHC) engines to double-overhead camshaft (DOHC) configurations really got rolling (Fig. 10-5). Some are easy to visualize (Fig. 10-6), others aren't (Fig. 10-7), but both enable variable valve timing (VVT).

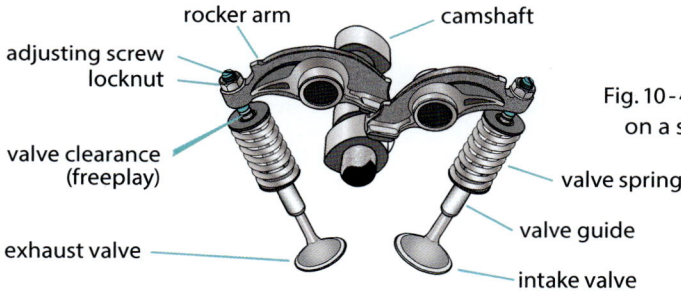

rocker arm   camshaft

adjusting screw
locknut

valve clearance
(freeplay)

exhaust valve

Fig. 10-4. Valves requiring adjustment on a single overhead cam engine
© PT

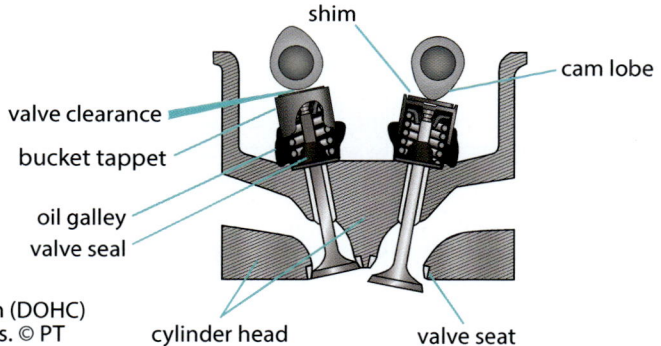

valve spring

valve guide

intake valve

shim

cam lobe

valve clearance

bucket tappet

oil galley
valve seal

Fig. 10-5. A double-overhead cam (DOHC) configuration with bucket tappets. © PT

cylinder head   valve seat

But first, some basics. You own a four-stroke engine, which means that for every cycle of each cylinder, its piston will go up twice and down twice. That requires the crankshaft to turn through two complete revolutions. Its (or each) camshaft will rotate just once, (Fig. 2-1, p.43).

Pick a cylinder, any cylinder: During its compression stroke, its piston is moving upward while both the intake and exhaust valves are closed, squeezing the air and fuel in that chamber up and around its spark plug. Just as the piston starts to round the "corner" (changes direction) the

10

ECU signals that spark plug's coil to fire and KABOOM, an explosion of atomized fuel suspended in air drives the piston downward, initiating the power stroke. As the piston bottoms out, the exhaust valve(s)[2] begin to open; enabling the rising piston to drive out the cylinder's exhaust gases. The exhaust valve(s) close at the top of the exhaust stroke, the intake valve(s) open, and during that final, downward stroke, air and fuel are sucked into the combustion chamber (intake or suction stroke).

Got that? OK, on to variable valve timing: For an engine sitting at idle, just a sip of gas mixed with oxygen–approximately 14.7 times the volume of fuel–is all that's needed to keep it running. The intake valves open, then close, and all is well. But at higher rpms, the fuel injectors are open longer, spraying considerably more fuel into the cylinders. What about the intake valves? The volume of air being sucked into each cylinder is essentially the same both at idle and at higher rpms. To create a clean burn, you need considerably more oxygen, which means that the intake valves need to open sooner. If they don't, unburned fuel remains, reducing mileage and increasing emissions.

So those are the theoreticals. The concept has been around for more than a hundred years, but designing a workable, reliable solution to the multitude of hurdles involved took nearly that long. Figure 10-6 shows an electromagnetic solution to the problem. The cut-away portion centered within the photograph exposes the genius. As long as the white, outer gear remains in the position shown, there is no advance along the camshaft. But as it gets pulled into its collar, the curvature of its teeth shift the relative position of the camshaft clockwise, thereby advancing intake valve timing. Complex, but simple.

With both basic and complex designs, one truth applies to all: If the timing belt strips or the timing chain loses its tension, you've got one helluva repair bill on your hands, because at least one valve will be occupying the space where its piston needs to be. Valve stems bend or snap, pistons end up with holes punched through them . . . ugly! Naturally, those engines blessed with an adequate supply of consistently clean, high-quality oil throughout their lives are far less likely to require internal engine work.

So what about valve adjustment? Unless your manufacturer recommends it in their Scheduled Maintenance, I wouldn't even think about it unless your mechanic recommends it to address one of the following problems:

- A rough idle: There are many possible causes for a rough idle (§6, p. 236), a tight valve being just one of them. Exhaust valves are more likely to lose their clearance because of the intense heat moving past them. Tight valves don't make noise.

- A pronounced valve tap: Likely causes include a low oil level or wear brought on by age. If your oil's fine and the engine is running smoothly, I'd leave well enough alone unless your mechanic gives you good cause to dive in.

- Noise associated with a worn or collapsed tappet. Here you have little to no choice: The engine will be running poorly, the noise will be alarming.

- Serious head gasket leakage (either oil or coolant). See (§5A, p.133) and (§8B, p.158).Valve adjustment would be one of the last steps during reassembly.

10

---

2   Many of today's engines employ two intake and two exhaust valves per cylinder.

camshaft position sensor

Fig. 10-6. Intake camshaft advance mechanism
Courtesy of General Motors

1 intake camshaft
2 exhaust camshaft
3 cam follower
4 barrel roller
5 hydraulic tappet
6 intake valve open
7 exhaust valve closed
8 combustion chamber
9 piston
10 connecting rod
11 intake manifold
12 exhaust manifold
13 ignition coil

Fig. 10-7.   Variable valve timing (roller-follower design)
Courtesy of Mercedes Benz

As always, generalizations like this must be evaluated on a case-by-case basis, relying on the advice of a knowledgeable mechanic.

1D. Valve Control – Roller and Follower Designs

As discussed in §1C, valve adjustment has been supplanted by–in most cases–an extra camshaft and set of shims, Fig. 10-5, p.185. And that extra cam has enabled variable valve timing. But there's another form of valve control: the roller-follower design shown in Fig. 10-7. The cam lobe interacts with a barrel roller that rotates freely within the camshaft follower's frame. Virtually frictionless. The follower moves up and down with the roller, pivoting on a hydraulic tappet at one end, opposite the top of the valve stem. The free end of the follower is "free" to move the valve down before returning to its neutral position as the high point of the camshaft rolls away. Complicated; beautiful.

10

throttle position sensor

throttle linkage

air inlet

electronic throttle positioner

idle control unit

throttle plate

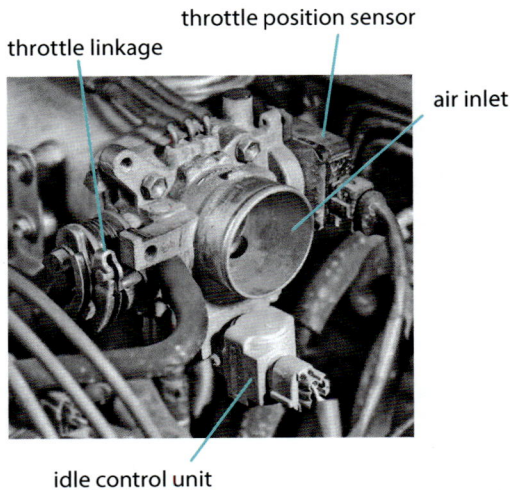

Fig. 10-8. Throttle body with air inlet tubing removed; throttle cable on the left.

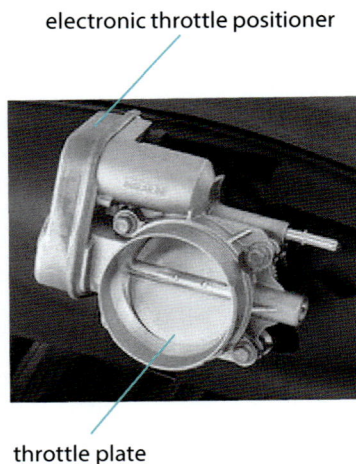

Fig. 10-9. Modern era throttle body
Courtesy of General Motors

1E. Improper Valve Adjustment

There are two reasons why a valve can go out of adjustment. The first–wear–creates noise. Valve tap can come from one or a bunch of valves and sounds just like the word–rhythmic tapping from the top of the engine. The other cause–heat–produces exhaust valves that swell a bit, thereby removing clearance. Tight valves don't make noise, but they do produce a rough idle because that cylinder has lost some compression. So, incorrect valve clearance can produce an engine that's noisier than it was to begin with, or quieter. Too much clearance won't affect compression, but too little will.

2. Maintenance Basics: Evaluations

To determine whether your mechanic has actually performed the work he billed you for, you will need to look under the hood . . .

Until you know what lives where, I'd suggest that you visit the engine compartment both before you go in for servicing and after. The maintenance section of your owner's manual is a good place to start: Most illustrate where the most basic components are located and offer instructions for replacing the air filter, etc.

Chapter 7 contains two relevant sections: For the Novice Interested in Tinkering, *§*2, p.113, and Safety Considerations, *§*3, p.114.

2A. Air Filter and Throttle Body

The air filter is usually housed in an assembly shaped like a shallow box (Fig. 10-2, p.183). Spring clips typically secure the air cleaner assembly's upper half. Good quality air filters have a tufted or cottony feel to them – thick like a good paper towel. The least desirable filters appear to be made from kid's construction paper – coarse and porous.

10

Fig. 10-11. A single unit coil over spark plug design
Courtesy of Bosch Auto Parts

Fig. 10-10. A coil over spark plug design
Courtesy of Honda Motor Corporation

The air filter assembly is connected to the engine's air inlet chamber by tubing that's approximately three inches in diameter. The opening to the air inlet chamber is called the throttle body. The throttle body is connected to the accelerator pedal by a throttle cable (Fig. 10 -8) or electronic throttle positioner (Fig. 10 -9). As the the throttle plate rotates (opens), air rushes past, pulled in by each cylinder's intake (suction) stroke.

Gaining access to the throttle plate for visual inspection requires removal of the air inlet tube. In most cases, that's no big deal, but it does require a few hand tools. Naturally, some manufacturers insist on complicating matters. On these, one can infer whether any cleaning was performed by using one finger to rotate the throttle linkage the first few degrees off idle (engine off and stone-cold). If it feels sticky, it was probably not cleaned or lubed. Note that this test is only possible with a mechanical linkage such as that shown in Fig. 10-8.

In order to access the throttle plate on an electronic throttle body, you need to remove the air inlet tubing. A sticky throttle plate is not necessarily a problem unless you're having problems with idling or until the deposits build to the point that the accelerator pedal sticks when first depressed in the morning.

2B. Spark Plugs and Coils

Figure 10-10 illustrates a row of coils sitting directly on top of a row of spark plugs. The coils are often concealed beneath a cover plate. Don't remove more than one coil at a time unless you first take note of which wire goes where: Reconnecting spark plug wires in the wrong sequence will produce a severe rough idle with the potential for engine damage.

Figure 10-11 illustrates a modular coil pack on certain four-cylinder engines. These units are almost always concealed beneath a cover plate. Inspection or replacement of the spark plugs requires that both the cover and coil module be removed first.

10

So how can you tell whether the plugs have been replaced without actually diving in there yourself? The short answer is: You can't. But an engine with a cover plate makes it significantly easier if you do your homework before taking the car in for servicing. First, take a look at what you've got. You'll find that the cover plate is secured by a series of cap-nuts on studs, threaded washers, or bolts.

A discreet scratch mark across the shoulder of just one nut or bolt and a complementary scratch out on the cover plate will suffice, simply because the odds of that bolt being reinstalled in the same hole, then tightened to the exact same degree are next to nil. A screwdriver blade is hard enough to make the scratch – use its corner. Cover plates that are held down by phillips-head bolts can be addressed simply by filling up a cross slot with grime. When the mechanic inserts his screwdriver, he will immediately drive some out and compress the rest. I've seen nail polish and touch-up paint used before . . . Enough said, it's too noticeable.

A pair of scratches can be used to verify removal of just about any part – as long as you're subtle about it. A heavy hand or overuse can kill a good relationship.

New plugs are obvious because their ceramic insulators are bright white; used plugs carry a ring of burned-on dirt located right below the end of the spark plug boot, Fig. 10-12. A small telescoping mirror can be very helpful for discerning any differences.

A significant number of V-block and flat-block (opposed head) engines have at least a few spark plugs placed in tortured locations. That can occasionally lead to a cross-threaded spark plug – one that's been installed at an angle slightly different from what was intended. A cross-threaded plug becomes increasingly difficult to install to its proper depth as one thread after the other is "customized." Cross-threaded plugs typically bottom out a few turns before their correctly-installed brethren, giving away their presence by sitting farther out, usually at a slightly different angle than the rest. The problem they present is that the threads within the cylinder head are stripped in the process of installation (and removal), making subsequent removal and installation increasingly difficult. What's worse

Fig. 10-12. This plug has been firing for 18,236 miles.

is that small pieces of aluminum can fall into the combustion chamber, scoring cylinder walls and creating all sorts of havoc. Should you encounter a spark plug that you suspect might be stripped, don't give the butcher shop another chance to ruin your engine – find a good machine shop for the repair.

10

2C. Spark Plug Tips

Whenever you have your spark plugs replaced, ask your mechanic to return your old ones. Compare them to the spark plug tips shown below. You can learn a lot about your engine.

Normal – Most of you will never see anything other than a Normal plug. No deposits, no abnormal coloration – just a dirty tan.

Too Lean – A bright white insulator can indicate one of two things: an engine that's been running too lean (§3D, p.196), or an engine that has overheated and blown its head gasket (§13, p.163). Your Check Engine light should have come on in either case , §4, p.199.

ash deposits

normal

oil fouling

too lean

carbon deposits

detonation

mechanical damage

Fig. 10-13. Spark plug tips clearly demonstrate the relative health of an engine. Photos courtesy of *Beru AG*.

Deposits – Black, sooty deposits can indicate one of two things: If all of your plugs are affected, then the engine is running too rich. Excess hydrocarbons aren't getting completely burned; the intake valves and cylinder head walls are loading up with the residuals. See §12, p247. If only one plug is sooty, then either that plug was misfiring or its coil is bad. The Check Engine light should have come on in either case, §4, p.199.

Mechanical Damage – You already know that your engine is suffering from mechanical damage. That whap-whap-whapping would have unnerved anyone within a one block radius . . . The engine will require rebuilding or replacement. See §6, p.138.

Detonation to the extent shown indicates a problem that's been going on for way too long. The knock sensor needs to be replaced. Hopefully, the engine held together.

Oil Fouling – If all of your plugs look the same, first verify that the PCV valve rattles when shaken and that your PCV system is clear, §2E, p.193. If they are, then it's time to replace either the engine or the car, §6, p.138. If the problem is limited to just one cylinder, then you have one totally shot valve seal.

Ash Deposits usually imply that you need to improve the fuel you're buying – either move to a higher octane or a different brand.

10

2D. Compression Testing

Compression testing is one measure of an engine's health. A consistently smooth-running engine will invariably have good compression readings, closely similar cylinder to cylinder. If you're considering a used car for purchase and you don't know a smooth idle from a rough one, or if the intake manifold vacuum is below 20 in-Hg (§6B1, p.237), then compression testing makes a lot of sense. Otherwise, I consider it a waste of time.

Good compression readings range from 145 psi to 185 psi. Compression readings above 185 are generally due to carbon buildup within the combustion chambers. Compression readings below 130 indicate a burned exhaust valve (Fig. 9-8, p.166) or worn piston rings. A certain amount of variation is acceptable between cylinders. As an example, the readings 180 – 185 – 160 – 180 (four-cylinder engine) should be considered satisfactory, but the readings 180 – 185 – 135 – 180 are problematic: That imbalance in compression would cause a rough idle. Just one cylinder with compression lower than 90 psi makes an engine hard to start, particularly in cold weather. Time to start considering your options.

If the engine was designed for regularly scheduled valve adjustments, then that adjustment should be made prior to the compression test. Just one tight valve is all it takes to drop compression by 30 – 40 psi within that cylinder.

The presence of carbon deposits within the combustion chambers is much more likely to produce a rough-idling engine than a tight valve is. Furthermore, since carbon deposits can directly affect compression readings – raising them and lowering them – every rough-idler should be treated to a can of fuel-injector cleaner and a tank of jet fuel (name brand gasoline, 91 octane or higher) prior to compression testing. Give it a chance to work: Schedule the test toward the end of that tank.

Compression loss produced by a worn or burned valve requires cylinder head removal plus valve work – a manageable expense on simpler engines. But repair costs can double on engines with two cylinder heads–the V block (V–6's, V–8's) and flat block (opposed-head) engines–unless the problem is confined to just one head. Ask for the compression readings; compare the two banks. If one side shows normal readings, leave it alone! (With the possible exception of valve adjustment.) Personally, I'd skip that too if the adjustment would require shim replacement, (Fig. 10-5, p.185) Most late-model engines have more than one intake and one exhaust valve per cylinder, a significant number of these now have variable valve timing as well. Naturally, this type of complexity translates into significantly higher repair costs.

Compression loss produced by defective piston rings requires removal of the cylinder head(s) *and* disassembly of the lower end, or block. To assist with that determination, see §9A, p.203.

There comes a point at which it becomes cheaper to replace the entire engine, either with a low-mileage used engine or a remanufactured one, than it is to pay a mechanical shop to completely rebuild what you have. However, if either the cylinder head or the entire engine is to be rebuilt, the valve work / block rebuilding should be performed by a machine shop – not a mechanical shop. It's customary for the good mechanical shops to limit themselves to removal and reinstallation of the cylinder head or engine, subletting the machine work to others more qualified. For further information, see §6, p.138.

10

2E. PCV Valve

The letters PCV stand for positive crankcase ventilation. This valve provides a pathway to pull both hot oil vapors and compression blow-by into the intake manifold where it will be sucked into the combustion chambers to be burned away.

This is a closed loop system: Inlet tubing connects the air inlet chamber to the valve cover, providing filtered air to the engine's interior. The PCV valve either sits in a rubber grommet atop the valve cover, or is threaded into it. Its return line joins the PCV's outer end to the intake manifold on the engine's side of the throttle plate, Fig. 10-14.

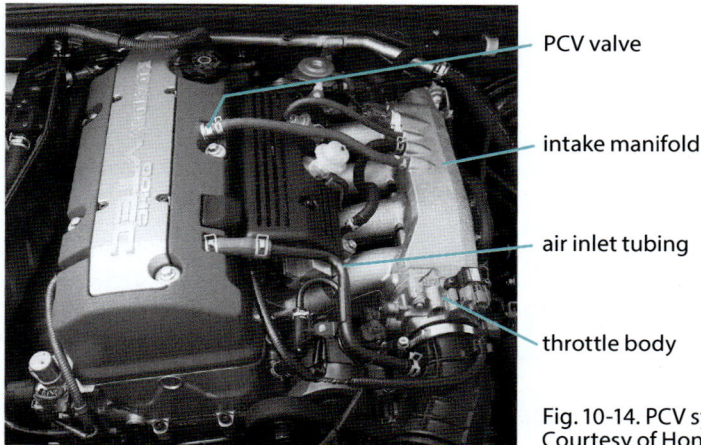

PCV valve

intake manifold

air inlet tubing

throttle body

Fig. 10-14. PCV system
Courtesy of Honda Motor Corporation

At idle, intake manifold vacuum pulls the PCV valve closed, preventing any crankcase vapors from circulating into the engine's combustion chambers. As rpms build, the valve opens, enabling both crankcase pressure and oil vapors to be sucked away into the fire.

The PCV system must be clear to prevent excessively high pressures from building within the engine's crankcase. Where's that pressure coming from? Oil expansion for one (as the engine moves from cold to hot); but also from blow-by past the piston rings: Even the most finely crafted engines eventually start to lose some compression past their valve stems and rings. The PCV valve should be inspected every 30 thousand miles to ensure that it doesn't become blocked, more frequently on engines that burn oil. A plugged PCV system will eventually blow out an engine seal, resulting in a hemorrhage of oil.

As always, a new PCV valve will be clean. On older models, check the grommet that it mounts into for a nice snug fit (grommets have been superceded by threaded valves). If the grommet is cracked, it should be replaced. A PCV valve that has been glued into place is very poor form.

10

As engines age, all that heat dries out both the grommet and the PCV return hose. Of the two, the grommet is more likely to need replacement – it's that much closer to the engine and unlike the PCV hose, its ends can't be trimmed to eliminate the cracks. (Most of the cracking in these hoses occurs right at the ends.)

The dealership parts department is probably your only source for the PCV hoses. Don't allow fuel line or coolant hose to be substituted, because oil vapors attack most kinds of rubber, rapidly softening and eventually collapsing the line. Excessively high pressures will build: You know the result – that hemorrhage mentioned above.

To test an aging PCV valve, remove the valve and shake it. There's a ball inside that should rattle as it rolls from end to end. Replace the valve if the sound is muffled – it's filling with sludge. If the ball won't roll at all, the valve has been plugged for some time.

2F. Fuel Filter

Most of the owner's manuals that I've studied make no mention of fuel filter replacement. While I did find several that recommend a 30,000 mile interval and a few more that recommend replacement at 60,000 miles, most failed to mention that they even exist, much less that they might need to be changed . . . Frankly, I'm not sure what conclusion to draw from all that. Having spent the better part of two decades replacing the clear plastic fuel filters found on carbureted cars of yesteryear, I can assure you that the dirt, water, and rust I've seen in them makes a strong case for observing some interval. The back-pressure produced by a partially plugged filter will dramatically increase the workload on the fuel pump – a far pricier proposition than filter replacement. Dealer's choice? I suggest once every 60 – 90 thousand miles if it's not mentioned in your owners manual.

Fig. 10-15. An assortment of fuel filters
Courtesy of Bosch and Toyota

If the parts list on your repair order includes a fuel filter, then you need to check that it was actually replaced. Neglecting to change these filters is a very common flat-rate ploy – in part because it can take a considerable amount of time to change them, in part because very few owners have even the slightest idea what they look like – or where they're located. EFI filters are often tucked away in the most obscure places – in or near the gas tank, for example. If your owner's manual won't show you, ask the shop about it.

Beyond verifying that it was replaced, you need only check that it's bolted securely in place and that it's not leaking. That assumes a lot, of course . . . That it's accessible! Check both the inlet and outlet fittings for wetness, preferably with the engine running – the system is not under pressure otherwise. Since most of these filters possess different line fittings at each end, they are seldom installed backwards.

2G. Vacuum Tubing, Vacuum Sources

If your engine has ever been apart for a valve job or head gasket work, if the intake manifold ever had to be removed, or if you know of a situation in which the engine gave some mechanic fits over a performance issue, it's quite possible that two or more vacuum lines got crossed. Consult your shop manual if need be.

As for the vacuum tubing itself, any line coming off the intake manifold or the engine itself gets very hot. This tubing is far more likely to fail there than, for example, along the firewall. Any line that's collapsed, broken, or extremely brittle should be either shortened or replaced. To avoid problems of misrouting, never remove more than one vacuum line at a time, and don't start moving lines around without first taking careful notes: Labelling lines with masking tape can help tremendously. Furthermore, use nothing other than factory replacement tubing; otherwise, over time you will likely encounter collapsed and broken lines due to the incompatibility of fuel vapors with the replacement rubber.

As for the vacuum ports themselves: As the miles spin away, these can gradually become plugged with deposits, rendering whole segments of the emission control system inoperable. Blockage to just one port can cause backfire, failed emissions tests, problems with driveability both cold and hot, lousy mileage, etc. And unfortunately, cleaning these ports is not in most mechanics' job description.

You can check these by disconnecting each vacuum hose in turn. Your fingers will thank you if you disconnect each line at its cooler end. Unless the line has collapsed or is badly cracked, the vacuum there should be equivalent to the port itself. Place your fingertip over the hose. Strong suction will be obvious at idle. These ports can be cleaned with throttle body cleaner, pinpoint sprayer attached, but *only* with the engine off and relatively cool. *Always* wear safety goggles!

3. Engine Sensors

The primary sensors involved in engine management are as follows:

- Mass Air Flow Sensor (MAF sensor)
- Mass Air Flow Temperature Sensor (MAFT sensor)
- Manifold Absolute Pressure Sensor (MAP sensor)
- Throttle Position Sensor (TP sensor)
- Oxygen (O_2) Sensors
- Crankshaft Position Sensor
- Camshaft Position Sensor
- Knock Sensor
- Temperature Sensor
- Wheel Speed Sensors

10

All of the data from each of these sensors passes through the CAN[3] bus–that maze of wires that you never want to venture into! Fortunately for all of us, neither the ECU nor its wiring harness is likely to fail. It's the sensors themselves and those wires leading to them–forever vibrating, getting super hot, cooling off to ambient temperature (whether that be 96 degrees or minus 10)–that cause virtually all of the problems and produce virtually all of the Check Engine codes.

There are more, sometimes many more, controllers that are wired through the CAN bus: Cruise Control, Traction Control and ABS Braking come immediately to mind. All three of these rely upon the four wheel speed sensors to feed data to their respective control modules. Other controllers include the air bag module, the door and window unit, the automatic transmission . . . Back to the engine's sensors . . .

3A. The **Mass Air Flow Sensor** (MAF) and **Mass Air Flow Temperature Sensor** (MAFT) are located adjacent to the air filter on the engine side of the air filter box, Fig. 10-2, p.183 The two share the same wiring harness connector and together transmit the "volume" of air heading toward the engine. A bit of physics creeps in here: The air's density varies with temperature and altitude, and turbocharged engines alter the pressure applied. All three affect how much oxygen the cylinders will see. So the engineers need to refer to air mass–a measure of weight–rather than air volume if they intend to calibrate their engines for peak efficiency. These calculations are done on the fly by the ECU.

3B. The **Manifold Absolute Pressure Sensor** (MAP sensor) monitors air pressure within the intake manifold when the throttle plate is open and air is roaring in. Being sucked in, actually, as each piston moves through its downstroke while that cylinder's intake valve is open.

Since air pressure and vacuum are opposite sides of the same coin, the MAP sensor doubles as a vacuum gauge. At idle, there is no manifold pressure, only a vacuum because while the cylinders are still doing their thing, the throttle plate is essentially closed. Actually, it is closed; the air required to keep the engine running is passing through the idle air control valve, Fig. 10-8, p.188. The MAP sensor is always connected to the intake manifold by a piece of vacuum tubing. It can be found mounted on the intake manifold or the firewall.

3C. The **Throttle Position Sensor** does exactly what you think it does: It relays to the ECU how wide open the throttle plate is, or isn't (at idle), Fig. 10-8, p.188.

3D. The **Oxygen Sensors** are located in the exhaust system. There are two: The first bolts into the the exhaust manifold or front pipe, just ahead of the catalytic converter, the other is mounted directly behind it. Their purpose is to measure the amount of oxygen remaining in the engine's exhaust gases so that the ECU can maintain the engine's optimal air-fuel ratio, and in so doing, protect the catalytic converter from overheating.

The fuel mixture can be too lean (too much air relative to the amount of fuel), too rich (too much fuel relative to the amount of air), or just about right to provide the proper balance between fuel economy, power, and emissions. Lean mixtures burn excessively hot,

10

3   Controller Area Network bus. The term "bus" is computer-speak for a distinct set of wires that carry data and control signals both within the components of a computer system and between systems.

producing nitrous oxides–terrible for the ozone layer. Over the long term, a consistently lean mix can cook the catastrophic converter. Overly rich mixtures will also overheat the converter by overwhelming it with unburnt hydrocarbons.

The trailing oxygen sensor monitors how well the catalytic converter is performing; its inputs also alter the air-fuel mixture to protect both the converter and the atmosphere.

All of this transpires instantaneously within the ECU: Data is fed in from every relevant sensor, the computer "maps" the appropriate fuel mixture and spark advance to correspond to the engine's speed and load, and delivers signals to the fuel injectors, ignition coils and other relevant controllers. A whole lotta science and even more genius . . .

3E. The **Crankshaft Position Sensor** is usually located at the rear of the engine; it points directly at the teeth of the flywheel, Fig.10-3. p.186 and Fig. 14-6, p.365. It's sole purpose is to feed engine timing cues to the ECU, specifically when top dead center (TDC) on cylinder #1 comes roaring past. (At TDC#1 the piston in cylinder #1 is at the top of its compression stroke, with both intake and exhaust valves closed. Ignition timing references that marker to set the spark plug firing sequence for every cylinder).

3F. The **Camshaft Position Sensor** is located at the front of the camshaft; it points directly at the timing chain gear, Fig. 10-6. p.187. It, too, is intimately involved with engine timing, but in this case, it exists to monitor camshaft position. Before the emergence of variable valve timing (VVT), this sensor was unnecessary, but VVT has become commonplace across the industry. Under hard acceleration, opening the intake valves sooner allows more air to get sucked into the combustion chambers. This makes for a leaner mix which in turn provides both increased power and fuel economy by offering a more thorough burn of the extra fuel being pumped in through the fuel injectors. Conversely, in low throttle modes the ECU can dial back the VVT: The lower the volume of fuel being injected , the less air is required, particularly if you're seeking a more optimal burn for the catalytic converter.

There are a number of ways to achieve this: Assuming a double-overhead camshaft configuration, the entire intake camshaft can be advanced or retarded with one controller, typically manipulated with an electromagnet, Fig. 10-6, p.187. That one camshaft can drive either one or two intake valves per cylinder.

Because the camshafts are arrayed across the top of the engine, they are the first components to suffer from oil starvation. I've hammered on this throughout the book, but this time . . . Let's say that one of these actuators suffers from wear and decides to bind. I've mentioned before that many modern engines are designed so that each valve fleetingly occupies the same space as its corresponding piston will. If that camshaft gets stuck–even momentarily–one or more of the valves associated with it might find themselves being crushed by their piston. The cost to repair that type of damage is astronomical. Over and over I've decried "What price power". But really, that's just shorthand; the purpose of VVT is to maintain healthy emissions, to increase fuel economy, and to derive more power from a smaller package.

3G. The **Knock Sensor** can be located anywhere along the side of the block. It operates much like a microphone, converting sound waves into an electric signal.

10

Knocking can occur under three conditions:

- When the combination of fuel and air are not thoroughly mixed. When fuel is not ignited during the initial spark plug firing, it can still ignite off the residual burn spreading outward through the combustion chamber. This secondary explosion creates a knocking sound.

- With low-octane fuel – same song, different tune

- When the engine's timing is off (pre-1996 – the distributor era)

These late explosions can be extremely destructive to the engine in that they set up shock waves running out of sync with the initial blast. Complete combustion simply drives the piston "south"; the added explosions stress every component in the combustion chamber, Fig. 10-7, p.187.

Fuel injectors can be mounted in the intake manifold, Fig. 10-2, p.183, or they can be inserted directly into the cylinder itself, adjacent to the spark plug. Gasoline direct injection (GDI) increases engine efficiency and power output. But not all technological advances are as good as they sound. Intake manifold injection (IMI) has been regaining market share because it offers the distance and time to produce a homogeneous mix of fuel and air prior to entering the combustion chamber. When the spark hits that charge, its uniformity permits a rapid, even burn throughout the chamber, with no knock.

The latest oil specification, API SP, was developed in part to help with engine knock.

3H. The **Temperature Sensor** is a critical component of engine control. Without it, the car is hard to start and tends to stall. Typically threaded into the thermostat housing, it has two primary function:

- It controls warm-up–providing a richer fuel mixture to prevent cold-start stalling–then dialing that back to enable improved fuel economy, and

- It triggers the Check Engine light if the car overheats.

Because the engine requires that richer mix when it's cold, with every intake stroke the fuel injectors fire for a longer interval. A temperature sensor that can only signal that the engine is warm would not cue the ECU to call for that. By the same token, a temperature sensor that can only signal that the engine is cold will cause the ECU to flood the engine once it's warmed sufficiently to require a leaner mix.

There are several other temperature sensors on board: One monitors radiator temperature, switching the radiator cooling fans on and off as needed to maintain steady engine temperatures. Air conditioning systems have their own temperature sensor, controlling radiator fan functioning when the AC switches on and off. These two are typically threaded into the radiator.

3I. One **Wheel Speed Sensor** is located at each wheel, bolted into the wheel hubs, Fig. 13-10, p.337. (That assumes the car has anti-lock brakes.)

These sensors feed data to a number of controllers:

- The ECU – It needs to know how fast you're going . . .

10

- The Anti-Lock Brake System (ABS) Controller: When a wheel locks up on ice, oil or water, the ABS controller dials back the hydraulic pressure being delivered to that particular brake assembly.

- The Traction Control Unit operates in identical fashion, except in this case it increases hydraulic pressure to slow the rotation of any wheel that's spinning.

- The Transmission Control Unit (TCU) adjusts shift points, taking its cues from the ECU. So if you floor the throttle to pass that laggard who's been messing with your serenity, the trans will kick down to the appropriate gear, where it will linger until you back off the throttle, then shift up to conform with the TCU's signal. All of that used to be controlled hydraulically within the transmission itself.

  The TCU also comes into play on four-wheel and all-wheel drivetrains, specifically with traction control.

4. The Check Engine Warning Light

The Check Engine Warning Lamp is one of many on the dashboard; it's usually framed by your steering wheel. You'll see it every time the key is turned to its first (On) position. And as with all the rest of them, it will turn off when the car starts–unless there's a problem. The light has two modes. Both are serious, both indicate that something isn't right with the engine or transmission, that something needs attention or repair.

Fig. 10-16. Check Engine light

> If you see the Check Engine light flashing, pull over and turn off the engine, NOW! Serious damage can result from continued driving. Have the car towed.

Some of the reasons why the Check Engine light would be flashing are:

- Engine overheating
- Catalytic converter overheating
- Spark plug misfire
- Fuel injector failure
- Problems with the ECU or its network

10

5. The OBD-II Access Plug

Whenever the Check Engine light comes on, a code is stored in the ECU. These codes can be read by an OBD-II reader or scanner that connects through an access plug beneath the dashboard on the driver's side. They've all looked the same since 1996.

Fig. 10-17. OBD-II access port
Courtesy of DHG

6. OBD-II, Readers and Scanners

The term OBD stands for On Board Diagnostics. Back in the dark ages, various electronic diagnostic standards existed across the automotive world, from none to a few manufacturer-specific attempts to simplify engine troubleshooting. OBD-I was the first standardized offering; it came on the scene in 1988 to assist in diagnosing the early fuel-injected products being produced for the American market. By 1991, all engines produced for sale here in America were fuel-injected. (Carbureted engines, with their own assorted grab bag of problems, had no such diagnostic platform; they, after all, had been around since forever.) But those early EFI models–the rough idling, the failed emissions tests–OMG! They could be inscrutable! OBD-I was not equal to the task . . .

As of 1996, the OBD-II standard became mandatory for all passenger vehicles in the United States. It has enabled tremendous leaps forward in automotive diagnostics. The problems we used to scratch our heads over– occasionally throwing a hammer at some innocent wall–became *so* much easier to diagnose. The evolution of OBD-II has not been exactly straightforward, but each in their own way, the automotive manufacturers worked to improve their products. At this point, most any problem you may encounter will have a diagnostic code for you to "read". The question is, how far do you want to go with this?

You can buy a code reader or a scanner. The difference is one of cost and value: Readers cost twenty bucks; scanners start at about $75. From there, the sky's the limit. Here are a few of the issues you should consider:

- The Check Engine Light will only come on when a powertrain sensor (engine or transmission, or the ECU itself) registers outside its normal range. Thankfully, it will flash if there is a serious chance of costly failure.

- No other problem with your car will cause the Check Engine light to come on, and, to my knowledge, no reader will enable you to look for anything else, such as an airbag or anti-lock braking system malfunction. But a reader will enable you to look for other P codes that have yet to illuminate the Check Engine light. These "temporary codes" will hang around for a while, then disappear if the problem doesn't recur.

- If you want to, a reader will enable you to turn off the light. You might, for example, just want to see if the problem recurs, and if it does, what exactly is going on with the car when it does light up.

- Now it may be that's as far as you want to go with this, but if you have an itch to delve in a bit deeper, a reader just won't cut it.

- Most of the better scanners will save what's in the car's memory chips should you need to disconnect the battery.

10

Every Check Engine code begins with the letter P. It is followed by four numbers.

- The first number indicates whether the code that follows is an industry standard (0) or is specific to a single manufacturer (1). To my knowledge, readers don't offer information regarding manufacturer-specific codes.

- The next number indicates the system that's affected: air metering, fuel metering, ignition control, exhaust monitoring, engine speed control, transmission troubles, and problems with the ECU. See §7, p.202.

- The third and fourth numbers attempt to detail exactly what part of that system is producing the problem.

A reader will translate all that information into English for you, but nothing else. That information coupled with the factory manual's diagnostic procedures can take you a long way, but not nearly far enough if, for example:

- Your ABS system (anti-lock braking system) is malfunctioning, or one of your tire pressure sensors is faulty and you'd like to know which one.

- You need to see a scan of the data that was collected by the ECU when the problem first occurred.

The OBD-II standard encompasses three additional areas:

- B (body–airbags, power windows, mirrors, door locks, etc.)

- C (chassis–ABS, tire pressure monitoring system (TPMS), traction control, etc.), and

- U (messaging troubles between the ECU and other controllers) I have no idea why they picked U, particularly since the term Controller Area Network (CAN) refers to all of that wiring and the system at large. Perhaps it means Unbelievably Expensive . . . This is an extremely important aspect of every modern car; think airbags, anti-lock braking–all of which needs to respond instantaneously to those sensors scattered all around the car.

You'll need a scanner to investigate every aspect of the B and C protocols because, among other things, no malfunction in either area will illuminate the Check Engine Light. But scanners will signal problems in those areas and the good ones will offer recommendations on repair procedures.

For what it's worth, there are plenty of scanners available for purchase that will list every manufacturer-specific trouble code–those beginning with C1, for example–but not offer any further information about what the last two numbers in that code mean. Another limitation: The manufacturers are not required to offer up B or C codes at all! That's why owning your factory's repair manual is invaluable. Whenever it is unavailable for purchase, AllDataDIY.com is a worthy substitute; the only real difference being that you'll want to keep the grease off of your computer keys . . .

So let's review: The Check Engine Light will only come on if a sensor within the engine, transmission, or CAN bus offers up an abnormal reading. Concurrently, the ECU stores in memory a snapshot, or "freeze-frame" of the powertrain's sensor readings at that moment

10

in time. That data will remain in memory until the problem is repaired or an OBD-II reader/scanner clears the code. A scanner can be used to investigate those readings and to watch live data of the engine/transmission in operation. The better the scanner, the more information will be available to you through its software. *Car and Driver* regularly reviews OBD-II scanners. This year (2023), their costs range from $68 to $330.

7. Metering and Monitoring

I'm no expert on OBD-II coding, so I won't even begin to try. Most any scanner would be a better resource. But I can offer some measure of understanding to the terms metering and monitoring.

Air and fuel metering refer to the ECU's responsibilities regarding the ratio of air and fuel entering the combustion chambers at various speeds, altitudes, humidity levels, load levels (up or down a hill, for example), etc. The catastrophic converter is a sensitive beast: It will overheat if the fuel to air mix is too rich (too much fuel) or too lean (too much air). And that's dependent upon every variable mentioned above. Chemistry uses the word stoichiometry to refer to the ratio between–in this case–air and fuel: To enable complete combustion, their ideal mixture needs to be in the vicinity of 14.7 parts oxygen to one part fuel. That's one of the ECU's primary functions.

Like every computer I've ever known, the ECU is fundamentally stupid: It needs engineers to tell it what to do. So it's "mapped" to respond to the input from sensors. Every one of the sensors listed in §3, pp.195-199 is involved with that decision, which takes place instantaneously and repeatedly for as long as the engine is running. That's the metering part.

As for monitoring, the primary controllers are the two oxygen sensors, but the crankshaft and camshaft position sensors get in the act with regard to engine timing: Not just the point at which the spark plugs fire, but also the degree of VVT (camshaft advancement) involved.

8. Engine Timing – Injectors, Intake Valves, and Spark Advance

The term engine timing used to refer solely to the point at which each spark plug fired relative to the position of its piston. This had everything to do with points and distributor settings. But since the advent of electronic fuel injection (§1A, p.182), spark generation controlled by the ECU (§1B, p.183), and variable valve timing (§1C, p.185), that term has taken on a far more sophisticated meaning that encompasses all three.

`First, a bit of mechanics. An engine under load – for example, one that's just begun moving up a hill – requires more fuel to maintain the same speed. That requires that the fuel injectors open a split second sooner and remain open for a beat longer, and that the variable valve timing (where present) needs to advance to accommodate that earlier injector firing. To translate the added fuel into increased engine performance, the point at which the spark plug fires relative to the upward motion of its piston must advance by a fraction of a second. The rate of this change must vary quite precisely in response to how aggressively your foot meets the accelerator pedal – this to keep the engine from pinging (knocking). Complicated, but beautiful!

10

9. Engine Tuning: DIY Detail

9A. Replacing Your Own Spark Plugs

Before you buy anything, take a good look at your engine, especially if you're working on a V-type or an opposed head (flat-block) engine: Can you see every plug? Study your line of access to the ones in the most troubling locations. Can you visualize getting a ratchet, the spark plug socket, perhaps an extension, possibly a swivel, *and* your hand into the vicinity of that plug? What about getting the new spark plug back in without cross-threading the cylinder head? All good?

If so, be sure to have everything on hand before you begin. Most owners' manuals offer several specific spark plug recommendations – both brand name and model number. Because these plugs are manufactured specifically for your engine, they'll be correctly gapped right out of the box. Having the correct gap between the spark plug's two electrodes is critical: improperly gapped plugs reduce fuel economy and performance. Installing spark plugs of the proper heat range is even more important, .

Fig. 10–18. Spark plug gap. Beru AG.

Always wait for the engine to cool sufficiently so that you won't complicate matters by burning yourself. If you will be dealing with "awkward" plugs, expect the job to take at least twice as long as you could possibly imagine and be delightfully surprised when it doesn't.

If you're going to need a swivel, spring for a one-piece spark-plug-socket-and-swivel combination (A). If your plugs live at the bottom of wells that reach into the engine's depths, you'll need to be equipped with tools that won't leave your socket stuck down in one of those holes. Losing a socket to the darkness of an engine's innards is not conducive to one's serenity . . . If you are concerned your tools won't handle it, the cheapest solution is to replace your extension with a brand new one. Or you can spring for a one-piece socket-and-extension combination (B), or a locking extension that will stick to that socket like glue (C), Fig. 17-24, p. 460. All are available through a *Craftsman, MAC* or *Snap-On* dealer.

Maybe your engine is a simple four-cylinder job that offers smooth sailing . . . So much the better! You'll still need a standard issue spark plug socket (D) so you won't crack an insulator: All spark plug sockets have a rubber insert to prevent just that.

Each coil's wiring runs to a specific cylinder corresponding to the engine's firing order. So never remove more than one coil at a time unless (a) the manufacturer designed their wiring harness to prevent any confusion as to which coil wire goes where or (b) you label their wiring first. Reconnecting spark plug wires in the wrong sequence can create a severe rough idle with the potential for engine damage.

Never tug at the wires: Always unclip the wire terminal connector that joins the coil and pull them apart there. Tugging on the wires can rip them right out of their plastic terminal; tedious to fix, expensive if you have to pay someone else to do it for you.

10

■

Make sure that your socket has fully engaged the spark plug's metal hex-nut base. (Those ceramic insulators love to crack.) Spark plugs unscrew counter-clockwise.

For the sake of comparison, lay the plugs out in the order they were removed so that if one of them is oil-fouled, drenched in raw gas, or cracked, you'll be able to communicate that information accurately to the man who'll be taking over the reins. See Fig. 10-13, p.191 for comparisons.

To reduce the chances of cross-threading, start the new plugs into place using your fingertips. Avoid using a swivel in the early stages of installation – it can greatly reduce your ability to control the spark plugs' entry angle.

If you feel increasing resistance, back the plug out to inspect its threads (and the threads in the cylinder head if possible). "Butching" a plug into place – forcing it in as far as it will go against ever-increasing resistance – will always cost more to repair than a prudent tow bill over to a machine shop.

Plugs should be turned in tighter than snug; but grunts and bulging veins are not only poor form, they are also ill-advised. A thin coating of *Anti-Sieze* applied to the plugs' threads is certainly acceptable practice, but is totally unnecessary unless the plugs were bound up during removal. Use no more than a dab, twirl it into the new plug's threads. Keep the goo off of the spark plug's electrodes! It will foul the plugs.

9B. Compression Testing

Accurate compression readings require that all spark plugs be removed.

To eliminate the possibility of any damage to the ECU, remove its relay, fuse or both from the fuse box (typically the one in the engine compartment), or disconnect each coil from the wiring harness. Note that the downside to fuse removal is that any trouble codes stored within the computer's memory will be lost. The upside to disconnecting the coil(s) from the wiring harness is that you won't get the shock of your life when you start to crank the engine.

Thread a compression gauge into each cylinder, one by one. Crank the engine for 5 to 10 seconds, just long enough for the gauge to reach its maximum reading. An assistant can do the cranking for you simply by turning the key, or you can set up a jumper wire from the starter's solenoid terminal that you can use to "bump" the engine yourself, §26A, p. 440.

Once recorded, the pressure within the gauge (and cylinder) is released by depressing the button that's located directly below the gauge, Fig. 10-19.

If you have low compression in a particular cylinder, the simplest way to distinguish between a burned exhaust valve and a bad set of rings is to insert the tip of a small oil can through the spark plug opening, preferably far enough in to approximate the center of the cylinder, and then to gently squeeze off two shots of oil – not more than a teaspoon – onto the top of the piston. Avoid splashing the oil! Give it a minute to spread, then crank the engine very briefly, 2 to 6 revolutions. Redo the compression test. If the readings are essentially unchanged, the problem involves a valve. If the readings are dramatically higher, the rings are at fault.

10

Fig. 10-19. Compression tester
Courtesy of Harbor Freight Tools

9C. PCV Valves

These are typically threaded into the valve cover. Use a six-point socket or wrench whenever possible. Put a bit of oil on the new valve's threads before inserting it, use your fingers to start it into place. Gentle torque here; no reason for bulging veins.

9D. Fuel Filters

EFI fuel filters are either bolted to the firewall or are beneath the car, adjacent to or inside the fuel tank. Replacement can be a dangerous mess with anything other than a near empty tank of gas unless the filter resides in the tank and the designers had the good taste to provide access through a plate beneath the trunk or rear seat. In any event, you'll need the factory manual to guide you through the process.

The reason I used the word dangerous is that once upon a time I nearly burned a garage to the ground. It was winter, I used a propane heater to keep the shop warm. I did (almost) all the right things: I turned off the heater, raised the garage door about a foot for ventiation, waited about a half hour, then proceeded to drain the gas tank into an open drain pan. Not long after, a flash of flame came shooting across the floor – right at me. It launched my career: I left the shade-tree world of auto mechanics behind, I opened up a real shop with all the expenses and regulations that entailed. But the whole ordeal was terrifying and could have led to real disaster. The lesson here? If you use a heater, get it the hell out of your shop before you begin. Have a near empty tank. And use a funnel to drain the fuel tank into a five gallon gas can. Cap it!

If you need to "drop" the tank to gain access, wait until the tank is nearly empty, then support it with your jack as you lower it to the ground.

Many of these filters use compression fittings. To avoid stripping anything, use a six-point socket or wrench on the bolt itself, a line wrench on the fuel filter's fitting, Fig. 10-20.

10

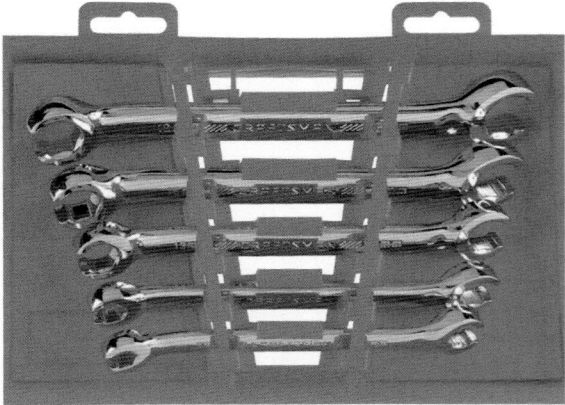

Fig. 10-20. Line (flare nut) wrenches

Chapter 11. Engine Troubles

Chapter 11. Engine Troubles

Notes

Engine Troubles

1. OBD-II

The OBD-II system records numerous codes that never show up through the Check Engine light. Whenever a problem develops with driveability, check your codes first. You may save yourself a considerable amount time and frustration by doing so.

If you're not familiar with the OBD-II system, read §4 through §6, pp. 199-201 before you start here. Once you've done that, you'll know exactly what I mean by "Check your codes".

2. Hard Starting Cold

Fuel-injected engines are designed to start with your foot off the accelerator. An automatic cold-start-fuel-enrichment system is built in to provide the additional fuel needed to overcome those cold combustion chamber walls. On all of the older EFI systems, you'll find a cold-start time switch housed in the cooling system's water jacket, and a cold-start injector mounted in the air plenum, directly behind the throttle body, Fig.11-3, p.218. The injector sprays fuel into the air plenum (Fig. 11-1) in response to a signal from the time switch. The length of that signal is dependent upon temperature.

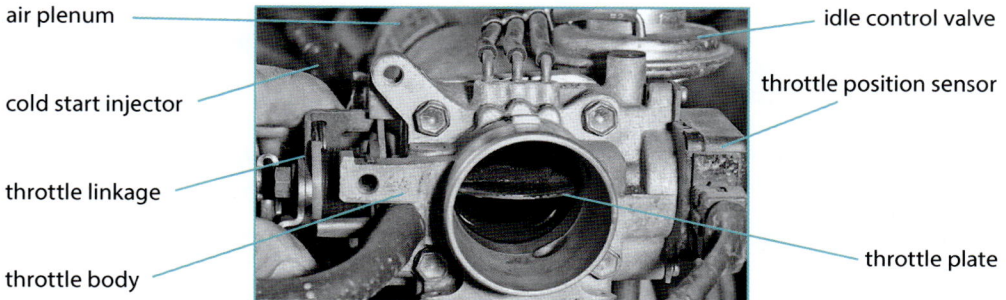

air plenum — idle control valve

cold start injector — throttle position sensor

throttle linkage

throttle body — throttle plate

Fig. 11-1. EFI throttle body. If your throttle plate is as gummy as that, imagine what your pin-hole injectors look like.

A separate cold-start injector is absent on many newer systems; they typically use the ECU to control cold-start fuel enrichment, turning on the individual fuel injectors for longer time intervals in order to inject more fuel. The ECU is prompted by the the cranking of the starter motor and the temperature sensor, §3H, p.198.

When these enrichment systems fail, the act of partially depressing the accelerator or pumping it while cranking the engine will usually turn the tide. The EFI system responds to the throttle's movement by providing more fuel through its standard pathway–the individual fuel injectors located above each combustion chamber.

An engine that cranks strongly but won't "hit" until you depress or pump the accelerator clearly has a problem with its cold-start system. Step one in any diagnosis is to consult with your OBD-II scanner. However, one thing that scanner won't tell you is whether your fuel injectors–cold start or otherwise–are dirty. If you've been buying off-brand gasolines, a plugged injector is a very real possibility. Try a can of fuel injector cleaner, §12, p. 247. Pour it into your tank, then fill up with a premium, name-brand fuel (Mobil or Sunoco, for example). Fit as many *cold* starts (3 to 4 hours between start-ups) into the use of that tank as you can. Even if this does not cure the cold-start problem, it will certainly improve your engine's performance.

One other aspect of the EFI cold-start system is an elevated idle speed during the first few minutes of driving. This is produced by bleeding additional air into the air inlet chamber, over and above the amount moving past the throttle plate. The path for this additional air runs through the idle control valve (ICV), Fig. 10-3 (#5) p. 184. Additional fuel is also provided through the individual fuel injectors: During the warm-up period, the individual injectors are fed a longer electric signal that yields a fuel pulse of longer duration.

If the idle speed is not elevated when cold, the problem is usually with the ICV. Where this valve offers access to its innards, lubricating the sliding plate within often cures the problem. Use CRC 5-56 or WD-40 spray. As with all fuel-injection components, the cold-start injector, its time switch, and the idle control valve are covered by your emission control warranty for two years.

If the car still won't start, the engine is suffering from a lack of fuel, spark, or air. As always, **the** safest way to sort through the possibilities is to consult first with your OBD-II scanner. Note that I keep using the phrase OBD-II scanner rather than OBD-II reader. I don't mean to be a snob; a reader is a perfectly satisfactory tool for many purposes, but if you're debating which one to buy, the scanner offers far more depth to engine diagnostics–if you need it–and right now, you don't know whether you need it or not.

One other essential tool is the factory's service manual, specific to your car, to provide you with accurate diagnostic procedures tied to whatever code(s) you may find.

If you're stuck in a parking lot somewhere with a limited number of tools, you need to first eliminate the possibility of a leaking gas line, because testing for spark in the presence of fuel vapors can be downright explosive!

- If gasoline odors are present, read on.
- If gasoline odors are not present, you can test for spark as outlined in §2B, p. 216
- If you find that a spark is present, check next for adequate fuel pump pressure, §2C, p. 216.

2A. The Smell of Gas

The most likely cause is a missing or defective gas cap. Problems include a brittle or broken seal or a defective (usually rusted) pressure relief valve, but in most cases the problem is simply an ill-fitting aftermarket cap. When the tank is full, it leaks; when half empty, it just spills fumes.

❑ **Fuel Line Leakage** usually results from a cracked flex hose or a loose line fitting. Gas line leaks are easy to locate—just follow your nose.

You will find a pair of flex lines in two areas on every car. One set joins the gas tank (usually located below or behind the rear seat, usually accessible only from beneath the car) to metal piping that runs to the engine compartment. The other set spans the space between those pipes and the engine. These lines are paired–one hose feeds fuel to the engine, the other returns any excess fuel back to the tank. Because of the engine's vibrations and heat, the lines within the engine compartment are the most likely to crack and/or seep. Check each line fitting: one will lead from the firewall or fuel filter to one end of the fuel rail, the other end of the rail will have a line leading back to the return piping. Unless the fuel filter is located near the fuel tank (consult your shop manual), the lines back there seldom leak, living out their lives undisturbed.

The gas lines on fuel-injected cars are under very high pressure (30 psi and higher). Consequently, compression fittings join the flex line to its piping. Even the slightest leak can produce a strong dribble of gas oozing through that crack (engine running). Even with the engine off, the line will be either wet ot stained.

Naturally, gas tanks can leak as a result of rust (old cars) or cracks (cars that have suffered through collisions). These usually puddle onto the pavement behind the rear seat.

❑ **Flooding** results from the combustion chambers getting too much fuel, drowning the plugs in raw gas. It is almost always accompanied by the unmistakably strong odor of gas coming from under the hood.

Flooding can be caused by a problem with the fuel pressure regulator Fig. 11-3, p.218 or a defective EFI temperature sensor (*S*3H, p.198). A bad temperature sensor casignal the EFI computer that the engine is cold when in fact it is fully warmed. This causes the computer to richen the fuel mixture (more fuel, less air) when it needs to be leaned. It can also be caused by a malfunctioning ECU, so check your codes!

The fastest way to restart a flooded engine requires removing the spark plugs, then cranking the engine to eject the raw fuel. New plugs are a nice idea if the electrodes on the originals are crusty. If you have no tools, let the car rest for several hours before attempting to restart.

11

Fig. 11-2. Lisle 2700 ignition tester and
spark plug ignition coil

2B. Testing for Spark

To verify that your ignition system is firing, you'll need an assistant and an ignition tester. The one made by Lisle is the best I've seen. Ignition testers offer the means to compare spark strength across each coil, a significant plus on older cars. More importantly, properly grounded, you won't get shocked!

- Unbolt any one of your spark plug coils and pull it off its spark plug. Leave both the spark plug and the coil's wiring harness connection alone.

- Attach the ignition tester's ground wire to any conveniently placed projection of metal–a bolt head works perfectly well.

- Adjust the tester's spark plug gap according to its manufacturer's recommended setting. Insert the ignition tester's tip into the spark plug side of the coil, Fig. 11-2.

- Have your assistant put the transmission into Park or Neutral with his foot on the brake. Then have him crank the engine.

You should see a pale blue spark between the tester's contacts which–if present–will arc for a brief instant, routinely once per engine revolution. In the absence of a spark, try reducing the gap within the tester. No dice? You either have a bad coil or the ignition sytem is not functioning. Try another coil.

If you find that your engine has a strong, even spark across every coil you try, then your next step is to pursue fuel feed as outlined in §2C, p.216.

❑ **Likely Causes for a No-Spark Condition**

- Blown ignition fuse or engine fuse: The only practical way to determine that a fuse is both intact *and* receiving power is to employ a test light, §10, p.421. Note that the ignition switch must be turned to the On position, that the test light lead must be properly grounded, and that *both* ends of the fuse must illuminate the test lamp.

- Defective ignition relay: ignition relay, engine relay, or main relay: Various names but the same function–delivering power to the ECU and primary engine components. Relays that power the engine's principal circuits are typically housed in the engine compartment's fuse block.

  A quick—though possibly misleading–test is to unplug the relay with the ignition switch in the On position, then to reconnect it. You should hear the relay click both when it is pulled and also when it is reinserted. While this test does not provide absolute verification of proper functioning, it does indicate that at least a portion of the relay is receiving power and that it's responding to that power. A silent relay could be defective, but it can also mean that the relay is not getting power from the ignition switch or that its ground circuit is no longer intact. See §10, p.421.

- A fried fusible link, §11, p.422.

- Defective coil or spark plug: While this only affects a single cylinder, missing one cylinder on a four-cylinder engine can totally disrupt starting the engine.

> The following is not relevant to engines that employ a timing chain.

- **Stripped Timing Belt:** Many engines drive the camshaft with a fiber belt that links it to the crankshaft. Unlike a fan belt's smooth or ribbed drive surface, the timing belt employs a series of notched teeth that lock the two shafts into synchronous motion, with the crankshaft turning through exactly two revolutions for every one complete revolution of the camshaft (Fig. 2-1, p.43). Not only must these shafts turn with a 2:1 ratio, they must also maintain their rotation in lock-step with one another, beginning their synchronicity upon engine assembly at a point called top dead center (TDC) for cylinder #1. Every one revolution of the camshaft returns the engine to this point unless or until the timing belt strips. Even if just one tooth is torn off, the engine will go "out of time", suffering a noticeable loss of power. Most engines won't even start if they're timing belt loses two teeth. If the engine is running while this destruction takes place, other teeth will shortly follow. The degeneration is quick, the result is an engine that won't restart.

Verification of a stripped belt usually begins with two observations:

- That the sound the engine produces as it cranks has changed to a hollow timbre. When the timing belt strips, the camshaft immediately stops turning. Since at least one valve is open at every position of the camshaft, at least one cylinder will have absolutely no compression. This lack of compression provides that telltale change in pitch, and

- That there is no spark.

Due to problems of accessibility, inspection of the belt itself is usually not possible without removing the engine's upper timing cover, but the fact that the belt has stripped can be verified with an OBD-II scanner: Either the camshaft position sensor or the crankshaft position sensor or both will register because the two will be out of sync with each other.

Certain engines provide a window in the timing belt's upper cover (typically a large hole with a rubber dust cover) for seeing the camshaft's TDC timing marks (Fig. 2-1, p.43). Have an assistant crank the engine while you observe the camshaft sprocket. If the timing belt is stripped or broken, the engine's lower end (crankshaft and pistons, crankshaft pulley and fan belts) will be merrily spinning while the camshaft gear stands motionless. You'll hear that telltale empty sound of an engine lacking compression in one cylinder.

Timing belt tension and overall condition should be inspected in accordance with the manufacturer's service schedule.

> There are far more possibilities for a no-spark condition than are listed here, so don't gamble on purchasing parts without first consulting your OBD-II scanner and factory manual, then carefully thinking this through, because virtually all suppliers of automotive parts refuse to accept returns of electrical components.

2C. Testing for No Fuel Supply

The most likely causes for a no-fuel condition are:

- Running out of gas
- Defective fuel pump or fuel pump relay
- Blockage within the fuel filter

Clearly, your first step should be to check your fuel gauge . . .

> I feel compelled to express my concerns regarding the dangers involved in handling gasoline. I have struggled with the wisdom of teaching those who might not be mechanically inclined how to create a situation that could lead to fire or injury. I have concluded that good information is better than none; hopefully, the following will enable you to make a sensible decision.
>
> Although gasoline will usually not ignite on contact with the engine, fuel vapors certainly will–particularly if they come in contact with an extremely hot surface such as the exhaust manifold, front exhaust pipe, or turbocharger piping. The following tests should ONLY be performed on a cold engine. Always have a fire extinguisher close at hand, and a rag to capture/mop up any fuel.

The fuel gauge sending unit and the fuel pump are typically mounted within the gas tank as a single unit. The tank is located directly below the fuel filler door. When the ignition switch is turned to the On position, the fuel pump should emit a distinct hum–not noticeable on a noisy street, not really noticeable from the driver's seat. Ask a friend to turn the ignition switch to the On position while you listen from the rear of the car. A humming sound indicates that the pump assembly is getting power: Its circuit from the ignition switch through the ECU to the fuel pump relay and from there to the pump is working as it should. The hum indicates that the fuel pump is at least trying, but that does not imply that it is actually producing sufficient pressure to drive fuel forward to the engine. For that you need a fuel pump pressure tester or, if you're lucky, just your OBD-II scanner.

Connect your OBD-II scanner and look for codes. Next, study your factory manual. Many of you will have already found your answer: The onboard diagnostics included with your car will tell you whether the engine is not getting sufficient fuel pump pressure and many will tell you why. Others will only find a code indicating insufficient fuel pump pressure and little else. On these cars, typically, you'll need to manually check the status of your fuel pump using a fuel pump pressure tester. They can be rented through most of the larger parts jobbers. Your shop manual will describe where to hook it up and whether you'll need any special service tools.

> Be certain that all of your spark plug coils are securely connected to eliminate them as a source of arcing (spark). Avoid getting any electrical components wet during this test (in particular the battery and coil wiring), by keeping the test as short as possible. Be certain that your assistant understands the words CUT IT OFF!

> EFI fuel pumps deliver fuel at very high pressures (30-60 psi), which is why all fuel line couplings employ compression fittings. Cracking one of these open while cranking the engine can produce a strong stream of gas shooting through the crack. Do not attempt to loosen these fittings!

Since these measurements are typically taken downstream from the fuel filter, a partially blocked fuel filter can produce low readings as well. So make sure to replace the fuel filter and run the test again before you jump into replacing the fuel pump (unless the fuel filter and pump are part of a single self-contained unit.)

In conclusion, don't attempt these procedures unless you feel comfortable with what you're doing. If you're feeling rattled by being stranded, let a professional take over.

A typical fuel rail is illustrated in Fig. 11-3. Fuel flows to the rail from the fuel filter; any excess exits from the rail's other end via the fuel return line. Fuel supply to the rail can be observed at any of the following; choosing the best site depends on accessibility:

- At the fuel filter's outlet
- At the rail's inlet
- At the line that joins the fuel rail and cold start injector (Fig. 11–3, p. 218)

11

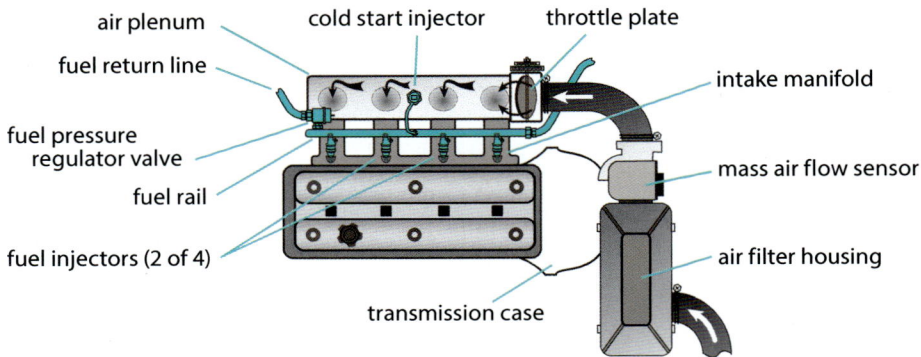

air plenum — cold start injector — throttle plate — intake manifold — fuel return line — fuel pressure regulator valve — mass air flow sensor — fuel rail — air filter housing — fuel injectors (2 of 4) — transmission case

Fig. 11-3. Fuel rail and related parts © PT

- The best spot of all—if your car has one—is at the fuel line's schrader valve. It has a cap and valve much like a car's tire valve. (Fig. 11–4, p. 218) A fuel pump pressure tester can be connected directly to it.

It's impossible to distinguish visually between the stream of gas produced by a healthy fuel pump and one that has weakened sufficiently to produce stalling. Evaluating whether your fuel pump pressure is weak requires a scanner or a pressure tester.

No fuel pump pressure is usually due to a bad fuel pump, but there are other possibilities:

- The fuel pump relay
- A blown fuse or circuit breaker
- A problem with the wiring—the pump's ground circuit is just as likely as its power supply wiring
- Possibly the EFI computer

Fig. 11-4. Schrader valve

None of the above possibilities is possible if the fuel pump hums when the ignition switch is turned to the On position. Relays, fuse and wiring are all taken up in Chapter 16.

2D. So What Else Is There?

Engines with timing belts occasionally "jump time." A stripped tooth on the timing belt alters the moment at which the spark plugs fire relative to piston position. Although the spark plugs are still firing, they're doing so at the wrong time to accomplish anything. A stripped timing belt degenerates rapidly into a no-spark condition, §2B, p. 214.

Another possibility is a bit more complex: Most engines these days vary their valve timing (VVT) in response to engine RPM and engine load (up or down hills, for example.) The only way to determine that an engine's VVT has malfunctioned is with an OBD-II scanner.

Air supply can be cut off by the air flow meter, typically located adjacent to the air filter between the filter and throttle body, Fig. 11-3. Possible causes include electrical disconnection, a jammed door, a problem with the unit itself, or failure of the EFI computer.

Air filters get incredibly dirty with neglect. They can also be filled with seeds and nuts if some hapless squirrel chooses to nest there. In either case, the engine will let you know of its ills through lousy performance and horrible gas mileage long before the air filter could actually choke off air flow.

Fig. 11-5. The dirtiest air filter I've ever seen! (The car looked about the same ...)

3. Hard Starting Hot

Engines almost always start easier when they're already warmed up–their cylinder walls are hot, oil is evenly dispersed. If your engine won't behave, suspect the fuel mixture first. The most likely culprit is the EFI temperature sensor, mistakenly registering that the engine is cold. In response, the EFI computer richens the fuel mixture by lengthening the pulse of each injector, the result being too much fuel. This can lead to flooding the engine in a very short period of time. Check your codes.

4. The Engine That Won't Start, Period

Two broad avenues of possibility lead away from this problem. The first includes all problems that result in a dead battery (or that mimic the characteristic silence of one) when the ignition key is turned to the Start position. The second includes every problem that can produce an engine that cranks strongly, but just won't "hit." It is essential to clarify this distinction between spirited cranking and the silence or slow cranking of a weakened battery, because the list of possible causes for the one diverge immediately and radically from the other, dependent upon that initial symptom. Furthermore, these problems are often intermittent, particularly in their early stages. The car is towed, the mechanic goes out to check it over, and the thing fires off the first time he turns the key. At this point he is left with little more than your input and whatever the OBD-II system is telling him. Your careful observation and communication may save him time and you money . . .

4A. The Sounds of Silence

Case 1 – Sluggish Cranking, or Click-Click, Click-Click-Click, Click

One of the most common misperceptions about batteries is that if the headlights (radio, etc.) still work, the battery can't possibly be dead. Wrong!!! Lights and frills require no more than a trickle of current in comparison to the demands of starting an engine. In my book, a battery is dead when it lacks the gumption to crank over the motor. An engine that turns over with ever-decreasing vigor–one that is especially slow cranking first thing in the morning–will degenerate over a period of days or months into stone-dead silence or a repetitive click-click, click-click-clicking sound when the ignition switch is turned to

11

the Start position. The clicking you hear is the engagement/disengagement of the starter solenoid or starter drive, caused by a lack of sufficient current to engage the starter drive and hold it there.

There is another type of clicking, typically much fainter, often much more rapid, that requires no key at all. These sounds will be coming from a fuse box or junction block either inside the car or under the hood. The clicking originates from one or more relays (electric switches) which, like the starter drive, are losing the current they need to maintain them in their "excited" state. Since more and more cars are being engineered with features that function while the owner is away (anti-theft devices, automatic headlamps, driver's seat motors), these relays will increasingly foretell battery discharge from a distance, even with no key in the ignition.

Typically, the cause of either slow cranking or repetitive clicking is a discharged battery. The five most common causes for battery discharge are:

- Leaving a door ajar overnight, or headlights on for several hours
- Corrosion on one or both battery terminals, or a loose terminal, *§*A3b, p.227
- An old battery
- A battery that's low on "water" (electrolyte), *§*A3a, p. 226.
- A defective alternator: The charge warning light usually alerts the driver to this predicament 30 miles (+/-) before the battery goes dead, plenty of time to investigate before you get stranded.

Let's say you wake up to a dead battery with no prior warning signs of a weakening battery (sluggish cranking). Start by investigating whether any doors have been left ajar, whether the interior lights are on, etc. Remember that trunk lights, underhood lights, etc. also draw down the battery, as will some owner-installed radios and other equipment that can play without a key. If that's what you find, then a simple jump-start should help you on your way, *§*A1, below.

> Never use the battery as a tool tray. If a wrench, screwdriver, or other metallic object accidentally bridges the gap between the battery's positive terminal and any adjoining metal surface, it will create a short circuit, producing an intensely hot spark with sufficient energy to melt (weld) the offending steel right into place. The most likely result is a severe burn to your hand as you fumble to remove it. Other possibilities include fire or a meltdown of the wiring harness. An exploding battery is not out of the question.

Battery terminal corrosion develops gradually over time and you won't experience that first no-start for quite some time. That's not to say that sluggish cranking won't be a part of its evolution, but if you're seldom under the hood and that drawdown is subtle, you might well be caught totally off guard. Cleaning battery terminals takes time and a few tools, *§*A3b, p.227; jump starting might be all you need to do until you've got the time for repairs. But the smart money is on taking those jumper cables along for the ride.

A1. Jump Starting a Dead Battery

For your own safety, please read the following section *in its entirety* before going anywhere near your car.

> **Jumper cables must be connected positive terminal to positive terminal, negative to negative. If you are unsure which terminals are positive, do not proceed until you do.** By connecting them backwards (positive to negative, negative to positive) you will at best fry the battery's fusible links or burn yourself as you try to pull the jumpers off. Other *likely* possibilities include frying the alternator and/or one or more of the car's computers. If your battery needs a jump-start, *get it right the first time!*

All batteries mark their terminals in some fashion, usually by stamping a + sign into the case, adjacent to the positive terminal. The positive terminal is usually covered by a rubber boot, the negative terminal isn't. The cable that leads away from the negative terminal always connects to the chassis (body) of the car. The positive terminal has more stuff hanging off it: fusible links, possibly a fuse block.

Examine the cables you intend to use. If they appear flimsy, cheezy, or cheap, get a better set: The amount of current that's going to flow through those things might surprise you. **A dinky set of cables can easily become hot enough to burn you.** If you intend to purchase a set, invest in #8 gauge cabling or better (#6 gauge).

❑ **If your jumper cables are two distinct wires, not joined in any fashion:**

- Use one of your cables to connect the positive terminal of the dead battery to the positive terminal of the other. Rock the jaws of each clamp as it's installed in order to ensure a good, firm bite.

- Use your second cable to clamp the negative terminal on the good battery, rocking the clamp as described above.

- The one remaining negative jumper end is secured to the car with the dead battery. The least preferred point of contact is the battery's negative terminal, because a spark can jump between the cable and terminal as the two meet. This spark has the potential to ignite fumes coming off the battery, although this is far less likely to occur these days because most modern batteries are totally sealed units. Choose either of the engine's two hoisting hooks or the car's negative cable where it bolts to the chassis. Again, rock the jaws of the clamp to scratch your way into fresh metal.

> Take care that the one remaining free end does not accidentally come in contact with either of the positive terminals or with any uninsulated portion of the cabling that joins them. Should you mistakenly connect a battery's positive and negative terminals together, even at a distance of eight feet away at the far end of your set of cables, that accidental contact–a short circuit–can conceivably fuse those clamps together.

❑ **If you are using a set of cables that are joined together as one unit:**

- Paired jumper cables are always color-coded to prevent accidental crossovers; the red clamps are always used on positive terminals.

- To reduce the likelihood of sparks, lay your cabling out on the ground with the positive and negative clamps laying apart from one another–and not in a puddle of water! Install one pair of clamps–both the positive and negative–onto the dead battery before moving on to the "hot" battery. Make your third connection to the hot battery's positive terminal, your last connection to a chassis ground. Rock the jaws of each clamp as it's installed in order to ensure a good, firm bite.

In forbidding temperatures, you will want to maximize the amount of current available. Turn off all accessories on both cars. Start the car with the good battery and, if possible, have an assistant raise its engine speed to 1500-2500 RPM (gearshift in Park or Neutral) and hold it there–this is to get its alternator humming. In temperate climates, these steps are usually unnecessary. Regardless, all accessories should be switched off on the car with the dead battery. Naturally, any procedures that you would ordinarily follow in starting the car–pushing down on the brake pedal, for example–should be performed.

If you find that the cranking is too sluggish for start-up, recheck your clamps to be sure that each makes good contact. **But be careful–use gloves or a potholder– because the jumpers might be griddle hot. If they are, the dead battery or some other major component on that car could have a dead short. DO NOT CONTINUE. Remove the jumpers without delay, negative cable first. Have it towed to your mechanic's garage and be sure to tell him what happened.**

If the jumpers are just warm and your connections seem good, then either the cables are not heavy duty enough or the positioning of that last negative cable clamp is unsatisfactory. Try moving it to a body ground rather than the engine's hoisting hook. If all else fails, there's always the negative terminal. Assuming a successful startup, remove your jumpers without delay, starting with one of the negative clamps.

Reposition the batteries' positive terminals inside their rubber boots.

Some batteries are so dead that they just won't sustain the engine when the battery cables are removed. This can be immediate, but it might also occur three blocks down the street. In this case, you'll need to:

- Charge the battery where it sits
- Remove the battery for recharging elsewhere
- Replace the battery
- Have your car towed

For those hapless souls who have a dead battery and no jumper cables, all is not lost as long as you happen to have a stick-shift transmission and a hill, the added good fortune to be parked at the top, and the good sense to be facing downhill.

- Turn your ignition key to the On position (not Accessory).

- If there are any procedures that you would ordinarily follow in starting the car–pumping the accelerator, for example–use them.

- Put the car in second gear with the clutch pedal to the floor. Release the handbrake and start to pray that you don't screw this up!

- When you hit 10-15 mph, let the clutch out as smoothly as possible while you rev that engine into life! Should you fail, immediately ram the clutch pedal back to the floor, leaving the thing in second, and pick up a bit more speed. The car is going to slow dramatically as you let out the clutch, in herky-jerky fashion.

You'll have as many attempts as the length of the hill allows.

A2. Battery Testing

The test that follows will differentiate between a good battery and a bad one, but only when used appropriately. It should not be considered indisputable without first verifying adequate electrolyte level and clean battery terminals, *§*A3, p.224.Also, keep in mind that the following test means nothing on a dead battery: Because it's already discharged the battery will always flunk.

❑ A battery load tester houses a heating coil (big resistor) and voltmeter. Attach it to the battery's positive and negative terminals (red to positive, black to negative). To read the battery's strength, flip the tester's load switch, hold it there for 10 to 15 seconds while you watch the voltmeter. The battery is considered defective if the needle dips too deeply into the weak range.

Fig. 11-5B
battery load tester.
Courtesy of
Battery Tender

A better measure is the actual voltage recorded under load: In our shop we consider 10.2 volts to be the lowest acceptable reading during that first 15 seconds, but ours is a temperate region. Naturally, the tougher your winters, the higher that number should be (10.4–10.6 volts). Heavy-duty batteries carry higher CCA ratings (cold-cranking amps) and should therefore read higher on the scale.

❑ A nice corollary to the above test is the battery sulfation test, in which that same voltmeter is used in conjunction with a high-output battery charger. (No need to go out and buy one–they're expensive. But if you do have access to one . . .) With the charger cranked to its maximum output (40 amps +/-), observe the voltmeter's reading. If it climbs above 15.5 volts within the first 30 to 60 seconds, that battery is toast. However, be aware that when this test is performed immediately following the fill-up of a battery that initially arrived low on water, roughly nine out of ten will flunk.

A battery that's low on electrolyte should be topped off with distilled water, *§*A3a below, then slowly charged (trickle charging). Wait for several hours before load testing the battery. If it flunks at that point, get yourself a new battery. But assuming it passes muster, don't consider it reliable until you test it the following morning.

Do-it-yourselfers with a limited tool kit: Top off the battery (*§*A3a, pp.226), then jump-start the car (*§*A1, p.221). Recharge the battery by driving for 30 to 60 minutes–preferably non-stop on a highway with all nonessential accessories turned off: the air conditioner especially,

11

but also the heater fan motor or rear window defroster draw heavily on the alternator, reducing the amount of charge available to the battery. End your road test where you can buy a battery if need be.

Should your car need a new battery, be sure that corrosion rings are installed as part of the job. If corrosion is evident on the terminals, have them replaced. Corrosion on the battery holddown strap speaks strongly for its replacement as well.

Fig. 11-6. Installing corrosion rings

❑ The load tester described above can also be used to test the alternator (assuming that the engine will actually start and run). The top end of the scale is devoted to alternator output.Raise the engine speed to 1500-2000 rpms. The load tester should show the alternator putting out something between 13.8 and 15 volts. Anything below that indicates a weak to useless alternator; anything above 15 volts indicates that the battery is overcharging. See 𝓢3 p.413

In a perfect world, the terminal in Fig. 11-6 would be replaced. That white powder encompassing the terminal's base is corrosion. Same for what's visible on the locknut inside the wrench.

As with anything else, you get what you pay for. Since there is tremendous variance in battery quality and value, a bit of research makes sense. But be advised: Most shops carry only one brand; if you want something different, you'll probably have to supply your own.

A3. Battery Servicing

> For your own safety, please read the following section in its entirety before going anywhere near your car. Begin with 𝓢 3C, pp. 102-106.

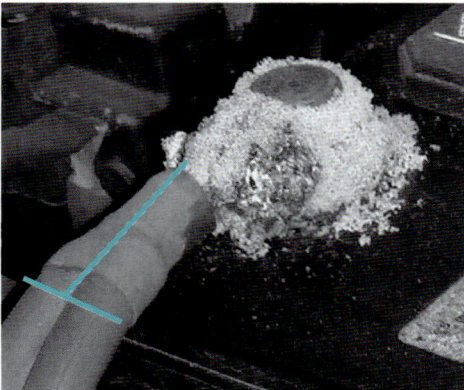

Fig. 11-7. The extent of this corrosion requires that the wiring harness cover be cut back until clean copper cable is visible. A universal terminal can then be installed. Attribution requested.

The purpose of battery servicing is twofold: to assure that the battery has enough electrolyte–the liquid contained within the battery: a mixture of sulfuric acid and water–and to remove corrosive salts that have been deposited on its terminals and other nearby parts.

If your battery is a totally sealed, maintenance free unit (Fig. 6-8, p.103), then skip to the next paragraph. Everyone else, stay with me. . . The continuous process of battery discharge and recharging leads to the gradual loss of electrolyte. Naturally, a battery that is run low on electrolyte will suffer for it. If the upper portion of its plates (Fig. 6-10, p.104) becomes exposed to air, a chemical process called sulfation sets in, reducing the battery's ability to accept and hold a charge.

Corrosion at either terminal can produce high resistance to current flow, reducing the battery's ability to deliver its punch to the starter, as well as the inverse problem of delivering the alternator's output to the battery, §3, p.413.

The word terminal refers to the clamps that actually slide down over top of the battery's two posts. Since these are either an integral part of the car's wiring harness or are bolted directly to it, corrosion in the terminal end will eventually spread into the harness, Fig. 11-7.

> Never permit a wrench, screwdriver, or other metallic object to bridge the gap between the battery's positive terminal and any adjoining metal surface of the car. Pay attention to the free end of any tool you might be using for removal and reinstallation of the positive terminal, and while cleaning the positive post, to assure that the tool always moves exclusively through clear space.
>
> Any metallic bridge from the battery's positive terminal to an adjoining metal surface will short out the battery to ground, creating an intensely hot spark with sufficient energy to melt (weld) that tool right into place. The most likely result is a severe burn to your hand. Other possibilities include fire, a meltdown of the wiring harness, or explosion of the battery.
>
> The same precautions hold true for every component that's tied to the battery's positive cable–the alternator, starter motor, the rear side of all fuse block assemblies, and the ignition switch. To avoid mishap, always disconnect the battery before removing any of these components.
>
> - Most manufacturers shroud their battery's positive terminal in a rubber cover; be certain that it has been reinstalled.
>
> - For obvious reasons, never use the battery as a tool tray. To avoid the possibility of explosion, never smoke in the vicinity of a battery.
>
> - To avoid the consequences of arcing, always disconnect the battery's negative cable first, then the positive. On reassembly, reconnect the positive terminal first, then the negative.

11

A badly corroded terminal end can be replaced either by unbolting it from the harness or by cutting it off, using a factory replacement or a universal terminal.Unfortunately, the complexity of most modern-era wiring harnesses means that their harness ends– particularly their + side–are not available as an individual part. You either buy the whole harness or clean up what you have.

Thousands of perfectly good batteries are thrown away each year because some less-than-thorough mechanic didn't bother to clean (with emery paper) the positive and negative posts prior to testing. Battery terminal cleaning or replacement should be performed whenever the white, dusty deposits of battery corrosion start to accumulate. You'll find generic terminals anywhere; the original factory parts are available from the dealer and, in many cases, from Amazon as well. Don't condemn a battery whose harness connections are buried under these deposits unless you see the problem recurring after a thorough battery service that includes a trickle (slow) recharge.

Those white, dusty deposits are a salt of sulfuric acid. Much like cancer, they eat away at both metal and plastic parts, blistering off any paint that might stand in the way. Dilution in water (including the moisture in your skin) produces sulfuric acid. Control these deposits early on with thorough servicing followed by installation of corrosion rings.

If your battery is going dead, the battery posts and terminals should be sanded whether corrosion is evident or not. Oxidation between terminal and battery post can greatly reduce current flow.

A3a. Inspection of Electrolyte Level

These levels should be checked once a year.

The typical battery has two rectangular plastic covers on its upper surface, Fig, 11-8. To access the six cells housed beneath, these covers must be pried up and off using a screwdriver. Variations on this include batteries with six threaded plugs as well as maintenance-free batteries that can *usually* be distinguished by the fact that their upper surface is flat, sealed, with no removable covers. Be advised that there are any number of batteries currently tooling around which, even though clearly marked as "maintenance-free," actually do have removable upper covers and benefit from annual replenishment.

> Electrolyte burns holes in clothes and paint; it's tough on cuts and excruciating in the eyes. Wear safety goggles and gloves; wash with soap and water as needed.

To prevent contamination of the battery's innards, first clean off any grunge that may have accumulated on the battery's upper surface. An aerosol can of battery terminal cleaner or baking soda chased with water and a brush work best. Brake cleaner or electrical parts cleaner also work nicely, but tend to dissolve any painted markings (name brand, cautionary statements, etc.). If the top surface of the battery and/or its terminals are buried in corrosion, remove the battery before cleaning: Once corrosion begins to take hold anywhere, it will continue to eat away at whatever surface it's attached to. If that seems excessive, remove what you can, then hose down the entire area until your satisfied.

On cars that carry replacement batteries, battery cap removal is often blocked by the holddown strap. Since removal of the strap often litters the top surface of the battery with corrosive dust, remove it before your cleanup.

Once the battery caps are removed, you will see six chambers (cells with a hollow plastic tube projecting downward into each cell, Fig. 6–10, p.104. Each cell must be inspected individually–their levels are independent. In a perfect world, you would top off your battery with distilled water. As a practical matter, more than 90% of the shops on this planet use tap water.

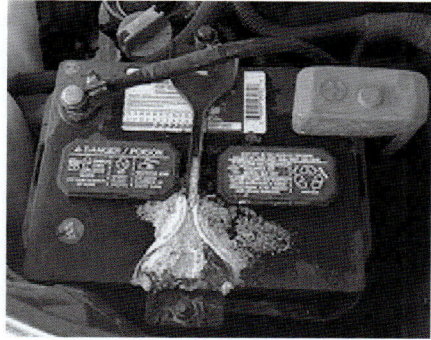

Fig. 11-8. An overfilled battery.
Attribution requested.

If you can see the upper edges of a series of thin plates stacked vertically beneath any one of these openings, then the electrolyte level in that cell has been too low for some time. Sulfation is the likely result, reducing the battery's ability to accept and hold a charge.

Following filling, the battery's upper surface needs to be dried with a *disposable* rag–standing water becomes dilute sulfuric acid in the presence of residual corrosion (on the battery strap, for example). Take care not to brush any crud into the cell's openings. Reinstall the cell covers, then the battery strap. If your battery terminals are clean, install a pair of corrosion rings now in order to avoid the hassles and cost of later repair.

Take a good look at Fig. 11-8. The corrosion on the battery strap is obvious. A bit more subtle is that wetness across the top of the battery. That moisture is what's left of the electrolyte that remained after all of the damage you see had already occurred. Whenever you leave a shop after servicing, check the top surface of your battery. If it's wet, return the car immediately, have the shop level out the the battery's electrolyte and remove every bit of moisture from its surface and all neighboring parts.

> **The electrolyte level should never be raised higher than the lower edge of each tube. Overfilling a battery can produce repeated boilovers of sulfuric acid until a proper level is attained. The resulting corrosion is far more costly to deal with than simply buying another battery.**

A3b. Terminal Cleaning

Since the real business of electron transfer takes place across the surfaces that are concealed when the terminal slips over its post, the only effective way to service a battery requires that both terminals be removed. You will probably need every one of the following:

- Wrenches or sockets of the proper size for (1) disconnecting the wiring harness ends from their terminals (where appropriate), (2) loosening the terminals, and (3) removing the battery hold-down strap.

11

- Slip-ring pliers to twist the terminals off their posts, or a battery terminal puller if the post wants to twist as well (A puller is a needless expense for most; typically the battery's posts don't twist unless the battery is junk to begin with.)

- Eye protection–preferably safety goggles–and disposable gloves.

- An aerosol can of foaming battery terminal cleaner.

- Coarse emery paper, a small wire brush and a pocket knife for cleaning/scraping. A battery post cleaning tool is nice, but frivolous for the weekender, particularly because they always leave some corrosion behind.

- Baking soda, water and rags; compressed air if you have access.

- A new set of corrosion rings

- New terminals, if needed. Positive and negative battery posts have different dimensions, the positive post being larger than the negative. Most universal terminals are made in only one size–the wrong one. Never try to beat a smaller terminal down overtop of a larger battery post. Doing so can destroy the battery.

- A heavy set of wire cutters if the terminal and wiring harness are formed as one piece and you have no option but to replace the terminal, Fig. 11-7, p. 224.

  Replacement of the negative terminal is usually no big deal because you are only cutting through cabling. However, if the positive terminal carries several undetachable fusible links, a fuse block, or some other complication, don't start chopping without first consulting a professional. Typically–and fortunately–the more complex the wiring harness end, the more likely it is that it will bolt up to a terminal rather than being formed as part of it.

- If needed, a new battery hold-down strap, available from your dealer. Universal hold-downs are available–a few are fine, most are difficult to fit. Whatever you choose, you might need new hardware (nuts, a bolt, a pair of washers, possibly a "J" bolt) to replace the original corroded stuff.

- An aerosol can of battery terminal coating (neat) or a layer of grease (yuk!)

- An aerosol can of plastic undercoating is needed whenever corrosion has worked its way into surrounding metal parts.

> The battery's cell openings must be closed throughout the following procedure, their covers in place and tightly secured to prevent cell contamination.

Let the terminal cleaner do its work for you. Spray it on, let it soak and bubble for 5-10 minutes, then initiate your cleanup with the wire brush. With terminals removed, use emery paper for the facing surfaces of post and terminal. Repeat this process until you are looking at clean metal. If the terminals are corroded and can be removed without butching the wiring harness, replace them–no sense wasting time trying to clean corrosion from a $5.00 part.

The upper surface of the battery case should be cleaned as well—oil and dirt can bleed down a battery. We use battery cleaner chased with water and compressed air. Brake cleaner and a disposable rag work fine, too, as long as you don't mind obliterating whatever information is painted on the battery's top surface.

On reassembly, be certain to install corrosion rings. Do not overtighten the clamps: Doing so can crack the terminal. Coat both terminals. If the battery strap is in bad shape, replace it–the corrosion it carries is guaranteed to spread to the terminals you just worked so hard to clean. If you need to reuse it, clean it well, then bury any corroded spots in plastic undercoating spray.

If you're dealing with a nightmare brought on by gross overfilling or a cracked battery case, one where the corrosion has taken hold below the battery within the body of the car, remove everything above it, then clean the area with a wire brush (wear those safety goggles!), terminal cleaner, and water. Once it's *bone* dry, bury the area in undercoating spray.

A4. Buying Batteries: A Note on Pro-Rata

Seems to me that every time somebody starts throwing Latin around, I'm about to get screwed. Would you expect anything less of the battery manufacturers?

When applied to batteries, the term *pro-rata* means that for every month you've owned it, the value of your battery has decreased, regardless of how long your warranty period lasts. Saying it another way, your "warranty" period simply means that at the end of its term, your battery will be worthless.

As an example, consider a battery that initially costs $150 and carries a warranty of 36 months. If it fails after 26 months, even your selling dealer won't replace it free and clear; instead, they'll give you (36–26)/36 = 10/36ths of its value *as a trade-in* against the price of a new battery. Such a deal!

A5. Sluggish Cranking Revisited

A small percentage of slow crankers are produced by excessive current draw through the starter, either because the starter is on its way out, or because of excessively high resistance through the positive or negative battery cable. This can be due to a loose terminal end at the battery, or the starter, or on any of the following three connections:

- negative terminal to chassis ground
- either end of the chassis-to-engine ground strap, or
- to deterioration of either the positive or negative battery cable. That deterioration is visible in its copper: It will appear brittle, kind of gray, and dusty. The "quick and dirty" test for all of these is to attach a voltmeter across the battery's positive and negative terminals and watch the voltage drop as the engine is cranked. Readings below 9.5 volts reflect a problem with high resistance through the starter or cabling.

11

Perish the thought, but slow cranking accompanied by low voltage readings can also mean excessively high resistance within the engine itself, as in the earliest phases of engine lockup. If the engine has been consistently run very low on oil, investing in a new starter and cabling may have no effect on your engine's startability. You can test this by removing all of the spark plugs, then turning the crankshaft with a breaker bar and socket. If the engine spins freely through 360°, then you're back to the starter or corroded cabling. If you experience binding at any point or if it's consistently tight while turning, see §6, p.138. If you're unsure, have a machinist test it for you.

Case 2 – The Single Clack

You turn the key and hear one distinct clack. This is not a timid click, but a solid slap of metal against metal. Nine times out of ten, a defective starter will be to blame. But cover your bases:

- Inspect the battery terminals for corrosion or looseness, §A3b, p.227
- Check the battery's electrolyte level, §A3a, p.226
- Load test the battery before condemning the starter (§A2, p.223)–a weakened flow of current can do strange things.

Are you stranded somewhere? Turn the key back to the Off position and try again. Quite often the car will start if you repeat this process two to twenty times. If this fails, try smacking the starter case once with feeling! Use a hammer and a long screwdriver or steel bar to focus the blow. (There is a set of carbon brushes (electrical contacts) inside the starter that are designed to slide inward toward the armature as they wear. Toward the end of their lives, they occasionally get stuck and prevent the starter from cranking.) Don't make a habit out of this procedure; the cure is temporary at best. **Don't miss and smack the wiring . . . It's electrified! Read the** Caution on p.225.

If the starter has to be replaced, ask your mechanic to comment on:

- The condition of the starter's B+ cable (the heavy cabling connecting the battery's positive terminal to the starter). The B+ cable can fry over time, producing ever-increasing resistance to current flow—the copper will appear grayish and "crinkly." Excessively high resistance produces sluggish cranking even with a new battery and starter.

- The condition of the flywheel's teeth. When energized, the starter's drive gear is thrown rearward to engage the teeth that encircle the flywheel, Fig.14-6, p.365. These teeth take quite a pounding–they can chip, crack, wear down, or suffer deep pock marks. Chewed-up teeth can produce either a single clack or a whirring, grinding sound.

Cable and flywheel problems are not particularly common, but it's far more cost-effective to know about related problems *before* the new starter goes on. Further, by asking the question you intimate to your mechanic that his careful observation has been requested and will be appreciated. Note that he'll be unable to comment on the condition of either one until the original starter is on the bench.

> **Never remove a starter without first disconnecting the battery. Read the** Caution on p.225.

Case 3 – Absolute Silence

Possibilities include:

- A stone-dead battery (read Case 1, p.219)
- Defective starter (*S*A5, p.229 and Case 2, p.230)
- Melted fusible link – If something real bad is happening electrically on your car, the fusible link responsible for supplying the ignition circuit can melt. Fusible links are typically located right at, or adjacent to the battery's positive terminal. They can look like wiring coming directly off the terminal, disappearing into the wiring harness; or they can be mounted in a junction block near the battery, taking the form of big fuses or short stretches of wire that plug into the junction block. To test for meltdown, inspect for an open circuit using a test light, *S*12, p.425.
- Defective ignition relay or starter relay
- Defective ignition switch (the electrical portion, attached to the rear of the cylinder and key)
- Defective neutral safety switch (automatic transmissions only) – The neutral safety switch prohibits the engine from being started in any gear other than Park or Neutral. Try jiggling the gearshift in both Neutral and Park, alternating between them and attempting a restart following each change in position. If the car now starts, the problem lies with the neutral safety switch, either because of worn contacts (requiring its replacement) or to misalignment (requiring an adjustment). If the car starts consistently in Neutral, there is no compelling reason to replace the neutral safety switch–as long as you consistantly remember to either engage the parking brake or keep your foot on the brake pedal.
- Defective clutch neutral switch (manual transmissions only) – Its purpose is to prevent the engine from being started unless the clutch pedal is mashed to the floor, thereby assuring that the car won't take off if it happens to be in gear. The clutch neutral switch is typically located adjacent to the clutch pedal arm, high up behind the dash. Don't confuse it with any cruise control paraphernalia–the neutral switch has two fat wires leading away from it. Power should pass through the switch only when the clutch pedal is depressed. Bypassing it with a jumper of equivalent thickness will either result in no change whatever or an engine that roars to life (when the ignition switch is turned to Start.)

With the exception of the last two possibilities, diagnostics should begin with the battery. Assuming that it passes muster, (*S*A2, p.223), you can differentiate between a bad starter and the electrical circuitry that feeds it by removing the starter motor from the equation: The starter has two wires leading to it–the heavy B+ cabling from the battery's positive terminal, and a thinner wire from the ignition switch by way of the neutral safety switch. Disconnect the thinner wire from the starter's solenoid. Put the probe from a 12V test lamp into that disconnected wire's terminal end. Connect the test lamp's ground wire to any convenient ground on the chassis. The lamp should not be lit. Now have an assistant turn the ignition key all the way over to the Start position. (Gearshift in Neutral or Park, or clutch pedal pressed to the floor).

11

- If the circuitry is sound, the test lamp will light, just as if power were being applied to the starter. The starter is defective.

- If the test lamp does not light, there's an open circuit between the battery and the end of that starter wire. You'll need the service manual to diagnose this problem; there's a lot of wire with numerous connector plugs involved, plus each of the components to investigate: ignition relay, ignition switch, neutral safety or clutch neutral switch, and the starter relay to name a few . . . Read §2, p. 410 and §10, p. 421 before you begin.

Before you begin that search, check to see if other systems are involved. With the ignition key turned to the On position, check the heater fan, headlights, wipers, etc. to determine whether the failure is limited to the starting system. If the problem encompasses more than just the starter, then the search proceeds from the battery's positive terminal through the fusible links and on to any individual relays or fuses that have been shown (by component failure) to be involved. Use a testlamp to search for power (§10, p. 421) at each point along the way from the battery to the dead spot. As a general rule, the more things that don't work, the easier the problem will be to locate, simply because the open circuit (meltdown) will be closer to the battery.

With multiple systems involved, I'd be looking at terminal ends within the junction blocks that contain some or all of the failed systems. The problem smells suspiciously like a melted harness plug. You'll need to remove the back cover of the junction block to gain access. See §23, p. 438.

4B. It Cranks Strongly, But It Just Won't "Hit"

An engine needs four things to run: fuel, air, a spark, and compression. An engine that lacks any one of these will simply not start no matter how well it cranks. Furthermore, the first three must be balanced in correct proportion and delivered at the proper time for startup to occur. So there are three broad groups of problems here:

- Those caused by an absence of fuel, spark, or air.
  - First, check for codes . . . (§6, pp. 200)
  - Testing for spark (§2B, p. 214)
  - Fuel pump pressure testing (§2C, p. 216)
  - Air filter and throttle body, (§2A, p. 188); MAF sensor (§3A, p. 196); TP sensor (§3C, p. 196)
- Those caused by an *imbalance* in mixture or timing
  - First, check for codes . . . (§6, pp. 200)
  - A defective fuel injector (§6C2, p. 238)
  - An excessively rich mixture (§6C3, p. 239)
  - Mechanical problem within a cylinder (§6C4, p. 239)
- A troubled engine:

      – Low or no compression (*S*2D, p.192); *S*9B, p.204)

      – Stripped timing belt (*S*6, pp.42-44),

      – Faulty timing chain tensioner (low or no oil pressure) (*S*4B, pp.131-132)

5. Poor Fuel Economy

5A. How to Check Your Mileage

To calculate your mileage, start with a full tank of gas. Fill the tank where you typically go and try to use the same pump for both this filling and the next. (The point at which the pump clicks off can vary from pump to pump). Don't try to add any extra, you want the pump to control when it wants to shut off. Write down your mileage.

There's no need to drive till you're almost out of gas; fill up again when your tank is about 1/4 full. Same pump, same click . . . Write down the mileage and the number of gallons it took to fill.

Subtract the mileage from your first fill-up from today's mileage. Divide that number by the number of gallons you just pumped.

To get a more accurate measure, average your usage over three tanks, (Mileage will vary drastically between your rush-hour-work-a-day world and a weekend road trip.) You don't need to fill the tank on that second fueling, nor do you need to write down your mileage; just keep track of the number of gallons you pump. When you do fill up the tank at stop number three, make note of your mileage. Subtract your mileage from fill-up number one and divide by the total number of gallons pumped over fill-ups two and three.

To maximize your fuel economy, never try to squeeze in that extra 19 cents worth so as to round off your purchase to the nearest dollar. Most of that gas ends up being lost to the vent tube atop the gas tank filler neck.

5B. Probable Causes

I can think of at least nine causes for lousy mileage, any one is just as likely as another:

- Low tire pressures
- Cheap gas
- Cold weather
- Hot weather
- A brake that's binding (That includes you left-footed folk who "ride" the brake pedal)
- A slipping clutch (manual transmissions only)
- Teeneage drivers and other "enthusiasts"
- Improper front- or rear-end alignment
- A problem with tuning

I've purposely placed tuning at the bottom of the list because of the incredible amount of money that's wasted each year on the assumption that "the timing must be off." In this era of electronic ignition and computer controls, the odds of engine tuning being the principal cause of poor fuel economy are constantly diminishing.

11

5C. Low Tire Pressures

A single tire, underinflated by 5 to 10 psi (pounds per square inch), can drop your mileage by 2 to 3 mpg (miles per gallon). Imagine what a whole set of underinflated tires will do! Studies have shown that at least half of our nation's tires are chronically underinflated. Tires deflate when they're cold, just as a balloon would if you parked it outside. Check those tires around Thanksgiving, and check 'em again regularly as you struggle through winter. Tire pressures should be checked about once a month, particularly when the weather turns cold. If you're thinking that your tire pressure warning light will monitor that for you, think again. Better yet, read p.94.

5D. Cheap Gas

You've heard the maxim, "Penny wise, pound foolish." I have never understood why otherwise intelligent people would pump low-octane, no-name gasolines into the tank of their $25,000 automobile and call that "economy." Let me give you three good reasons why you should insist on using name-brand fuels of the octane rating recommended by your manufacturer.

❏ The better grades of gasoline have better additive packages. These "detergent" additives continuously clean the pinholes within the fuel injector nozzles through which fuel passes on its way to combustion. A clean injector produces a nice parasol of atomized gas, whereas the gum and varnish deposits that settle in on a dirty injector produce a dribbling stream of largely unburnable fuel. The atomized parasol burns much faster, more powerfully, and cleaner.

❏ Premium fuels offer more consistent quality than their no-name cousins. And just as their detergents work to clean fuel-injector passageways, they also scour the engine's combustion chambers, valve stems and seats, and spark plug electrodes, producing a greatly diminished rate of carbon buildup. Carbon deposits cause rough idle, hard starting (especially cold), hesitation under acceleration (cold and hot), and power loss on hills.

❏ The EGR valve (exhaust gas recirculation) has been around in one form or another since 1974 on virtually every engine in this country. Its purpose is to draw off a portion of the gases that exit the combustion chamber and return them to the intake manifold to lower combustion temperatures, thereby decreasing nitrogen oxide emissions. Just as the unburned hydrocarbons of lower octane fuels deposit themselves on the interior walls of your combustion chambers, they also take up residence within the EGR's plumbing, gradually choking off the system and defeating its purpose. Plugging the EGR system reduces both mileage and air quality.

As if that were not enough, these deposits occasionally adhere to the valve seat within the EGR valve, preventing it from closing at idle. The result is a large intake manifold vacuum leak, severe roughness at idle, and stalling. Just another trip to the repair shop at a time in the car's life (70K and up) when you might want to forego it! Fortunately, EGR valves are electronically controlled by the ECU and will signal their discomfort with a code.

5E. Cold Weather

With the onset of winter, at least three gremlins come out to play:

❑ Engines are much harder to start in the cold and a lot tougher to keep running – at least until their cylinder walls reach operating temperature. Cold start enrichment becomes a necessity: Fuel injectors begin to fire sooner and for longer intervals, engines with variable valve timing (§1C, p.185) accommodate this by opening sooner, and an idle compensation (IC) valve maintains idle speeds at several hundred RPM higher than when the engine has warmed up. All told, these cold start features dump significantly larger quantities of fuel into the cauldron relative to summer's needs, and this goes on for roughly five minutes every time you fire it up.

❑ Brakes love to tighten up in winter. Front brake calipers get so cold that they actually shrink a bit. It might take three or four blocks – in some cases three or four miles – before they warm sufficiently to stop dragging (binding). And how 'bout those handbrakes? Folks in Minnesota and Maine know all about not using their handbrakes during the winter months; otherwise, their handbrake cables could easily "ice" the car till spring! But even in more southern latitudes, handbrake release can be fairly slow, particularly when those cables get to be four and five years old, because their return springs are puny to begin with and weaken with age. Most handbrake cables are sealed within a sheath, making them virtually impossible to lubricate.

❑ We've already discussed tire deflation, §5C, p.233.

5F. Hot Weather

Those of you who have blessed yourselves with air conditioning know all too well that your mileage drops off about as precipitously as your engine's performance when you turn that baby on.

5G. Binding Brakes

I mentioned above the effect of cold on brakes, both front and rear. But calipers can bind just as easily in hot weather. All it takes is inadequate lubrication of the caliper guides, a sticking brake pad, or a bit of rust in the caliper bore, §6, p.268 and following. A badly bound caliper can literally cut your mileage in half.

Virtually every service interval that involves drum brakes includes a rear brake adjustment. It's easy to overtighten them if you're in a hurry, and quite often in the heat of a high-pressure day, it happens. A little drag to one side or the other will drop your mileage by 2-5 mpg. Replacement of the rear brakes can lead to the same problem. For more on rear brake adjustment, see §6C, p.270 .

5H. The Clutch

A car that's equipped with manual transmission, still running the original clutch at 70,000 miles, could easily be costing you a few miles to the gallon without your ever noticing that the clutch has started to slip. Clutch wear is so gradual that it's hard to detect until it becomes too noticeable to miss. See §2C and following, p.365.

5I. Improper Alignment

Improper alignment will also rob you of mileage due to the increased friction of "scrubbing" your tires down the road. Badly chopped or worn-out tires reduce mileage as well, even if the misalignment that caused the abnormal wear has been corrected.

11

5J. Problems Related to Tuning

And now, finally, we come to tuning . . . O.K., I'll admit it, there are certain aspects of tuning that will rob you of mileage:

- A plugged air filter, *S*2D, p.219
- Dirty fuel injectors, *S*6C2, p.238 and *S*12, p.247
- Excessive fuel flow (an overly rich mixture), *S*6C3, p.239
- The wrong spark plugs (incorrect heat range), *S*2B and *S*2C, pp.189-191
- A vacuum line that's been left dangling, a pair of them that have been reversed or a plugged vacuum source, *S*2G, p.195
- Other, more improbable issues, mostly having to due with the human comedy

I can only hope that those of you who have made it to this page will refrain from your self-congratulatory "I knew it!" just long enough to acknowledge that the BIG picture might just save you a bundle of money over a lifetime of driving.

6. Rough Idle

6A. Definitions

Any time an engine is fully warmed and just sitting there running – waiting at a traffic light for example, or in line at a toll booth – that engine is idling. Any time you give it gas you're either revving the engine (the car is standing still) or you're accelerating.

Vibrations at idle can range from the whole car visibly rocking down to a light trembling of the rear-view mirror. If you raise the hood, you can see what it takes from the engine – the extent of roughness required – to produce that vibration. If you correct the problem that's making the engine vibrate, you will usually eliminate the other symptoms as well. Every running engine produces a certain amount of vibration – it's inherent to the beast. What's typically viewed by mechanics as a rough idle is an excessive amount – an imbalance – that is not being dampened sufficiently by the engine's motor and transmission mounts (big chunks of rubber and an occasional small shock absorber used to insulate the engine/transmission from the body of the car).

Cars equipped with automatic transmission will always vibrate more at idle than the stick-shifts, because an automatic is continuously coupled to the engine – as long as it's in Drive or Reverse – whether you're moving or not.[1] Notice the change in engine speed and the differences in vibration level when the gearshift lever is moved from Drive to Neutral and then back again. This change in load on the engine drops the idle speed by 150 rpm or more because of the additional energy required to spin the transmission's innards. Slight vibrations to the steering wheel or rear-view mirror that disappear as soon as you shift into Neutral are common and normal.

Fig. 11-9. EGR valve. Courtesy of Hella

---

1   On a stick-shift automobile, the transmission must be in Neutral whenever you stop, otherwise the engine would stall. Getting into Neutral either requires clutch engagement (pushing the pedal to the floor) or that plus moving the gear shift into the Neutral position.

6B. Causes

The following discussion is aimed primarily at engines that have reached normal operating temperature. Start by seeing what an OBD-II scanner has to say . . .

6B1. Higher Mileage Engines

- A buildup of carbon deposits within the combustion chambers due perhaps to the age of the engine or low octane fuel, §5D, p.234. Add a can of fuel injector cleaner (§12, p.248) to a full tank of premium fuel.

- A buildup of carbon deposits within the EGR (exhaust gas recirculation) valve, §5D, ❏3, p.234. EGR valves can be hard to find, but every one bolts up to the intake manifold and has a pipe leading to it from the exhaust manifold. Use your shop manual as your guide.

- Leaks from a broken or disconnected vacuum line make a hissing sound. A cracked vacuum line may only act up intermittently. If the car is getting old (100,000 miles+), check the lines for brittleness; move them from side to side while the engine is running to see if you can produce any change.

- The intake manifold gasket may be leaking. Same for the gasket that connects the throttle body to the manifold. Like any other part on an engine, the bolts securing the IM to the cylinder head may loosen with age, the gasket may dry up for similar reasons. Fuel injector seals can crack as well. It's impossible to actually see any one of these problems. You could always go ahead and tighten every bolt you see–and all those you can't!–but in doing so, you may mask the problem. See §13, p.247 for testing procedures.

- Cracked intake manifold

- Low engine compression

All of the above reduce intake manifold vacuum. A vacuum gauge attached to any pure vacuum source on the manifold[2] might give readings as low as 17 in–Hg (+/-); healthy engines provide vacuum readings of 20 – 22 in–Hg (inches of mercury, a measure of vacuum strength).

6B2. Worn or the Wrong Temperature Spark Plugs

Check your owners' manual for recommended spark plug intervals, brand and model number (heat range).

6B3. Ignition Coil

If the problem occurs only when it's raining, damp, or humid, suspect one or more of them, §2B, p.214.

---

2 Vacuum lines that originate on or in the vicinity of the throttle body are typically ported lines, meaning that their suction varies in response to throttle plate position. They are useless for our purposes. But the MAP sensor attaches directly to the manifold and so qualifies as an ideal vacuum source. It can be temporarily disconnected for testing purposes, using its port to connect a vacuum gauge. There's a downside to that: It will trigger the check engine light. Fortunately, most engines have an accessory port adjacent to the MAP sensor's port, capped to maintain vacuum until you need to swap in a vacuum gauge.

11

6B4. A Tank of Bad Gas (Water or Dirt)

The car will begin to run poorly within a few blocks of the station, certainly within the first several miles. Intermittent or continuous stalling is likely whenever you take your foot off the gas. In most cases, you'll also experience problems with acceleration.

If only a few gallons were added, you might be able to avoid a repair bill by filling the tank at another station with premium fuel. Add a bottle of dry gas and a can of fuel injector cleaner. If your tank is full, it should be drained – preferably for no charge by the station that sold you the gas! In any case, notify them as soon as possible to reduce the number of motorists that will have to suffer the same fate. When the tank is refilled be sure to add dry gas, possibly a can of fuel injector cleaner.

6B5. Defective Idle Control Valve

Check your codes. For more, see §2, p.211 and Fig. 10-8, p.188)

6B6. Improper Valve Adjustment

Exhaust valves are more likely to lose their clearance because of the intense heat moving past them. Tight valves don't make noise. See §1E, p.188.

6B7. Torn or Weak Engine/Transmission Mount

More severe vibrations in gear are commonly due to a weak or torn engine or trans mount.

6C. A Dead Cylinder

6C1. Cylinder Balance Testing

An engine that shakes violently has lost a cylinder. A scanner is built for problems like this! A cylinder balance or cylinder performance test can not only isolate cylinders that are simply not functioning, it can also pick out cylinders on the margin between healthy and weak. Another approach is to unbolt each coil in turn, remove it from its spark plug, then start the engine. Four cylinder engines won't even start if you disconnect a healthy cylinder because you now have only two operable cylinders. If you pick right, you'll disconnect the offending cylinder first; the engine will behave identically whether that coil is connected or not.

So . . . Bad coil? Bad spark plug? A problem with the wiring to and from the coil? The odds are strongly in your favor that the ECU is not at fault. See §2B, p.214 for coil testing, §2C, p.191 for spark plug inspection. **Don't forget to inspect the wiring!** But first, read on . . .

6C2. Lack of Fuel Flow to a Particular Cylinder

Caused by a physically plugged or electrically malfunctioning fuel injector. The backyard approach to diagnostics is to lay a fingertip against the side of each injector, one at a time with the engine running. You'll feel a rhythmic vibration to the working injectors; an injector that's not firing (electrically) will feel dead. Note that a (partially) plugged injector can't be identified like this because the electrical aspect can be perfectly functional.

> Don't try this on an engine with direct fuel injection: Those injectors are way too hot to touch!

Before you let anyone talk you into injector replacement, check the OBD-II codes and read §12, p.247.

6C3. Excessive Fuel Flow (An Overly Rich Mixture) Delivered to a Particular Cylinder

A cylinder can't fire when it's flooded with fuel because gasoline won't burn unless it has first been atomized (dispersed into a fine mist) in air. The spark is literally drowned in fuel. A leaking fuel injector is about all that can cause this.

An engine–or a single cylinder–that's running on an excessively rich mixture can usually be identified by the smell of rotten eggs (H_2S). This smell is particularly noticeable as you slow to a stop after accelerating hard over the course of several blocks. Hydrogen sulfide (H_2S) results from an overworked catalytic converter. An overly rich mixture can literally melt the "catastrophic" converter by forcing it to operate at excessively high temperatures.

6C4. Mechanical Problem within a Cylinder

- A burned exhaust valve that reduces the amount of compression that can be generated, §1E, p.188 and Fig. 9-8, p.166.
- A broken valve spring that eliminates compression altogether, Fig. 10-4, p.185.
- A piece of carbon stuck to the head of a valve which yields a temporary loss of cylinder compression, §2C, p.191, §2D, p.192 and §12, p.247.
- A blown head gasket that is adding coolant to one or more combustion chambers, thereby reducing their ability to fire, §13, p.163.

Of these, all but the last can be verified by compression testing, §2D, p.192 and §9B, p.204. All spark plugs must be removed, all cylinders should be tested for the sake of comparison. A blown head gasket will reveal itself through that squeaky clean spark plug.

To eliminate the possibility of a piece of carbon yielding temporary loss of compression, try a tank of name-brand fuel and a can of fuel injector cleaner before embarking down the path of engine disassembly – as long as the engine isn't making ugly noises.

On certain engines, replacement of a broken valve spring can be accomplished with the cylinder head in place. Get two opinions on this, one from a machinist. A burned exhaust valve requires a valve job (cylinder head removal). Section 13, p.163 addresses means to verify a blown head gasket.

7. Engines with a Loping Idle

(Varying Up and Down As If Hunting for a Steady Idle Speed)

7A. Dirt and Water Contamination

Dirt and water can accumulate in gas station storage tanks over time. Dirt can easily become lodged in the pinpoint openings of fuel injectors producing an irregular stream of fuel. Water produces similar results through a different mechanism: The stuff just won't burn!

Engines that suffer from encountering the dregs of a near-empty storage tank are far more likely to suffer these problems, because dirt and water are both heavier than gasoline and become concentrated toward the bottom of the tank.

11

The exact same statement applies to your car's gas tank – running on empty can be an expensive proposition.

Begin by adding a can of fuel injector cleaner and a can of dry gas (particularly in winter) to the gas tank. Don't fill the tank just yet because you might need to drain it, but don't go on this run with less than a quarter of a tank of fuel. Drive about 5 to 10 miles, maintaining the RPMs above 3,500 for as much of the drive as possible, using D2 or second gear. This remedy can work wonders. If it does, fill the tank with premium, high octane fuel. For those unlucky enough to need more than that, repairs might encompass some or all of the following, listed in rough order of probability:

- Take a close look at your gas cap. Is the seal cracked, squashed or showing signs of rust? Is the vent valve – that central disc of steel – rusted or frozen in place? Any one of those problems should send you packing to the parts jobber for a new OEM cap.

- Fuel filter replacement, primarily to determine whether the fuel is contaminated with water or crud: Shake out its contents through the filter's inlet side into a clear glass jar.

- If the fuel filter provides evidence of excessive amounts of water, dirt or rust in the gas tank, you'll need to drain it, §15, p.249 If severe internal rusting seems evident, replace the tank.

- When all else fails, the electric fuel pump might have to be removed to clear its pick-up screen. This becomes increasingly relevant on those engines that cut out and won't restart due to fuel starvation, §2C, p.216. This would also be a very good time to clear both the fuel supply and return lines with compressed air.

7B. Other Possibilities

If dirt or water contamination are not involved, several more possibilities surface:

- The throttle body has an idle control valve bolted close by the throttle plate. It alters the engine's idle speed in response to certain conditions – a cold engine, turning on the air conditioning, etc. A problem with the signal that feeds the control valve can cause erratic variations in engine idle. How would you know? Watch the throttle linkage, Fig. 11-1, p.211: If you find that it's moving slightly whenever the engine speed changes, you may have found the culprit. The diagnostic work required will be to uncover the reason for that movement; it may be controlled electrically or with vacuum (older models in particular).

- Most engines with a loping idle are likely to have low intake manifold vacuum. A vacuum gauge attached to any pure vacuum source on the manifold[3] might give readings as low as 17 in–Hg; healthy engines provide vacuum readings of 20 – 22 in–Hg (inches of mercury, a measure of vacuum strength). See §6B1, p.237.

---

3   See Footnote 3, p.237

- A problem with the signals being sent to the ECU can result in erratic signals being fed back to the injectors, altering the amount of fuel being delivered as well as its timing. This problem could be caused by any of the following:

  – Any one of a number of defective EFI sensors: Check your scanner!

  – A problem with an electrical connection, either within the plugs that connect a sensor to the wiring harness, or a poor ground connection, *S*14, p. 427. The primary ground strap connecting engine to body is another possibility.

  – A defective EFI computer: Unlikely.

Be advised: If you are unlucky enough to find that your problem is not easily remedied with cleaner fuel, then you may need a specialist. The sooner you get there, the cheaper your bill will be. Remember that all components of the fuel injection system are covered by warranties valid only through the the factory's dealership network, most for only its first two years. See *S*2B, p.6.

8. Engine Misfire

8A. A Rapid Succession of Missed Beats

These are most pronounced under hard acceleration up a hill. The beats can be erratic or very regular; each misfire is a momentary loss of power. Possibilities include:

- A bad spark plug – its electrodes buried in carbon deposits, a cracked insulator, or oil fouling, *S*2B and *S*2C, pp.189-191. Replace them all.

- A defective coil, *S*2B, p.214.

8B. A Single Misfire

A single misfire produces what feels like a stumble, as if your heart just skipped a beat. This condition typically occurs at idle, and can be caused by a problem in the:

- Fuel injection system

- Ignition system

- Emission control system

The cause for this type of misfire is often quite difficult to isolate, and usually ends up as a process of elimination that begins with the ignition system. Many turn out to have a bad fuel injector or deteriorated injector seal; the problem is – which one? Electric fuel pumps have been known to skip a beat on their way to the trash can. Other possibilities include a loose electrical connection or bad ground.

I used to feel a sense of dread whenever I'd see one of these coming through the door, but a sophisticated scanner has made diagnosing these problems immeasurably simpler. Check your codes!

11

9. Losing Power

Power loss on a hill can be caused by:

- A slipping clutch
- Low fluid level within an automatic transmission
- Ignition breakdown
- Carbon buildup within the combustion chambers
- Fuel starvation
- Problems with timing
- A plugged exhaust system
- An emission control system malfunction

Not enough choices, eh? Fortunately, four are easy to differentiate from the rest of the pack. If you can rev the engine while you are losing power, then fuel starvation and a plugged exhaust are definitely ruled out. Furthermore, the fact that engine RPMs can increase while the car is losing speed suggests a slipping clutch on the manual transmissions, a slipping transmission on the automatics.

9A. To Test for a Slipping Clutch

Approach a long, fairly steep hill in second gear at 25 mph. As you get into the steeper portion of the incline, shift quickly to third gear, flooring the accelerator as soon as you do so.

- Clutches in good condition might permit the engine to race momentarily as the disc begins to "bite," but almost immediately the engine's RPMs will drop into sync with the rear wheels, producing the low RPMs of a car that's straining uphill in the wrong gear.
- If the clutch is slipping, the engine's RPMs will just race away while the car loses power.

For those of you who live in flat terrain, there's a variation on this test that requires nothing more than a slight incline. Traveling in third gear at 50 mph, shift quickly into fifth gear while flooring the accelerator. A slipping clutch will produce the same result – an engine that's screaming, yet producing no power.

Clutches usually last 80,000 to 100,000 miles. Refer to §2, p.365 for a deeper discussion.

9B. An Automatic Transmission Will Slip at Much Lower Speeds

Slippage is usually most noticeable from a standing start, particularly when the car is cold. You give the engine gas, the RPMs go up, but the car barely moves. Check your transmission fluid level immediately to prevent destroying the transmission's clutch packs, §1A, p.357.

9C. A Breakdown in the Ignition System

This typically occurs at higher RPMs, 3,500 and up. A single misfire will feel like a stumble, as if your heart just skipped a beat. The more common scenario is a rapid succession of misfires – staccato fashion. The more the thing misfires, the more power is lost. See *§*8A and *§*8B, p.241.

9D. Carbon Buildup on Combustion Chamber Walls

Carbon deposits accumulate within engines that are consistently fed low-octane, unbranded fuel, *§*5D, p.234. This process is enhanced by a lot of stop-and-go city driving at low RPMs. Carbon also builds within engines that burn a lot of oil, particularly when the source of the problem is worn valve guides and valve seals, Fig. 10-4 and 10-5, p.185. Regardless of cause, the valve heads accumulate deposits that eventually reduce engine compression as the deposits begin to interfere with valve seating (closing). Routine cleansing with a good fuel-injector cleaner can forestall the need for valve work; a can is added at the first sign of power loss on hills, *§*12, p.247.

9E. Power Loss in Wet Weather

An engine that's hard starting in the rain, won't accept gas without bogging down, and acts as if it will cut out any second has a problem with at least one of its coils. For more on this, see *§*2B, p.214.

9F. Fuel Starvation

Your first hint of trouble is usually when climbing a long hill. Just as you need to give the car gas in order to maintain constant speed, the engine bogs down – you'll simply lose power and with it, speed. The RPMs drop off, the engine might even stall and may not restart.

Fuel starvation is caused by a worn-out fuel pump or by blockage within the fuel system. Blockage is more likely to occur to those who consistently run their cars low on fuel because this concentrates sediments within the gas tank before feeding them to the engine. See *§*2C, p.216 for fuel pump testing procedures.

If blockage is the cause of your fuel starvation, the point at which that blockage occurs will always be at the point of greatest restriction to flow:

- the screening within the fuel filter, and
- within the pinhole passageways of the fuel injectors themselves.
- the fuel-pump's pick-up screen

Since none of these are visible or easily accessed, you might want to invest in a can of fuel injector cleaner (*§*12, p.247). Unless your tank is quite low, I'd skip adding gas: Because the fuel pump is located in the gas tank, you might have to drain it sooner than later. Run the car hard in first and second gear, maintaining the RPMs above 3,500 as much as possible for 5 to 10 miles. This will clear the pinhole passageways through your fuel injectors, enabling a stronger flow of fuel and a more clearly defined "parasol" spray pattern. (Aerated fuel burns much more readily and powerfully than a dribble or droplet). If that doesn't cure the problem, move directly to fuel filter replacement. Don't buy into fuel rail[4] and injector cleaning without:

---

4   The piping that actually supplies the fuel injectors, it runs alongside the intake manifold, Fig. 17-20, p.459.

11

- A new fuel filter in place, and
- Strong fuel pump pressure readings within the fuel rail.

Since the only way to identify a failing fuel pump is with a pressure tester or a scanner, and since pressure tester measurements are commonly taken at the fuel rail, a partially blocked fuel filter can produce low readings. So make sure that the fuel filter gets replaced and that new readings are taken before you authorize fuel pump replacement.

The current supplied to electric fuel pumps is greater while the engine is cranking than once it's running. This requires current flow through two different paths involving both a relay and a connection to the ECU. As with any electrical problem, an uninterrupted path to ground is just as important as its source, *§*15, p.429.

9G. Engine Timing: Injectors, Intake Valves, and Spark Advance

First, a bit of mechanics. An engine under load – for example, one that's just begun moving up a hill – requires more fuel to maintain the same speed. That requires that the fuel injectors open a split second sooner and remain open for a beat longer, and that the variable valve timing (where present) needs to advance to accommodate that earlier injector firing. To translate the added fuel into increased engine performance, the point at which the spark plug fires relative to the upward motion of its piston must advance by a fraction of a second. The rate of this change must vary quite precisely in response to how aggressively your foot meets the accelerator pedal – this to keep the engine from pinging (knocking). Complicated, but beautiful! (*§*1C, p.185, *§*3F, p.197)

9H. Blockage Within the Exhaust System

Blockage usually develops so gradually that it's hard for an owner to detect until it's well advanced. There is one exception to this: A muffler that self-destructs over a bump – its interior baffling coming loose, blocking the passageway out.

A plugged exhaust can change the sound of an engine as exhaust gases struggle to find a way out. The exhaust fumes exiting the tailpipe can feel like a light breeze even under high revs – as if the car were still idling. In more advanced stages of exhaust pipe clogging, the engine won't be able to rev past 1,500-2,000 RPMs.

> To avoid burning yourself, place your hand at least 18 inches away from the pipe. Have an assistant rev the engine for you.

The only way to know with certainty that exhaust blockage is present requires disconnecting the catalytic converter from the front exhaust pipe. This results in a frightful amount of noise, plus a hot blue flame shooting out of the pipe with every rev. Witnessing this will forever impress upon you the accuracy of the term "muffler."

The converter is far and away the most likely cause of blockage, particularly if the engine has been running poorly for some time: A consistent diet of unburned (overly rich) exhaust fumes feeding the converter can cause it to melt. If unbolting the converter restores power to the engine, reconnect it temporarily, then unbolt the next pipe down the line. If you've restored the problem, you'll know beyond a doubt that you'll need to buy a new converter. But if the engine is running fine with the converter back in place, breath a sigh of relief: You've just saved yourself hundreds of dollars.

> The exhaust manifold, front pipe, and catalytic converter are blisteringly hot almost immediately after starting an engine. Give them at least a half hour to cool before initiating this work.

For those of you with the problem still unsolved, unbolt other pieces down the line until you find the source of the blockage. The resonator and muffler are the two most likely candidates (Fig. 17-5, p. 447). Remember that every piece you unbolt will need a new exhaust gasket and probably, new hardware–preferably stainless steel.

10. If Your Car Has Stalled, Proceed as Follows

❑ Inspect your instruments immediately. What is the temperature gauge reading? If it's pegged or running close to the top of its scale, read $1, p. 147 immediately!

❑ All of your warning lights should display as the engine stalls, regardless of cause, and will remain on while the engine is being restarted. Check to see if a warning light is out when it should be on. It could be quite important as far as diagnosing your problem.

❑ If the temperature gauge is within the normal range, try to restart the engine. Assuming that it does restart, check your oil pressure readings. If the oil light stays on or your oil pressure gauge is reading low, shut down the engine immediately and read $4, p. 130.

- If the engine cranks slowly or won't turn over at all, see $A5, p. 229.

- If the engine cranks strongly but won't restart, see $4B, p. 232.

- If the car starts but the charge warning light remains on following startup, see $3, p. 413.

- If you're out of gas, start walking . . .

❑ If any warning lights remain on while the engine is running, shut it down and consult your owner's manual. Equally as important, write your observations down – your notes could save you a bunch of money in diagnostic time. Include exactly which warning light came on plus the sequence of events leading up to the stall – whether you were in heavy rain, on a long hill, engine temperature, etc.

❑ Check your oil level, just to be safe. While you're under the hood, check to see if anything is wet or smoking.

❑ If engine temperature, oil pressure, and oil level seem OK, see if you can drive a few blocks, preferably along quiet streets where you needn't worry about traffic. Watch your gauges and listen for noise. If the engine continues to stall whenever you take

> Don't touch! Do Not Open the Radiator Cap!

your foot off the gas but exhibits no other problems, you can probably save a tow bill in exchange for a dose of anxiety. (Read the next paragraph.) On the other hand, if the engine has overheated or lost oil pressure, call in the hook: a tow bill is always cheaper than a new engine!

11

To Keep an Engine from Stalling at Idle

"Feather" the gas pedal – providing just enough gas to maintain the engine above idle. This requires shifting the transmission into Neutral as you approach every stop. On cars equipped with automatic transmission, that requires manually moving the shift lever into Neutral – NOT Park. Putting an automatic transmission into Park while the car is still moving can be catastrophically expensive.

+There is a distinct difference between an engine that stalls while it's sitting motionless at a traffic light and one that quits while driving along with your foot on the gas. The first will usually restart and continue to run as long as fuel is supplied. (Reread the paragraph immediately above.) Conditions that cause an engine to quit while in motion usually prohibit restarting. Time to pull out your code scanner and factory manual.

Stalling at idle can actually occur while the car is moving. As soon as you take your foot off the accelerator, engine speed drops off toward idle. Having your foot off the accelerator for more than several seconds will bring it all the way down.

If this is a problem that's limited to the first several minutes following start-up on a car that's been sitting for at least four hours, that's a cold stall. See §2, p.211. If the problem only occurs after the temperature gauge has reached the middle of its scale (typically 10 minutes or more following start-up), that engine is hot.

If the problem only occurs during those 3 to 10 minutes following start-up, you'll definitely need to consult your scanner and factory manual: The EFI system is switching over its functions from cold to hot.

The most common causes of stalling are:

- Bad gas – contaminated with water, dirt, or rust
- Old tune-up gear: worn spark plugs, clogged air filter, plugged fuel filter
- Blockage within the injector nozzles
- Weak fuel pump
- An intermittent no-spark condition or other problem within the EFI/ECU system: a failing sensor, loose or corroded terminal connection, etc. Check your scanner for codes, use your manual's diagnostic flow charts to narrow down the choices.

11. Backfire

A leak in the exhaust system, particularly when it's close to the engine, can "backfire" heartily because there is nothing to muffle the roar. Exhaust manifold and front exhaust pipe leaks should be repaired as quickly as possible, in part for the noise reduction, but more importantly to keep you from breathing those fumes – they're loaded with carbon monoxide.

12. Decarbonizing Fuel Injector Cleaners: Their Many Uses

I know of two excellent fuel tank additives – BG44K and Slick 50 Fuel System Formula – that clean dirt, varnish deposits, and carbon from fuel injection systems, the combustion chambers, and carburetors, for that matter. They can be used as often as every 3,000 to 4,000 miles for any of the following driveability problems:

- Resists taking gas, cold or hot, *§*2 and *§*3, pp.211-219
- Rough idle, hot, *§*6, pp.236-238
- Hard to start, either cold or hot, *§*2, and *§*3 pp.211-219
- Losing power on hills, *§*9, pp.242-244
- Surging, cold or hot, *§*7 and *§*8, pp.239-241

Either product can be used to advantage on engines that burn oil, particularly where the source of the problem is worn valve stems and valve seals. The valve heads continuously accumulate burned oil deposits, reducing engine compression as the deposits begin to interfere with valve seating (closing). Routine cleansing with Slick 50 or BG44K can forestall the need for valve work; add a can at the first sign of power loss on hills.

Both products are added to a full tank of gas, preferably jet fuel (a name-brand super premium). To help blow out carbon deposits, run your car at higher RPMs than you ordinarily would for periods of two to three miles at a time throughout the course of that tank – keeping the revs above 3,00 RPM is ideal. On an automatic, this can be accomplished by city driving in second gear (D1) or highway driving in third (one down from overdrive).

With the exception of carbon removal, these products are seldom needed by those who consistently use name-brand fuels. If you compare the differential cost of a can of fuel system additive every few months to the expense of burning a name brand fuel instead of some schlock no-name, you'll find that the costs balance out.

13. DIY – To Test for a Leaking Intake Manifold Gasket

Either dried out, cracked, or adjoining a loose intake manifold, a leaking intake manifold gasket provides a path for additional, unregulated air to be drawn into the engine, producing a dramatic drop in intake manifold (IM) vacuum as well as leaving one or more cylinders burning way too lean. The symptoms of a bad intake manifold gasket are usually less noticeable on cold engines; they become much more pronounced on the tail end of a long drive on a hot summer's day. A vacuum gauge attached to any pure vacuum source on the manifold might give readings as low as 17 in–Hg (+/-) on a troubled engine; healthy engines provide vacuum readings of 20 – 22 in–Hg (inches of mercury, a measure of vacuum strength).

To test, proceed as follows:

- Start by putting a fire extinguisher next to you: You'll be using an aerosol can of flammable injector cleaner (with a pinpoint sprayer attached) on a hot engine . . .

11

- With the engine running in Neutral or Park, offer a very brief (1 second) shot of cleaner along the seam between each pair of bolts at the top of the cylinder head/IM flange, listening for an instantaneous increase in engine RPM. Should that happen, you've found your leak. **A longer than brief shot is unwarranted and potentially dangerous, particularly on a hot engine due to the risk of fire.**[5]

Rest assured that the intake manifold is the coolest part of the engine, constantly being replenished with fresh air.

> **If the sound of this makes you uncomfortable, let your mechanic diagnose the problem for you.**

If the intake manifold will have to be removed, the gasket must be scraped off both the manifold and the cylinder head. Wait till the engine is cool before you begin; you'll need a considerable amount of patience when addressing the hidden bolts along the lower edge of the manifold. Remove them first.

Use a steel straight-edge to inspect both the block and manifold for warpage. A slight warp can be finessed with high-temp gasket cement applied to both sides of the new gasket. A significant warp implies new parts or a trip the the machinist.

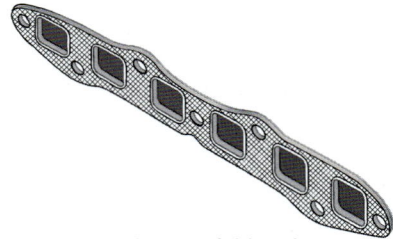

Fig. 11-10. Intake manifold gasket, six cylinder engine

The cylinder head should have pins on which to hang the manifold while you start each bolt into place. Put a couple in finger tight to make the others easier to install.

You'll be tightening from the center out, but you need to do this in two or three passes to gradually bring the manifold into contact with its gasket and the cylinder head. For the final pass, you'll need a 3/8th inch torque wrench: Space can be hard to find.

14. DIY – To Test for a Leaking Fuel Injector Seal

Read the section directly above . . . Same dance, different tune.

---

5   If you think I'm crazy, well, that's debateable; but I probably had to run this test a hundred times over the course of my career with only one flash of flame, gone before I could reach for the fire extinguisher. The good thing about extremely flammable aerosols is that, unfed, they flash away in an instant.

15. DIY – Fuel Pump Replacement, Fuel Tank Removal

Some of you will be lucky enough to discover that your fuel pump is accessible by removing the rear seat cushion and the metal cover below it. The rest of you should be grumbling, because for you, the tank will have to be drained, then lowered to the floor. Then and only then will the pump be accessible.

In either case, the battery needs to be disconnected. Some scanners offer a program to save all of the car's memory once it loses its power source. Use it!

Fuel Tank Removal – Tanks are either bolted or strapped in place, or both. Bring the tank down about a foot to enable you to access the fuel lines and electrical connections. Disconnect them all before lowering the tank to the ground. The pump, its filter, and the fuel gauge will all come out as a unit. Your shop manual will provide the details.

CAUTION! I once nearly burned down my garage while draining a gas tank. Wintertime, a propane heater, every precaution I could think of . . . Turned off the heater, moved it to the far end of the shop, made myself some lunch, *then* started draining the tank. About five minutes later, a flash – no, a roar – emanated from the heater, at least 30 yards away, shot toward me and that open pan of gas. My customer's car? Toast. My clothes? Singed. My ego? You can only imagine . . . The only bright spot in this tale is that the fire launched my career, leaving me with a much larger shop, commercially zoned, with all of the regulation and expense that implies.

Gasoline vapors are what engines run on. Not liquid fuel, but atomized spray. The vapors that emanated from that open metal drain pan gradually filled the shop. The wick in that heater had just enough residual warmth in it to ignite those fumes. I should have moved the heater outside; far, far away.

Use one or more five gallons gas cans and a large funnel to keep the vapors corralled, and ventilate! May my folly save your shop!

Notes

Chapter 12. The Braking System

12

Chapter 12. The Braking System

Notes

Notes

The Braking System

1. Brake Anatomy

Most economy cars and trucks employ a pair of disc brakes for the front two wheels, a pair of drum brakes on the rear. Higher-end models run discs both front and rear.

Drum brake designs are blessed with tremendous mechanical advantage. Unfortunately, this same leverage results in the undesirable tendency to lock up under panic braking, creating the unsettling sensation of the car skidding out of control. By contrast, disc brakes will never lock up on dry pavement because from the standpoint of mechanical advantage, they don't have any. But what they lose in leverage, they make up in brute force: The additional "standing power" required to slow the car is supplied by way of a vacuum-assisted booster betwen the brake pedal and the brake master cylinder.

Disc brakes are unadjustable; as their pads wear, the caliper simply repositions itself to compensate for that wear. This offers a distinct advantage over drum brakes, in that they require routine adjustment. Disc brakes have the additional advantage of fading less than drum brakes under hard braking. On the down side, it's hard to design an effective parking brake assembly into disc brakes due to their lack of leverage. Consequently, many cars with rear disc brakes have a small drum brake assembly mounted inside their rear brake rotors, controlled solely by the parking brake.

Since wet pavement offers virtually no traction for tires, even disc brakes will lock up in the rain when slammed on in a panic. Anti-lock braking systems (ABS) were developed to prevent this. These systems precisely meter the amount of hydraulic (brake fluid) pressure being applied to each wheel. Whenever a wheel approaches lock-up, the ABS modulator drops the fluid pressure applied to that particular brake to permit the wheel's continued rotation, then increases it just enough to bring that wheel's rotation into sync with its mates. This process repeats itself continuously at each wheel until the slippage is controlled. Whenever this occurs, you'll feel a light pulse in the brake pedal as the modulator applies pressur≤e, then releases it, over and over in rapid succession.

When you press on the brake pedal, a small volume of brake fluid moves out of the brake master cylinder, transmitting hydraulic pressure to the ABS modulator assembly. There, the fluid pressure is distributed to each wheel, controlling for the car's weight distribution, speed of each wheel, etc. Wheel speed is monitored by an electronic sensor behind each wheel. Those signals are transmitted to an electronic control unit (ECU) mounted on the ABS modulator. The ECU governs valve action within the modulator, one valve per wheel. These valves can be wide open (brake fully engaged), fully closed (no pressure supplied) or anywhere in between. In order to maintain appropriate pressure to lines that are cycling between engagement and release, the modulator also houses a pump to restore pressure to those wheels that were deprived by valve closure.

ABS modulator   vacuum booster

master cylinder reservoir   master cylinder

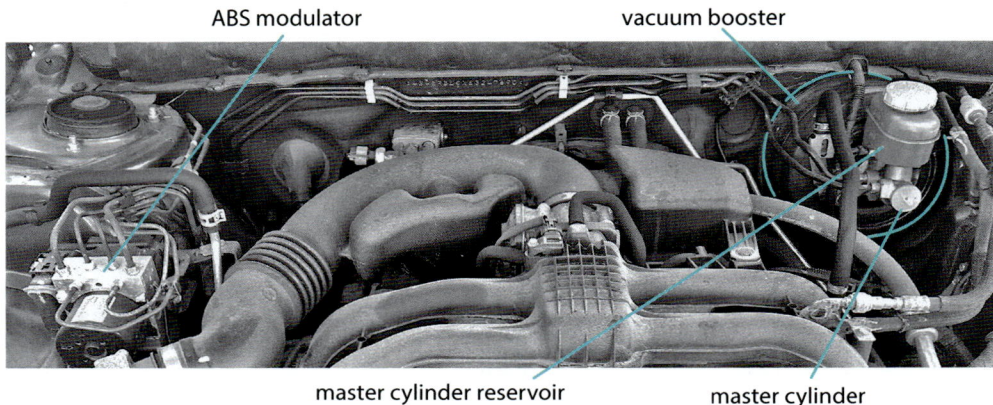

Fig. 12 – 1. Brake system, engine compartment

Steel lines are used to transfer hydraulic pressure from the brake master through the ABS unit and down to the vicinity of each wheel. Since both front wheels need to turn and all four wheels move up and down, flex lines are employed to span the gap from rigid chassis to caliper (or rear axle).

The brake caliper is like a lobster's claw, squeezing inward against a disc of steel – the brake rotor – when the brakes are applied, Fig.12–2. Since the rotor bolts up directly to the wheel, slowing the speed of its spinning slows the speed of the car. The caliper contains one or more pistons housed within it that are pushed toward the brake rotor whenever hydraulic pressure is applied, Fig. 12–3. In single-piston configurations, the caliper is suspended overtop the rotor on a pair of caliper guide pins that enable it to slide laterally relative to the rotor and the pad bracket, permitting the caliper to shift its position slightly. When the brakes are applied, the force that's produced by the piston pushing against its side of the disc literally pulls the opposing fingers of the claw toward the disc's other side. This transfer of pressure enables the two brake pads to squeeze evenly against the rotor, creating a balanced set of opposing forces.

caliper

flex line

guide pin, lower

rotor & rotor hub
(one piece)

pad bracket, lower end

Fig. 12 – 2. Brake caliper and rotor
Courtesy of Goodheart-Willcox

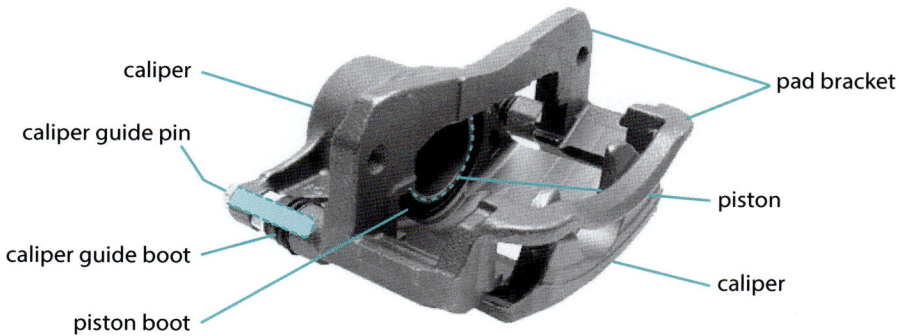

Fig. 12 – 3. Single-piston brake caliper, related parts
Courtesy of Raybestos

When the brake pedal is released or the ABS modulator closes a line, hydraulic pressure drops to zero. Consequently, the brake piston retracts ever so slightly, the caliper "unflexes," and the brake relinquishes its drag. Note that the brake pads remain in constant contact with the rotor even when the brakes are off. The only difference between braking and not is the amount of pressure that's applied.

Note that if the caliper guides are not properly lubricated, the caliper is unable to relax (unflex) when the brakes are released. This leads to excessive heat buildup and accelerated brake wear, often accompanied by noise. The heat can actually cause a brake rotor to warp, producing a pulsation to the brake pedal and/or steering column whenever the brakes are applied, *S* 4B, p. 264 and *S* 4C, p. 266 and *S* 5, p.266. Frozen caliper guides can also produce a caliper that can't distribute its "bite" evenly to both sides of the rotor. As a result, its brake pads wear unevenly and the car often pulls under hard braking (assuming the caliper on the other side of the car is working more efficiently).

Frozen caliper guides are just one of the five principal causes for a binding caliper. See *S* 6B, p.268 for further enlightenment.

In a two- or four-piston caliper, Fig. 12–4, the caliper itself is bolted rigidly in place; the term caliper guide has no meaning. The pistons are paired, one (or two) to either side of the rotor. Assuming unrestricted fluid flow and happy pistons–not rusted or mired in sludge–these calipers work great. But if the pistons stop working in consort, then brake pull results. Unfortunately, these calipers can be tricky to overhaul; re-establishing that harmony is sometimes easier said than done.

A typical drum brake on a front-wheel drive car is illustrated in Fig. 12–5. Fluid pressure works on a pair of small pistons that are housed within the wheel cylinder, driving them outward against the upper end

Fig. 12 – 4. Four-piston brake caliper
Courtesy of AC Delco

of each shoe. Since the lower end of each shoe is held stationary, this movement effectively pushes the shoes outward against the inside face of the brake drum. Friction slows the wheel. When the brake pedal is released, the return spring pulls the shoes inward toward one another, releasing the brake. Brake shoes ride in very close proximity to their drum, but should not be in contact the way disc brake pads are with their rotors. Should you be present during a brake inspection, you will probably notice the mechanic spinning your tires, looking for binding in the brakes. If so, you will notice that–all else being equal–your rear drums spin much more easily and for longer duration than your front discs.

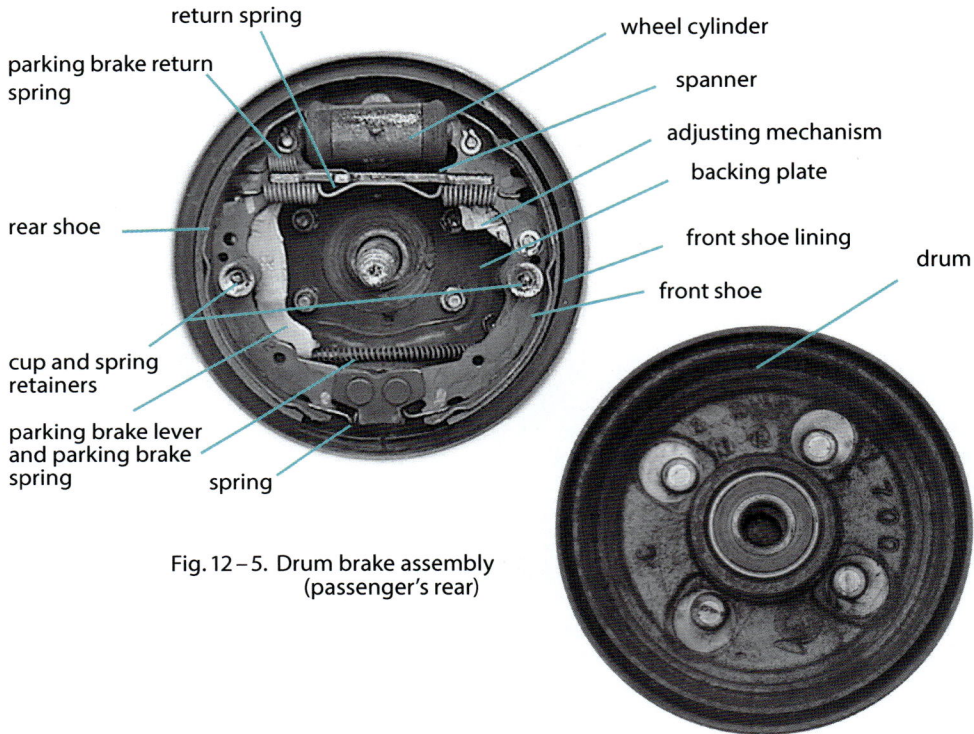

Fig. 12 – 5. Drum brake assembly
(passenger's rear)

Drum brakes require routine adjustment to compensate for wear, usually once every 15,000 miles (or annually, whichever comes first). The brake adjuster is rotated to alter the relative distance between front and rear shoes. Access is usually provided through a slot in the backing plate or through a cutout in the drum. Some adjusters can only be accessed with the brake drum removed. Drum brakes are typically equipped with a self-adjusting mechanism as well.

From an engineering standpoint, the real beauty of a drum brake is in the simplicity and nominal cost of its parking brake lever and spanner. Engaging the parking brake tightens the cable that attaches to the lever's bottom (free) end. This pulls the lever away from the rear shoe, driving the spanner forward, pressing the front shoe tightly against the drum. Since the lever pivots at its other end, the rear shoe is forced backwards against its portion of the drum as well.

2. Evaluating Your Brakes

2A. Stationary Tests

With the engine running, pump the brake pedal three to four times. Does the pedal height increase? If so, either your rear brakes are in need of adjustment (drum brakes only) or you have air in the brake lines, or both. Taking it one step further, pump the pedal again three to four times, then hold it on firmly. Over the next minute, does the brake pedal start to creep toward the floor? If so, you probably have a brake master cylinder that's on its way out, or a leak somewhere else in the system, *§* 18, p.307.

If you feel that increasingly you need to stand on the brakes to slow the car, you may have a problem with the brake vacuum booster. To test, turn off the engine. Wait a moment, then apply the brake pedal firmly. Start the car. If the vacuum booster is working properly, it will suck the pedal toward the floor, perhaps a half inch to an inch. If there has been no change, ask your mechanic to first check whether the booster is receiving an adequate supply of vacuum. If it is, then the booster becomes suspect.

The brake vacuum booster is seldom guilty of bad manners. The most frequent cause for a car that requires brute force applied to the brake pedal is a set of binding calipers, *§* 6, p.268. Another possibility: crud binding up the valving within the ABS modulator.

2B. Road Testing

I used the word "road" for lack of a better single-word term. A mega-mall parking lot is a much safer place to perform these tests–early in the morning while the lot is still empty. Choosing a location with excellent visibility and no traffic is essential both for your own well-being and the composure of others. Do your braking on an area that's as flat as possible–even a perfectly conditioned car will wander toward the low side of any road as soon as you let go of the wheel. The road needs to be dry.

> A car's direction can change very quickly when braking, particularly when there's a problem. Start with light, gradual braking, and move up to more moderate applications if appropriate.
>
> Always check your rear view mirror before applying the brakes!
>
> Never do these tests in traffic!

❑ Driving along a straight stretch, travelling at 35 to 40 mph, apply your brakes firmly to produce moderate deceleration. Don't slam 'em on, we're not looking for heroes here... Repeat this several times, noting any tendencies toward pulling[1] and pulsation (vibrations while braking) with a brief interval (2/12 mile) between each test. A brake pulsation is most easily felt when braking from 50 to 25 mph, especially if you are headed downhill.

You need not come to a complete stop. In fact, those last 20 yards–the distance you'd cover when slowing from 15 mph to 0–will be no more than an exaggeration of whatever takes place during the initial deceleration.

---

1 Movement toward left or right, consistent under consistent circumstances.

❏ Next, travel the same stretch of road at constant speed. Once your car is nicely centered in its lane, move your hands a hair's breadth off the steering wheel. In most cases, your fingers will be off the wheel for periods of no more than a few seconds at a time; very few cars travel in a perfectly straight line for any great distance.

Repeat the test several times until you are certain that you are seeing consistent behavior. This portion of the drive will identify for you any tendencies that the car might have independent of braking, such as a misalignment of the front end or a problem with one of your front tires.

❏ Finally, repeat the above run, this time combining parts one and two, applying your brakes with your fingertips just off the wheel.

While this might seem a bit dicey to some, rest assured that most mechanics perform this test several times daily. **The most dangerous part is forgetting to check your rear-view mirror prior to hitting the brakes.**

> Let me repeat the important points: a *dry* road with *no cars behind you, no traffic around you, reasonable* speed (35–40 mph), *moderate* deceleration. *Initiate the test with several trials* having your hands placed firmly on the wheel. After *all* of these conditions have been met, *then* proceed to the second and third parts of the test. **Your fingers should always be within tickling distance of that wheel!**

All the drama but none of the excitement–probably 90% of the cars out there will behave like perfect gentlemen: tracking straight, stopping straight. Perhaps 30% of you will experience some degree of pulsation when braking, either through the brake pedal, the steering wheel, or the seat of your pants. If so, see §5, p.266.

3. Brake Pull: Wandering Left or Right

1. **If the car tracks straight while driving, and tracks straight when braking,** then at least for this one road test, you don't have a demonstrable problem. Still, it wouldn't hurt to check your tire pressures and inspect the surface of your tires for wear, §2, p.317.

2. **If the car seems to stop straight, but wanders consistently in one direction while driving,** you are either experiencing a problem with one of your tires or a problem with the alignment, §2, §3, and §4, pp.317-327.

3. **If you find that your car has an intermittent pulling problem associated with either tracking or braking,** you may be experiencing an evolving brake problem or a problem related to temperature: Both your brakes and your tires take a mile or two to heat up; does the problem manifest itself only within that time frame? That could easily suggest:

    • A problem with binding in one of your front brake calipers: To start, try lubrication of the caliper guide pins (single-piston calipers only), §11A, p.287.

    • An evolving ply separation in one of your front tires, §3A, p.320.

    • On high-mileage cars, the need for front end lubrication, including the bearings at the top of each strut tower, Fig. 17–26 (p.461)

12

Problems in their early stages of development are the most difficult to isolate unless the customer can define precisely how to duplicate the problem. While a brake inspection/lubrication of caliper guides/tire inspection/lubrication and inspection of the front end/alignment check will certainly not hurt anything but your pocketbook, be advised that this expense might not turn up anything, particularly if the circumstances that produce the pull are too ephemeral to pin down.

4. **If your car pulls strongly in one direction while driving and pulls even more strongly in the same direction when braking:**

    - Binding front brake caliper, *§* 6A, p.268

    - Defective tire (severe ply separation)–replace immediately! – *§* 3A, p.320

    - Alignment problem, *§* 10, p.338

5. **If your car tracks reasonably well at constant speed but pulls when you apply the brakes,** one of your front brake calipers is probably binding. If the car is pulling right, the right brake is working more effectively than the left. That can mean that the left brake is frozen and barely working, or it can mean that the right brake has been binding to some extent. That friction produces heat. Combine that with the fact that the brake is already partially engaged creates a caliper that wants to grab. Overhaul or replacement of the calipers is the minimum fix. Unless the brake pads were replaced recently–like last week–they too will have to go. Rotors are a possibility as well. See *§* 6, p.268 and *§* 7F, p.276.

6. **If your car pulls in one direction while driving at constant speed, then pulls strongly in the opposite direction when the brakes are applied,** it's likely that you have a combination of issues that has degenerated to another level. See 4. and 5. directly above.

7. **The car pulls in one direction under hard acceleration, then heads back the opposite way under deceleration.** When travelling at constant speed, the thing will track fairly straight. In most of these cases, the car will stop straight as well.

This is fairly rare and happens only on very high mileage cars. Each end of the front stabilizer bar passes through its lower control arm. Those unions are cushioned by a fat rubber bushing. As these bushings wear out, the lower control arms become free to shift somewhat under acceleration and deceleration, moving forward and rearward relative to the stabilizer bar.

❏ Rear Brake Pull

The above road test will not tell you much, if anything, about the rear of your car, simply because the rear must follow wherever the front end leads. From the driver's perspective, the effects of a relatively minor problem up front will feel about the same as a potentially catastrophic problem emanating from the rear. It usually takes a wet road and a good scare for these to become manifest. You'll be driving in the rain, entering a curve when you hit the brakes and–cheese and rice!–the rear end breaks loose! The car might just start to spin.

Because rear braking problems are so well hidden under dry road conditions, and because of the very real dangers involved in trying to experience them on a wet road, be sure to have a thorough brake inspection once a year or every 15,000 miles, whichever occurs first.

What causes rear brake pull? On drum brake set-ups, any of the following:

- Frozen rear wheel cylinder piston
- Frozen handbrake cable
- Improper rear brake adjustment
- Brake linings soaked in brake fluid (from the rear wheel cylinders) or gear oil (due to a bad rear axle seal)
- Jammed or binding self-adjustment mechanism
- Rusted brake shoe retaining pin (or clip) or other hardware that has fallen out of its proper place and into "the works"

On rear disc brakes, possibilities include:

- Binding caliper, *§6, p.268*
- Binding pad, *§6, p.268*
- Parking brake issues:
  - Most rear caliper systems employ a small drum brake that's housed within the rear brake rotor, Fig. 12–6, Type 1. Any of the drum brake issues cited above could be involved.
  - On those that employ a lever-actuated brake housed within the caliper Fig. 12-6, Type 2, the only possibilities are a frozen handbrake cable or the caliper assembly itself.

Type 1   Type 2

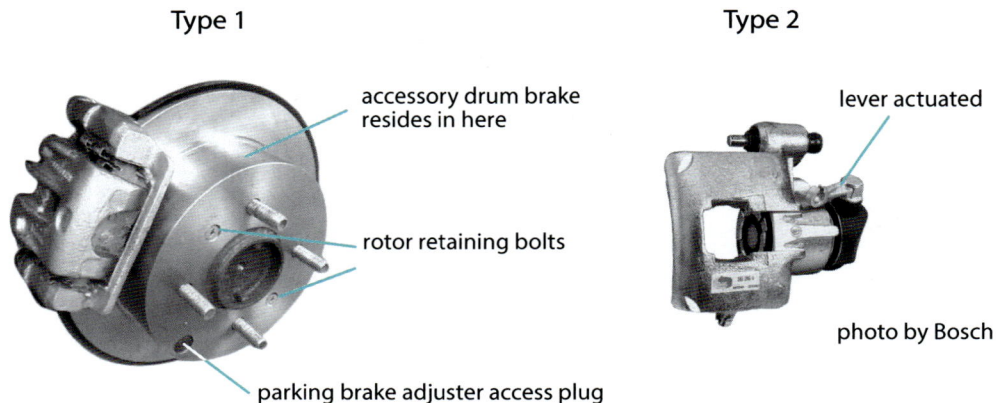

accessory drum brake resides in here

lever actuated

rotor retaining bolts

photo by Bosch

parking brake adjuster access plug

Fig. 12 – 6. Parking brake designs, rear calipers
Attribution for Type 1 caliper requested

4. Brake Noise

It's very important for you to pin down whether the noise is coming from the front or the rear of the car, not only because it helps the shop with their evaluation, but more to the point, because that knowledge will help you to avoid paying for rear brake work when you're certain that the noise is emanating from the left front!

To distinguish between front and rear, an assistant can be extremely helpful. A person who is sitting in the center of the back seat with his head held forward can position his ears about as centrally as you can get. Perform a portion of the road test with the windows rolled down, a portion with the windows completely closed.

4A. Scratching or Grinding

Most disc brake systems squeak from time to time, typically when they're applied, most commonly in the morning during the first few blocks of driving while the pads and rotors are warming to their task. Brakes that squeal more than that might benefit from some attention, *§* 4B, p.264.

Scratching or grinding sounds are far more consequential than that: Over time, that scratching becomes grinding, indicating that the friction material on at least one of your brake pads has worn away. Its metal backing is ripping one or more grooves into your brake rotor, Fig. 12-7.

To prevent that, most brake pads come equipped with a depth warning indicator–a thin metal projection designed to contact the rotor before the pad is actually spent, Fig. 12-8. As it does so, the depth indicator starts to cut a thin groove in the rotor; that scratching sound is the result. The damage is nowhere near as extensive as what the pad backing itself can do. The clock is ticking: Replace the pads *now* before the repairs cost twice as much: When caught early enough, that light scratch line is insufficient cause to replace the rotor.

The thinner the brake pads, the more heat is generated. That heat can ruin a rotor all by itself by producing hot spots in the steel or by warping the rotor. Both produce pulsations when the brakes are applied. In certain cases, both rotors can be turned–shaved thinner on a lathe to eliminate the problem. In most cases it's best to just replace them, *§* 7F, p.276.

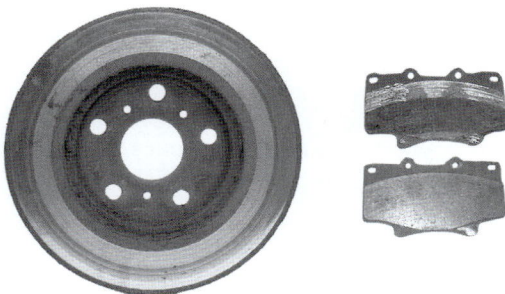

Fig. 12 – 8. Depth warning indicator clip

Fig. 12 – 7. Brake pads and rotor, overcooked

■

12

Variations

1. The earliest warning occurs intermittently: When turning left or right (usually not both), a high-pitched scratching sound emanates from the front end. That's the depth indicator clip on the thinnest pad beginning to contact the rotor.

2. The scratching sound may be evident whenever you are in motion –except when applying the brakes. Same symphony, different movement.

3. The scratching sound will ultimately become a crunching sound when the brakes are applied, becoming increasingly more noticeable at speeds of 5 mph and slower.

4. The grinding becomes continuous whenever the car is in motion.

4B. Squeaking, Squealing, or Chirping

The causes of brake squeal are listed below, along with some cures. To summarize, **OE or OEM replacement brake pads, properly installed, are always your best alternative in reducing these sounds to a minimum.**

❑ Brake pads of the wrong composition: The amount of friction inherent in the pad surface greatly affects the amount of heat that develops as the car rolls down the road. The material used must be hard enough that it will last a respectable number of miles, but soft enough that it won't squeal.

❑ Aftermarket pad kits often come with four pads and nothing else. Factory pad kits typically include anti-squeal hardware–spring clips, end plates and backing plates that cradle each pad, preventing direct contact with both the pad bracket and the caliper, Fig. 12-9. These clips are made of alloys that don't rust. Properly installed, they eliminate potential binding. If a full complement of new hardware is not supplied with OEM pads, they will be available under a separate part number.

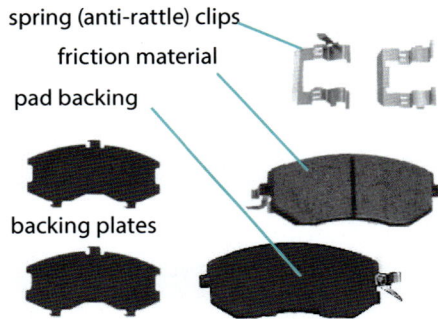

spring (anti-rattle) clips
friction material
pad backing
backing plates

Fig. 12 –9. Brake pads with associated clips

Over the course of 40,000 miles, these clips lose much of their tension and depth, so even the most conscientious mechanic in a shop selling junk will be forced to reassemble your brakes with aging clips. So right from the start, your pads can be loose in their brackets. This often produces a clack whenever the brakes are applied. A loose clip can work its way free, only to get jammed between pad bracket and rotor where it can destroy your rotor almost as well as a wasted brake pad.

❑ Frozen caliper guides/pins are a common cause of brake squeal. By not allowing the caliper to unflex when not in use, heat build-up is rapid and intense. See §6B, p.268 and §7E, p.274.

12

Whenever you have a caliper that's been binding due to a frozen guide or piston, the rotor surface will usually be glazed—the surface that contacts the pads will appear shiny as if highly polished. Glazing can result from aftermarket pads as well when their composition is too hard for the application. In these cases we use an orbital sander and emery paper (#120) to sand away the glaze. It can also be done by hand.

Fig. 12 – 10. Hot spots burned into a rotor
Attribution requested.

There are plenty of cases in which sanding is insufficient. If one of the calipers has been binding badly, the heat that's been generated will have discolored the rotor's steel from something near the color of pewter to a pale blue, much like you'd see in a watercolor. These "hot spots" will not sand out and often won't even turn out, Fig. 12-10. Hot spots cause a rotor to grab, creating brake action that feels comparable/identical to rotor pulsation.

There are a number of products available that can reduce or eliminate brake squeal. All are applied to the back side of the brake pads (the pad backing) where they perform as a vibration dampener between the brake pad and caliper piston (or pad bracket). The best ones I know of are Wurth's SBS Brake Treatment II, Disc Brake Quiet by Permatex, and Disc Brake Quiet by CRC. The first two come as an aerosol spray, the third is more of a goop.

The aerosols are akin to a spray adhesive that inhibit the formation of rust on the backing (anti-squeal) plates that separate pad backing from caliper and caliper piston, Fig. 12–9. A generous coat is sprayed across the pad backing (not so much that it starts running down the side of the pad). Once that's dried (ten minutes), the anti-squeal plates are installed over top, then another coat is applied. Allow another ten minutes for drying before installing the pads. Don't try using these compounds on any other surfaces, they'll just gum up the works.

When anti-squeal plates must be thrown out due to rust, use a thicker-bodied compound such as CRC's Disc Brake Quiet. We use the CRC version whenever aftermarket pads must be reused due to their remaining depth and the owner's slim resources; the added goopiness substitutes for that missing shim.

None of these compounds will transform junk into a good set of pads, nor can they be expected to eliminate the need for anti-squeal clips and plates in good condition.

Should you decide to replace your pads rather than live with what you've got, OE and OEM pad kits are often supplied with a packet of caliper grease for lubricating the caliper pins/guides. If it's not, pick some up or consider a different brand. Smear a thin film across both sides of the anti-squeal plates.

12

4C. Brakes That Moan or Groan

Get 'em checked; in all probability, you have a caliper that's binding badly. See *S* 7, p.270.

4D. Noises From the Rear

For rear brake caliper systems, see *S* 8A p.278 and *S* 7, p.270. What follows pertains only to drum brake issues.

The most common brake noise arising from the rear is the crunching sound that brake dust makes on a humid day. All that's required is an inspection to verify that your brake shoe linings are sufficiently deep, that each wheel cylinder is not leaking and that its pistons are freely moving, and that the axle seal is not leaking gear oil into the brake drum area. If that's all hunky-dory, a simple brake cleaning and adjustment should be all that's necessary.

Naturally, some linings won't let you off that easily. If the lining surface is glazed (shiny), they'll need a good sanding (emery cloth #120). If they've been soaked in brake fluid or gear oil, the shoes will have to be replaced. Crunching sounds can also be due to linings that have gone metal to metal. Metallic grinding sounds are of the same family. If you hear it only when braking, you'll probably find worn out linings and the need for drum refacing/replacing. If you hear it all the time, it will often be due to a rusted piece of brake hardware that's fallen into the works. For more on this, see *S* 8B, p. 279 and *S* 16, p.300.

A fluid leak of long standing, whether it be from a wheel cylinder or axle seal, will soak the brake dust as well as the linings. The result can be uneven braking or rear brake pull. Wet linings occasionally produce a moan or a crunching sound, or a kind of howling whenever the brakes are applied.

5. Vibration/Pulsation When Braking

If you experience a vibration that is felt only when braking, either through the brake pedal, the steering wheel, or the seat of your pants, you have a brake pulsation.

A brake pulsation is most easily felt when decelerating downhill from 50 to 25 mph: You'll experience the brake pedal vibrating beneath your foot. If the problem is due to a defective front brake rotor, you will feel the vibration coming through the steering wheel as well. Test this by braking with your fingertips placed lightly against the wheel.

If the problem is coming from the rear, the steering wheel will be steady while braking, but the body of the car will be vibrating in sync with the brake pedal–you'll feel it in the seat; in severe cases you'll see it in the dash.

With rear drum brake systems that are equipped with a hand brake, slow your car to 15 mph, then apply the handbrake *partially and gently,* **one click at a time** with your foot off the brake pedal, the fingers of your right hand placed lightly on the hand brake's handle. A warped drum will cause the hand brake's handle to vibrate strongly about 2–4 clicks up. The body of the car may shake as well.

> Applying the parking brake with too much vigor will lock up the rear brakes instantly. Certainly scary, potentially *very* dangerous!

On cars equipped with a foot-operated parking brake, vibrations in the body will probably be your only means of isolating rear drum vibrations, because your shoe-clad foot is nowhere near as sensitive as your fingertips. For the same reason, these systems are far more dangerous to test.

It is impossible to distinguish from the driver's seat whether it's the left or right rotor (or drum) that's warped; for that you need a brake inspection.

To eliminate *most* suspicions about driveline vibrations, shift the transmission into Neutral, foot off the gas, at 50 mph. The car is now coasting; the engine and transmission are no longer engaged with the wheels. (Unfortunately, the driveshafts and differential cannot be disengaged because they are continuously linked to the drive wheels.) Coasting for a short period prior to applying the brakes will establish a baseline from which to measure subsequent vibration when the brakes are applied. The transmission can be reengaged into Drive at any time to create further contrast.

Brake pulsations can range from very mild to frightening, depending upon the degree of warpage. Mild pulsations don't affect your stopping ability and need only be addressed if they affect your sense of security, but you should get your brakes checked anyway to prevent a minor problem from developing into something major. Severe vibrations should be repaired immediately because they markedly reduce stopping ability.

Brake pulsations are always due to one of the following:

- Warped rotor (like an old vinyl record that's been left in the sun)
- Rusted rotor
- Rotor with hot spots, Fig. 12-10, p. 265
- Rotor with non-parallel faces
- Brake drum that's out of round
- A rotor that is currently being torn up by a brake pad that has worn away to its steel backing, or one that suffered the indignity with a previous set of pads. See §7D, p. 272 through 7F, p. 276.
- Ditto for brake drums, §8B, p. 279 and §16, p. 300.

Because of the fact that disc brakes run so much hotter than drums, a rotor is much more likely to be at fault. The only way to get rid of a pulsation is to isolate the offending rotor (drum), then reface or replace it, §7F p. 276 .

It's been my experience that in most cases of rotor pulsation, both rotors are involved. One of them will be the chief offender, but the other rotor, if not addressed, will be the reason why the car comes back a second time with the comment, "You know, it's better, but it's still not right."

12

6. Brakes That Bind

6A. Symptoms

A brake that's binding will signal its predicament in one or more of the following ways:

- Poor gas mileage will be evident in all cases. After all, you're driving with your brake on!
- By a surprising amount of heat emanating from a particular wheel after a drive of 5-12 miles. You'll notice a marked difference between the amount of heat that's rising off that corner of your car as contrasted with the others, simply by walking nearby.

> **Do not touch the wheel, hub, or rotor of any brake that you suspect to be binding**–it could burn you badly.

- A lurid stench from one wheel well is another indicator.
- Pulling when you apply the brakes, §3, p.260.
- A brake that's binding fades as it gets hotter; that is, it loses some or all of its braking effectiveness. You'll notice an increasing need to stand on the pedal in order to slow the car. The more severe the binding, the more evident brake fade becomes, because fading is directly related to temperature. In extreme cases, the brake fluid hits such temperatures that it makes the pedal feel soft or "spongy."
- Moaning sounds, groaning sounds are not uncommon.

6B. Possible Causes of Disc Brake Binding

- ❑ Single-Piston Calipers only – Frozen caliper guide pins, Fig. 12–3, p.257 and Fig. 12–22, p.289: If the caliper guides are not adequately lubricated, the caliper will be unable to relax (unflex) when the brakes are released. This leads to excessive heat buildup and accelerated brake wear, often accompanied by noise. The heat can actually cause the brake rotor to warp, producing a pulsation to the brake pedal and/or steering column whenever the brakes are applied, §5, p.266. Frozen caliper guides will also produce a caliper that cannot distribute its "bite" evenly to both sides of the rotor. As a result, its brake pads will wear unevenly, and the car will often pull under hard braking (since the caliper on the other side of the car is working more efficiently).

  Caliper guide lubrication requires specially formulated high temperature grease that won't produce swelling or deterioration of the guide's rubber boots. Two good products are BG's 608BK and Permatex's Ultra Disc Brake Caliper Lube.

- ❑ Single-Piston Calipers only – Extensive rust within the pad bracket: Brake pad brackets see a lot of water. Those made of steel alloys develop rust that can render a once smooth surface cratered and rough. This is most likely to occur on older cars, those that live near the ocean, and those that sit a lot. Higher end cars employ aluminum brackets which suffer far less, though even those can develop some oxidation. The whole purpose of having a shim/anti-rattle clip between the pad

12

bracket and each end of the pad is to keep that rust away. (Shims are made of alloys that don't rust.) In most cases, just removing the old clip and installing a new one will knock down any rust that's been building beneath the clip, but in severe cases additional scraping is required. The edge of a screwdriver blade is all that's needed.

❑ All – Ill-fitting brake pads: Those that are too large to fit their pad bracket will bind right out of the box. Those that are too small usually wear funny; after a while they become cockeyed within the pad bracket. (Yes, they actually do exist . . .)

❑ All – Rust inside the caliper bore reduces the piston's ability to retract following brake application. Calipers seldom have trouble engaging, it's the retraction that makes for trouble. The only two forces at work are the brake pads' spreader clips (where present) and the piston seal's desire to return to its original (resting) state, Fig. 12–11. That's not much to work with when you consider the tremendous hydraulic pressure that's brought to bear against the piston's rear face when braking.

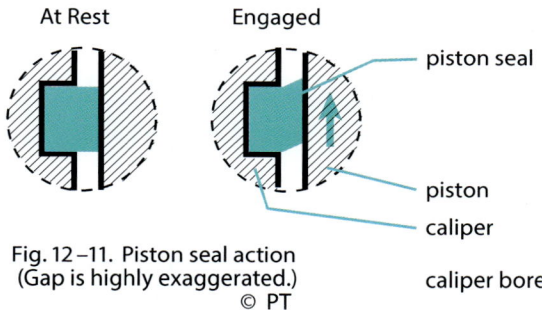

Fig. 12–11. Piston seal action
(Gap is highly exaggerated.)
© PT

piston seal

piston

caliper

caliper bore

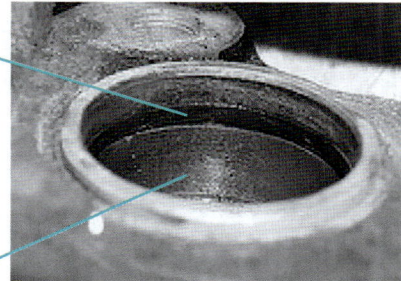

Fig. 12–12. Rusted caliper bore
Attribution requested

Rust only develops inside the caliper bore in response to water. It can enter through a torn or ill-fitting piston boot, or it can enter by way of the brake fluid. Brake fluid is hygroscopic –it literally takes moisture right out of the air.[2] For this reason, a thorough bleeding of your brake system once every four or five years makes a great deal of sense, particularly since virtually all of you have anti-lock braking systems. (ABS modulators are *very* expensive . . .) Check your owners' manual for recommended service intervals.

❑ All – A deteriorating flex line can cause a brake to bind, but its frequency of occurrence is way down the list. See § 19, p.308.

---

2   Synthetic brake fluids do not absorb water. DOT 5 was the first classification that is exclusively synthetic. DOT 3 and DOT 4 classifications are now available as synthetics. DOT is an acronym for the Department of Transportation.

12

6C. Binding in Drum Brake Systems

The most likely cause of rear brake binding involves human error – overtightening while adjusting the brakes. If your car has just been serviced and the handbrake no longer travels more than three or four clicks (from fully off to fully engaged), check for binding as follows:

With the brakes fully warmed–a drive of three to five miles will suffice –position the car on a slight incline (1-2 degrees) and bring it to a full stop. Engage the parking brake, then put the transmission into Neutral (not Park). Release the parking brake. The car should start to roll of its own accord. If it won't, try a slightly steeper incline; it's conceivable that the frictional drag of your front brake calipers could be holding it back. If the car simply won't roll, return to the shop and ask them to verify that each wheel spins freely. That will require their putting the car back on a lift.

Other possibilities for binding include:

- Frozen rear wheel cylinder (binding piston)
- Frozen handbrake cable
- Jammed or binding adjustment mechanism
- Rusted brake shoe retaining clip or other piece of hardware that has lost its mooring and dropped into "the works"

6D. When Both Front or Both Rear Brakes Are Binding

If you find that two or more brakes are binding, either the brake master cylinder or the ABS modulator is prohibiting the release of fluid pressure from those wheels. Possibilities include defective valving or just a piece of crud blocking a passageway. Sediment from the brake master cylinder reservoir can lead to all kinds of complications in systems that are not flushed on a routine basis – check your owners' manual for recommended intervals. Check the fluid in the master cylinder reservoir for clarity and the reservoir itself for sediment before anybody starts jumping to conclusions about the number of parts that will require replacement.

7. Front Brake Work

7A. Introduction

The extent of front brake work required varies with age and mileage. Assuming that you have at least some vestige of pad material remaining, read on. If your brakes have gone metal-to-metal – grinding every time you apply them – skip to §7D, p.272 and §7F, p.276.

The life expectancy of a set of pads varies with engine size, type of transmission,[3] whether you live or work in a hilly area (or use a parking garage), the car's age, teenage sons, etc. I've seen some front pad sets last 80,000 to 90,000 miles, others that didn't last six months. Most cars average 40,000 to 60,000 miles.

---

3 Brake pads last significantly longer on cars with manual transmissions.

If you come in for servicing at 30,000 miles and are told that your front brakes have roughly 25% of their pad depth remaining, what do you do? Is it too early to change them? That depends, in part, on how frequently you get your car serviced. Wearing pads down into their last 5–12% significantly increases the odds of warping a rotor. That remaining 15% of usable pad would carry you roughly 1/5th as far as you've driven so far (15% x 5 = 75%), except for the fact that thinner pads have less ability to dissipate heat. Consequently, what's left will wear out quicker.

If you've made it to 30K on 75%, you should be able to stretch those pads comfortably for another 6,000 miles. But what about the difference in cost between changing them now or waiting till your next oil change? That's a question for your service writer: Right now the wheels are off, the inspection has been made. Will the cost of the brake job reflect that fact? Or will the cost be the same now as it would be later?

Another issue: Can you rely on the accuracy of their percentage reading? What if the guy looking at your pads hasn't bothered to remove the caliper, so the 25% he's seeing would actually look more like 5–12% to a more thorough mechanic? Take a look at Fig. 12-21, p.287 and read *S* 11, p.287).

Most shops won't actually give you a percentage. They'll offer you a thickness in millimeters, which is valuable if you know (a) the factory's recommended minimum allowable thickness, (b) the pad depth of a new pad, or (c) can visualize the thickness of several millimeters. Obviously, if you're in the shop, you can see it for yourself . . .

7B. The First Two Sets of Pads

Low-mileage cars in need of their first set of brake pads should require no more than pad replacement, lubrication of the caliper guide pins (on single-piston calipers), and cleaning and adjustment of the rear brakes (if equipped with rear drum brakes).

In most cases, the same can be said for the second round of pad replacement, with one possible addition: replacing (flushing) the system's brake fluid. Check your owners' maintenance schedule to determine when it's recommended. Pad replacement does not require that the brakes be bled; however, having the front half of the system flushed will never be cheaper than it is right now.

Take a look at your brake fluid. It should be clear or nearly so. If it's brown or murky, ask the mechanic to bleed out the front half of the system before he removes the old pads. Replacement involves driving the calipers' pistons back into their bores (Fig. 12-22, p.289), this to enable the installation of thicker pads where thinner pads have been. As a piston is pressed into its caliper, brake fluid has to be displaced: It moves upward through the brake lines and ABS modulator toward the brake master cylinder. That can lead to crud becoming lodged in the passageways and valving within the ABS unit. Odds are that at this mileage the fluid will still be fairly clean, but if you're due for fluid replacement within the year, save another bill and that second trip. Get the rear fluid swapped out when the rear brakes need to be replaced.

This approach will suffice for roughly 80% of all brake jobs under 120,000 miles. Three exceptions follow:

- If you are experiencing a problem with pulling right or left when the brakes are applied, start with *§*3, p.260. Considering your mileage, suspect nothing more than a defective tire, or perhaps, a binding caliper guide pin.

- If you are experiencing a pulsation when braking, read *§*5, p.266 and *§*7F, p.276. Suspect that the reason the rotor warped was either a problem with rotor casting (dating back to its fabrication) or a binding caliper, *§*6, p.268.

- Some brake rotors just don't last. They'll have a roughed-up, gouged-out or pitted surface that's obvious to both eye and fingertip. Whether that's the fault of the pad, the rotor, or the manufacturer is impossible to determine unless there's been a factory recall, *§*2B, p.07. If the factory won't pay for it, find a better alternative than original equipment. There are plenty of OEM replacements.

7C. Above 80,000 Miles

Long about the third set of pads, you should be considering caliper overhaul. The timing of this decision rests largely upon whether the lifespan of this latest set of pads rivalled the longevity of the prior set. If so, perhaps you can put off the added expense for another round. But if they didn't last as well, ask your mechanic for his opinion on the amount of rust in the guides, whether the piston boots are showing signs of dry rot or cracking, and what the relative costs of caliper replacement vs. overhaul might be. OEM calipers are considerably less expensive than factory originals, remanufactured replacements are cheaper still. Naturally, a problem with brake pull or pulsation must enter into the equation, *§*3, p.260 and *§*5, p.266. Be sure to stay current on routine flushing of the brake's hydraulic system. The fact that you are about to embark upon the expense of caliper overhaul or replacement is not sufficient justification for doing anything to your brake rotors other than sanding their pad surfaces unless they're close to or thinner than legal spec[4] or if they're giving you problems with pulsation.

7D. Brakes That Have Gone Metal-to-Metal

If one of your pads has worn away to the point that its steel backing is now gouging a groove into a rotor (Fig. 12–7A, p.263), virtually every mechanic on the planet will recommend that both rotors be replaced. There's nothing wrong with that necessarily, except for the cost. And I wholeheartedly agree that if your brakes have been grinding for more than few days, at least that one rotor will have to be either refaced or replaced. If the damage is extensive, you will have no choice but to replace the rotor –the factory has a minimum allowable thickness for every rotor,[4] it's illegal to cut them thinner. This, by the way, is for your own protection: Rotors that are too thin will overheat–they just can't dump the heat produced by braking (friction) fast enough. The entire brake assembly gets too hot, resulting in a loss of braking effectiveness (brake fade). Rotors have been known to crack as well, producing potential loss of control.

Then the question becomes, what about the other rotor? If it's not damaged, I'd suggest you compare its thickness to a new rotor. If it's close (1-2 mm), I'd be tempted to save the money. "Conventional wisdom" states that if you turn one rotor and leave the other at its original thickness, the car will have a tendency to pull when braking. Personally, I've never

---

4   Every rotor is stamped with its minimum allowable thickness, either in it's outer circumference or centrally located overtop of the hub.

12

road-tested any car that would conclusively support that statement, but one of my technical proofreaders is convinced that *significant* differences in rotor thickness can produce a pull brought on by the resulting difference in the rate of heat dissipation between the two rotors. I guess it depends on how hard you drive your car . . .

In my opinion, a narrow groove, particularly if it's shallow, is not sufficient justification for either refacing or replacing a rotor as long as that rotor is thicker than minimum spec *and* you are not experiencing any pulsation while braking. In fact, the noise you've been hearing may well be just the sound of a depth indicator clip warning of impending destruction. Assuming low mileage and relatively even wear across all four pads, you may well get away with just replacing pads, cleaning and lubricating the caliper guides, and adjusting the rear brakes (drum brakes only). On the other hand, a wide swath of roughed-up metal requires that:

- The rotor be refaced or replaced; otherwise your brakes will both grab and pull when applied. It is typically cheaper to replace the rotor(s) with OEM parts than it is to pay a mechanic to reface what you've got.

- Caliper overhaul or replacement.

Go online and price out the parts that cost real money–rotors, pads, and calipers–before you head to the shop. That way, you have three purchase options: from the shop, from a local parts jobber, or from the internet. You will also need:

- A pair of new inner hub seals in every case that requires repacking the wheel bearings, Fig. 12-15, p.278, and §14, p.298.

- A pair of new cotter pins (stainless steel) or hub nuts (stake or compression).

Calipers come in four flavors: new from the factory (OE), remanufactured by the factory, new aftermarket calipers (OEM), and remanufactured OEM. Actually, there's a fifth variety–junk. **If you're looking at aftermarket new or remanufactured, be sure to verify that what you intend to buy is clearly marked as meeting OEM standards.**

If you're dealing with a tight budget, then be sure to evaluate all four brake pads (both sides of the car) and both calipers before deciding what needs doing.

- If, for example, one of your right pads has accomplished some real damage, but the left brake still has 20% or more of its pads remaining, then you will need caliper replacement or overhaul to correct the obvious binding that's been occurring on the right-hand side. This is in addition to pad replacement, replacement of at least that one rotor, and a rear brake adjustment (drum brakes only). See §7E, p.274.

- If both the left and right sides are evenly worn, you could possibly get away with skipping the caliper work –at least for now. But be advised: A whole lot of heat is generated by driving metal-on-metal, and the amount of damage that has occurred indicates that you've been driving like this for a while. That heat can produce cracks in the piston boots and guide boots, and it can cook the brake fluid. At a *minimum*:

  – the piston boots need to be inspected for cracking, Fig. 12–3, p.257

  – same for the guide pin boots (single-piston calipers), Fig. 12–3, p.257

12

– the caliper pistons must move freely within their bores, *§* 11C, p.292

– ditto for the caliper guide pins, *§*11, p.289.

If any one of those is a problem, the calipers–both of them–will need to be replaced or overhauled. In addition, those two calipers will have to be bled, *§* 20, p.309. If there is any suspicion that the brake fluid was cooked, the entire system will require bleeding.

7E. Caliper Overhaul, Caliper Replacement

Calipers are rebuilt in pairs, calipers are replaced in pairs: If one caliper requires attention, then both must be addressed. What's done to the one caliper *must* be done to the other. Rebuilding or replacing only one caliper can make your car's braking unpredictable, possibly dangerous.

Brake calipers should be overhauled or replaced in *all* cases that involve binding of a caliper piston. On single-piston configurations, finding widely differing rates of pad wear between the two calipers is a dead giveaway for one or both of the following:

- a frozen (binding) piston
- frozen guide pins (one or both) in single-piston calipers.

If the problem resides *solely* with binding guides, it may be possible to skip a complete overhaul, but the piston boots need to be sound (not cracked or torn, fairly supple) and both pistons have to move freely outward when the pedal is depressed. Furthermore, each must slide all the way back into its bore with little effort. That needs to be repeatable–two to three times on each side. The effort required to move the pistons back into their calipers should become easier with each squeeze. To prevent driving crud back up the brake lines, the bleeder screw should be cracked open prior to squeezing the piston back into the bore, then immediately closed. See *§* 11A, p.287 for inspection procedures.

On two- or four-piston calipers (*§* 11B, p.290), a frozen piston will always be evidenced by a strong pull when braking, as well as uneven pad wear within the same caliper. They have no caliper guides.

Caliper overhaul or replacement is a sensible approach to brake work on cars that are about to get their third round of brake pads and/or cars with more than 80,000 miles on them, particularly if the car wore through its last set of pads at a markedly faster clip than the previous set. An overhaul should also be considered if your mechanic advises it, but insist that his reasons be specific and clear.

For example:

- The caliper piston on the left side is frozen in place and won't retract (slide back into its bore).
- Both of the caliper piston boots are badly dry-rotted and cracking.
- The left caliper piston boot is torn.

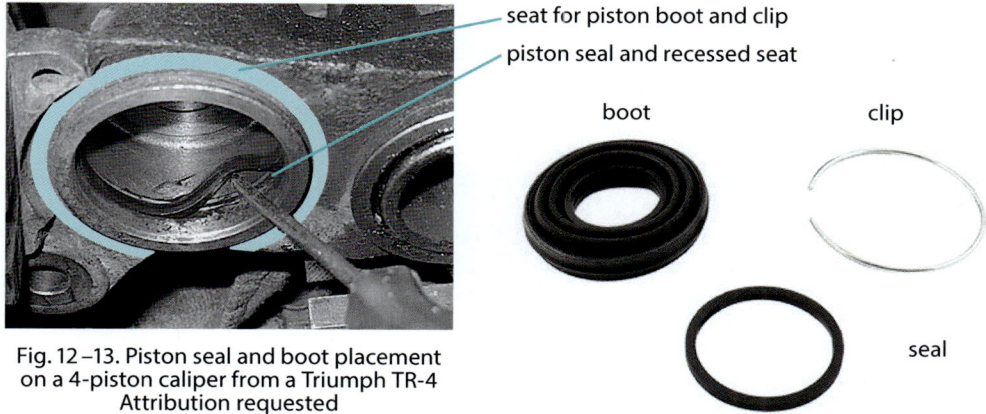

seat for piston boot and clip
piston seal and recessed seat

boot

clip

seal

Fig. 12 –13. Piston seal and boot placement
on a 4-piston caliper from a Triumph TR-4
Attribution requested

Better yet, ask him to show you. In all cases involving a torn piston boot, have that caliper disassembled to inspect for rust in the caliper bore–and on the piston for that matter, §12, p.294. Or save the cost of inspecting their innards and just replace them.

The question then becomes overhaul? Or replace? New OE calipers can be quite pricey, but they are pretty, and they are new. Factory remanufactured units can cost about the same as an overhaul, particularly if a piston and some guides will be needed. With today's labor rates, OEM calipers are usually less expensive, particularly if you supply them, but most are remanufactured units. (Nothing necessarily wrong with that; it all depends on who did the overhaul and the quality of the parts used.)

Remanufactured calipers do not necessarily have rust-free bores. No matter how well reconditioned they are, once rust has started in steel, it will always return. That being said, I don't want to leave you with the impression that they are unreliable or should never be used. Just be sure to verify that the ones you're looking at meet OEM standards. Compare their cost with factory rebuilt units.

Every overhaul should include external rust removal followed by attention to the interior. That includes scraping out all crud and rust within the piston seal recess and from the recessed ledge where the piston boot resides, Fig. 12–13. This is followed by honing the caliper bore and a thorough cleaning inside and out using brake cleaner.

Honing is a process that removes surface rust and crud from the caliper bore by polishing the surface with a set of three spring-loaded wetstones spinning at the end of a drill. Light rust responds nicely. Heavy rust that has left significant pockmarks in the caliper wall (Fig. 12–12, p.269) is best addressed by replacing the calipers. For judgment calls lying somewhere in between –small pockmarks scattered within the rust–most can pass muster with a good honing and cleanup. Note that the pocks are not removed; doing so would destroy the caliper by taking its bore way past maximum tolerance. While there is some leeway to the extent of rust allowable within the caliper bore, there is absolutely no leeway regarding its piston(s). Any piston whose highly polished chrome surface has become pitted with rust will have to be replaced, preferably with a factory original or OEM substitute. Some cars carry ceramic pistons; they obviously won't rust, but older ones do crumble.

12

Pistons don't develop rust unless the caliper bore has some significant rust already. If your caliper overhaul is going to require one or more pistons, consider caliper replacement.

7F. Refacing Rotors, Replacing Rotors

The terms *refacing* and *turning* both refer to the use of a brake lathe to remove metal from both sides of the rotor. Done right, it produces flat, smooth, and parallel braking surfaces.

A typical brake lathe can cut fast or slow. One or more fast cuts are made initially to get the rotor past its damage. This should always be followed by a slow cut to provide a smooth finish. Fast cuts produce a surface that has a very uniform, very repetitive series of concentric circles with distinct ridges that you can literally feel with your fingernail. By itself, this type of cut eliminates problems with rotor pulsation, but leaves ample cause for noise – a zinging sound when the brakes are applied. Most shops have only one lathe; in busy shops, most mechanics make just one pass.

I've never been a fan of refacing rotors. For one thing, they'll be thinner: less able to dissipate heat, more susceptible to brake fade, warping, and cracking. Also, the typical mechanic in a busy shop will not be able to produce a surface as fine and finished as on a new rotor. You'll be paying for labor, money that could be applied to the cost of a new part. Labor rates being what they are these days, you may find that a brand new OEM rotor will actually cost you less. But be advised: While they should work just as well as their factory cousins, some will not be as pretty. Particularly on higher end cars, the alloys used by the factory will be superior: The hubs of OEM rotors will be far more likely to rust.

One last consideration: Should you choose to reface, ask that your repair order include the thickness of your rotors following refacing, as well as the rotor's minimum legal thickness.[5]

Most shops want to sell rotor turning or replacement with every brake job. In my book, there are only four reasons to replace/reface rotors:

- If one of your brake pads has started to gouge the rotor
- If your rotors have been visibly roughed up by their brake pads
- If your car is suffering from brake pulsation
- If the rotors have hot spots, Fig. 12-10, p. 265.

For those of you being told that you really *have* to get this work done, ask them why.

- "To prevent brake noise" is not a good answer. Whether the brakes were noisy prior to this brake work is irrelevant unless it involved grinding sounds. Warning sounds from a depth indicator are not sufficient cause unless you let it go on too long, cutting a canyon into the disc. (Perhaps you should consider turning off your radio once a week . . .)

- "To prevent brake pulsation" is even more sketchy. Did you have a pulsation when braking prior to today?

- "The braking surfaces are no longer plane (flat)." If that's true, then the car had a pedal pulsation before you drove in.

---

5 Every rotor is stamped with its minimum allowable thickness, either in it's outer circumference or centrally located overtop of the hub.

12

- The rotor's surface is glazed. If so, why not just sand them?

So how can you tell if a rotor has been turned? If you had a rotor pulsation to begin with, then its absence should be sufficient proof for most. But if you need to take a peek, those cutouts in your wheels can occasionally provide decent access. Those of you with full wheel covers (hubcaps) have to pry them off first. A rotor that has been refaced should be smooth, flat, and shiny, as if it has just been polished. Metal that appears dark grey, particularly if it has darker lines running at random across the surface, and rotors with distinct, circular, concentric ridges (or grooves) have not been turned. On second thought, that's not entirely true: A fast cut produces distinct, circular, concentric ridges, but they will be absolutely uniform and the steel will be shiny, not grey.

Brake rotors come in two flavors:

1. By far the most common are those that slide over the outboard side of the wheel hub, Fig. 12-14. They are employed on most FWD sedans and SUVs. Outboard rotors are accessed simply by removing the caliper and pad bracket, Fig. 12-2, p.256. Because the hub stays home, its wheel bearing(s) require no attention.

2. Many older rear-wheel drive cars, most half-ton and larger trucks, and some FWD sedans and SUVs have their rotors mounted to the inboard side of the wheel hub, bolted to it with four or five stubby bolts, Fig. 12-15, p.278. Should the rotor require replacement, the hub must first be removed to access those retaining bolts. That can get expensive:

Fig. 12 – 14. Outboard rotor

The inner and outer wheel bearings are packed with grease, as is the hub (though not to the same extent). Since refacing a rotor creates lots of metal shards, and since grease is the equal of any magnet, it's always a good idea to request that your hubs be thoroughly cleaned in a parts washer following their trip to the lathe. Since all it takes is one piece of steel lodged in that old grease to completely trash a bearing, it is always sound practice. Note that the inner hub seal and bearings are removed prior to the hub's trip to the lathe.

Rotor replacement eliminates all concerns over getting steel shards in a wheel bearing, but specs of rust breaking loose as the rotor is separated from the hub are not uncommon. (Separating rotor from hub usually requires a solid whack with a hammer.) However, as long as the hub seal is in place and a rag is shoved into its center, problems are minimal. Even so, the bearings should be repacked, the inner seal must be replaced.

There's another approach to all this: Portable lathes that bolt right up to the car–typically to the pad bracket's bolt holes–enable rotor refacing to be performed in place without removing the hub. These lathes are expensive and not that common, so if you do need your rotors turned, see if you can find a shop that has one. Paying 2.5 to 3 hours' labor for a mechanic to disassemble each side is no small price when contrasted with the 1.5 hours per side that you might pay in a shop that's properly equipped. Minor to light-moderate damage[6] to an inboard front rotor on a FWD car or truck is the only circumstance that would lead me, personally, to refacing, and then only if I had access to a shop with a portable lathe. In all other cases, I'd pay to have the rotor(s) replaced.

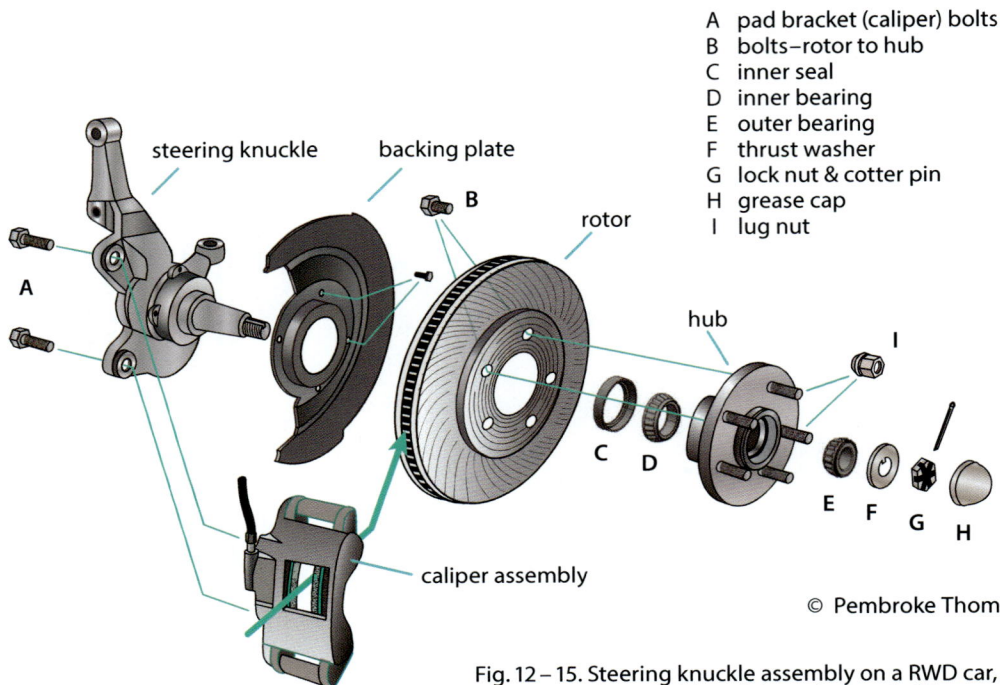

A pad bracket (caliper) bolts
B bolts–rotor to hub
C inner seal
D inner bearing
E outer bearing
F thrust washer
G lock nut & cotter pin
H grease cap
I lug nut

© Pembroke Thom

Fig. 12 – 15. Steering knuckle assembly on a RWD car, shown with an inboard rotor © PT

8. Rear Brake Work

8A. Disc Brakes

Everything that's been said about front disc brakes (*§* 7, pp. 270-276) applies to rear discs as well, with two exceptions–one liguistic, the other meaningful:

- Since rear wheels don't steer, the steering knuckle is replaced by a spindle. Figures 12-29, p.301 and Fig. 12-31, p. 303 show spindles used in conjunction with a drum brake.

- There's a parking brake involved. As illustrated in Fig. 12–6, p.262, your car might have a small drum brake housed within the brake rotor (Type 1), or the caliper itself might house a levered assembly (Type 2).

---

6 On second thought, I'd leave the rotor as is if the minor damage hasn't produced a pulsation while braking or pulling. Moderate damage, yes; a shallow scratch, probably not.

All Type 1 rotors are mounted outside the hub. Their drums house dinky little brake shoes with thin linings and no wheel cylinders–the parking brake cables are their sole activator. While it's conceivable that the parking brake shoes will wear out about the same time as the rear pads, it's unlikely–particularly the first time around–unless you have a habit of driving around with the parking brake engaged.

As for Type 2 calipers . . . Maybe they've gotten better over the years. We used to find a lot of rust in them, water seeping in around the parking brake lever's shaft. After attempting to repair a bunch of them, I ultimately decided to just replace them rather than attempt another overhaul. Replacing brake pads on calipers exhibiting no particular problem requires a special service tool: The piston must be rotated back into its bore to accommodate the new, thicker pads, *§* 15, pp. 300.

8B. Drum Brakes

Rear linings usually last 40,000 to 60,000 miles. The leading (forward) shoe always wears out much faster than the trailing shoe because the car's load will fall disproportionally toward the front.

Whenever linings need replacement, the rear wheel cylinders need to be evaluated. Are they presently leaking? By all means replace them. Are they seeping? If so, change them now if you can afford to. Leaving them in place until they've sprung a leak will cost you at least the price of a second brake inspection, an additional labor charge for installation, and in some cases, an accident. If need be, ask your service writer to break the costs down for you: doing it all in one shot vs. the shoes now, wheel cylinders later. If you have a frozen wheel cylinder, then you have no choice but to replace both.

If both wheel cylinders are dry and their pistons are freely moving, then there is really no good reason to replace the cylinders, particularly on a lower-mileage car. However, on older cars there's a tendency for wheel cylinders that have been disturbed to start leaking, particularly if the brake fluid has not been routinely replaced. When new shoes are installed against an original equipment cylinder, its pistons are driven inward toward one another to accommodate the new shoes' added thickness. Any crud that has accumulated within the cylinder bore that's directly behind the piston seal will tend to flip the seal's lip outward, providing a pathway for future leakage. So if you decide to pass on wheel cylinder replacement, monitor the brake master's reservoir for signs of fluid loss, and get your rear brakes reinspected 6,000 to 12,000 miles following rear shoe replacement.

Wheel cylinders must be replaced in pairs to assure even braking across the rear, even if only one cylinder is leaking. The other cylinder is experiencing much less fluid pressure than it will once that leak is eliminated. And don't forget that it's the same age as the leaky one, with just as much crud in it. There's no reason to believe that it will last once it's being stressed again. Change 'em both and live a little! They're not particularly expensive.

Some shops love to sell wheel cylinder overhauls. This approach provides the mechanic with a much bigger labor ticket than simple replacement. New OEM wheel cylinders cost less to install than overhauling your originals.

12

As for brake shoe replacement, there are several points that should be mentioned. The leading shoe and trailing shoe differ one from the other in the positioning of their lining material (relative to the shoe's metal backing), Fig. 12–32, p. 303. You would be absolutely amazed at how often we find these shoes reversed –perhaps 20% of the aftermarket shoes we see. And it's not that uncommon to find the two leading shoes on one side, the trailing shoes on the other (maybe another 12%).

Some shops, most notably the big chains, like to sell brake hardware: the springs, retaining clips, etc. that are used to secure the shoes to their backing plates and to each other, Fig. 12–5, p.258. This stuff really doesn't go bad that often, due to the fact that all of it is enclosed by the brake drum and backing plate. By contrast, the anti-rattle clips and anti-squeal plates found on disc brakes are continuously exposed to the elements. You might find that cars from the Northlands need a few pieces replaced due to rust by 120,000 miles, but you sure won't see that in Virginia at 50K. Again, ask for your old parts, or better yet, wait on the car until it's been checked out.

As for turning drums, I see no reason to do so unless:

- The drum is warped, in which case you've been experiencing brake pulsation, §5, p.266 and §16C, p.304.

- The drum has a lip that will prevent proper adjustment, §16, p.300

- The lining surface of the drum has been torn up by your going metal-to-metal, in which case you've been experiencing some rather intense grinding sounds back there whenever you hit the brakes. As with brake rotors, some drums are just too thin to turn. Once again, it's a good idea to insist on finding out the thickness of your drums following refacing, alongside the drum's minimum legal thickness.

9. Evaluating Brake Work

- ❑ If you've done your homework, then you road tested your car following the procedures of §2, p.259 before you dropped it off for servicing, providing you with a baseline against which to measure the work performed. Upon its return to you, the car should track straight and stop straight –assuming, of course, that your tires are in good shape and that the alignment is sound.

- ❑ The brakes should be quiet.

- ❑ The brake pedal should not pulsate under your foot when the brake is applied.

- ❑ The brake pedal should feel firm and solid. This assumes that the car has been adequately road tested following pad or shoe replacement: New linings at either front or rear usually take a few miles to "burn in." During that period, the brake pedal often feels somewhat spongy and less effective than it did prior to lining replacement. This can be particularly noticeable when both the front and rear linings have been replaced.

  Here's a classic example of why you should write down your odometer reading at the point of dropoff. If the mechanic drove the car for six miles and the brakes are still spongy, it's fair to assume that another few miles won't help. Even so, unless something

12

seems really wrong, I'd suggest that you evaluate the brakes for a day –nothing so unpleasant as returning the car three minutes after you've picked it up–level heads might consider you a bit hasty.

❑ All of the brake lights should work. While inspection of the tail lights is not necessarily within the domain of the brake work you payed for, failure to do so reflects a lack of attention to detail.

❑ Drum Brakes only: Front brake replacement does not necessarily require rear brake adjustment. However, just as with tail lights . . . enough said.

Adjusting rear drum brakes alters the point at which the brake pedal begins to slow the car, raising the point of engagement closer to the driver. Because the rears are now tighter than they were before, your foot should travel a shorter distance to the point at which the brakes start to engage. For the same reason, adjusting rear drum brakes reduces the number of clicks when pulling on the handbrake lever.

❑ Engaging the parking brake should encompass 3–6 clicks. One click more or less is not worth getting excited about as long as the car will pass the following two tests:

- Park on a steep grade with the parking brake applied and verify that the car won't budge with the transmission in Neutral (not Park).

- With the brakes fully warmed up –a drive of 3–5 miles will suffice –position the car on a slight incline (~ 2 degrees) and bring it to a full stop. Give your curbside tires enough clearance that they won't get hung up on the curb. Engage the handbrake and put the transmission into Neutral (not Park). Release the handbrake and see if the car will start to roll of its own accord, the weight of the car should overcome the frictional drag of your front calipers without a whole lot of help. If it won't, try a slightly steeper incline. Naturally, if the parking brake traverses no more than one or two clicks before it's fully engaged and the thing simply won't roll, you should return the car and ask them to verify that each wheel spins freely. That requires their putting the car back on a lift.

❑ Neither pad replacement nor rear shoe replacement requires that the brakes be bled. However, pad replacement does involve driving the calipers' pistons back into their bores, this to enable the installation of thicker pads where thinner pads have been. As a piston is pressed into its caliper, brake fluid has to be displaced: It moves upward through the brake lines and ABS modulator toward the brake master cylinder. That can lead to crud becoming lodged in the passageways and valving within the ABS unit, particularly on higher mileage cars that have not had their brake fluid flushed routinely. So neglected systems should have a thorough flush before the pads are replaced.

Assuming a clean system that's not being flushed, that brake fluid can produce master cylinder overflows. Brake fluid can discolor paint–particularly if it's not cleaned off. It can also mess with electrical junctions, coolant hoses, etc. If needed, an aerosol can of brake cleaner is the solvent of choice within the engine compartment; it can be used on any of the above **as long as the engine is off.** (Breathing the vapors is terrible, and a hot engine can light it up.)

Something far gentler should be used on fenders. Brake fluid is not water-soluble, so soap and water are typically a waste of time. Most household cleaners will do the trick, but also they will take off most anything else, particularly wax polish. Those with bleach can discolor paint if not rinsed away promptly.

Many master cylinder reservoirs are covered by a butyl rubber cap. Butyl rubber works great as a cover, but when continuously exposed to brake fluid, it gradually deteriorates, clouding the fluid and eventually turning it brown. Check your reservoir level: Most have two lines etched into the reservoir wall that indicate maximum and minimum fill levels. If overfilled, use a paper towel to soak up the excess so the cap won't be constantly exposed.

❑   Each of the following operations requires that the brakes be bled:

- Replacement of the brake master cylinder, brake vacuum booster, or ABS module

- Caliper replacement or overhaul – Remember, these should always be addressed in pairs.

- Rear wheel cylinder replacement or overhaul – Likewise, these should always be addressed in pairs.

- Flex line replacement

Of these, only the first group necessitates bleeding the entire system. It's not essential to address the entire system if the work is limited to one axle or the other. However, combining routine brake fluid flushing with either front or rear brake work is the least expensive way to accomplish that work. Having it done as a separate line item when you go in for an oil change, for example, is a total waste of money. Everyone needs brakes from time to time; do it then!

Three issues come to mind:

- You will almost always be able to tell whether bleeding has been performed by the residual wetness that brake fluid leaves behind. Brake fluid feels oily and darkens the areas it hits. It's fairly easy to examine the front end: Unlock the steering if necessary, then turn the steering wheel hard over. You'll now have access to one of the calipers, sometimes simply by squatting down next to the wheel, sometimes by standing on your head . . . You'll need to crawl partway beneath the car to gain access to the rear.

- Every wheel cylinder and caliper on this planet left the factory with a rubber cap overtop its bleeder screw, Fig. 12–22, p.289. And every butcher shop between Manhattan and Malibu leaves them to litter their floor. Verification of their presence is worth crawling under the car–they serve to protect the bleeder from dirt and corrosion.

- Drum brakes all have an adjuster, most are accessed through a slot in the backing plate. Every one of them should have a rubber plug.

10. Brake Inspections and Repairs – DIY Detail

> To protect both your safety and your pocketbook, please read each section in its entirety–including all other referenced sections–before you begin. Start with *§*2, p.112 and *§*3, p.113.

12

10A. Initial Steps

1. Start with a thorough road test, *§*2, p.259.

2. Crack the lug nuts loose while the tires are still on the ground, Fig. 12–16. Be sure to engage your parking brake first. Don't back off the lug nuts more than 1/2 to 3/4 turn; you want the wheels to be secure for Step 4 and *§*12B.[7]

3. Raise the car off the ground, support it on jack stands, *§*4C, p.118, *§*7A, p.139 and *§*7D, p.140. Be sure to engage the parking brake prior to raising the front end, then release it once the car is secure on all four stands. Re-engage the parking brake prior bringing the car back down.

Fig. 12 – 16. Loosening lug nuts without an air gun (parking brake ON) to

4. A worn wheel bearing or steering component can affect driveability, making it difficult to differentiate pulling/wandering problems associated with the brakes from those issues.

    Check for loose wheel bearings as follows: Grab each tire at top and bottom and attempt to rock the wheel by alternately pushing inward with one hand while pulling out with the other, then reversing those movements. Perform the same test with your hands at the front and rear of each tire. If you find any wobble at all, read *§*11, p.342.

    Front wheels only: If the wheel is rock steady when your hands are at top and bottom, but wobbles ever so slightly when they're positioned front and rear, it's likely that a tie-rod end is worn. See *§*13A, p.348. Note: You are not trying to turn the wheel; rather, you are looking for slight movements before any significant resistance is felt. Uneven tire wear almost always accompanies free-play in the steering.

---

7 It's virtually impossible to loosen fully-torqued lug nuts once the car is in the air–unless you've got a 1/2 inch-drive air gun handy. A live body standing on the brake pedal will suffice in a pinch, but it's much easier to accomplish this work solo on the ground.

10B. Testing for a Binding Brake

> **Binding brakes produce a tremendous amount of heat. If any wheel feels or smells hot, let it cool down for a half-hour before removing the wheel. Thick leather gloves are an alternative, but proceed with caution–a hot wheel foretells a red-hot rotor and brake assembly.**

1. Put the transmission in Neutral (not Park); release the parking brake.
2. Spin each wheel to inspect for binding. **Be sure to use the tire to spin the wheel to eliminate any chance of burning your hands**, while at the same time maximizing your leverage.

   Don't jump to any conclusions when you discover that differences exist in the amount of drag exhibited between front and rear. Disc brakes invariably exhibit more drag than drum brakes, and drive wheels always have more inertia[8] to overcome due to their driveshafts, differential gearing, etc. So, for example, front-wheel drive models with disc brakes at all four wheels will exhibit significantly more drag in the front than in the rear. You are comparing the relative amount of drag between left front and right front, or between left rear and right rear.

10C. Visual Inspection for Brake Pulsation

Remove all four wheels. Release the parking brake if you have not already done so.

10C1. Front Brake Rotors, Front Wheel Drive (FWD)

As discussed in *S* 7F, p.276, brake rotors come in two flavors:

- Outboard–To ensure that the rotor aligns flat against its hub, some manufacturers employ a pair of flat-headed bolts, centrally located on opposing sides of the hub (Fig. 12–6, p.262). The rest are secured solely by their wheel: As the lug nuts are tightened sequentially to their proper torque, the wheel squeezes the rotor up against the hub. Once the wheel comes off, you may find that the rotor is loose against the hub. If so, secure it by threading two or three lug nuts onto opposing lugs, just past finger-tight. Those that have flat-headed bolts securing rotor to hub are good to go as is.

- Inboard–Because these rotors are bolted to their hubs, they are ready to inspect as is. Should the rotor require replacement, the hub must first be removed to access those retaining bolts.

Start the engine. If necessary, let it warm sufficiently to achieve its normal idle speed (800 RPM +/-). With your foot on the brake, engage the transmission in low (automatics) or first gear (manual transmissions), then *slowly* release the brake (and clutch, if relevant) so that the front wheels *gradually* take on power. One or both of the front wheels will be slowly turning.

---

8   The tendency for a body at rest to remain at rest; or if moving, to continue moving in the same direction.

12

> Never throw an automatic transmission into Park without first bringing the wheels to a complete stop. Use the brake pedal; keep it depressed until the trans shifter is set in Park. Failure to heed this warning is guaranteed to create a horrifying sound, and on occasion, significant damage to the transmission.

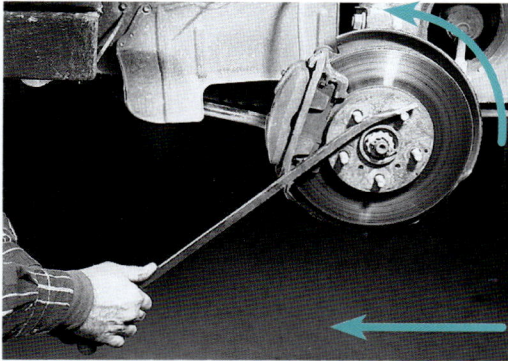

direction of rotation

Fig. 12–17. Proper placement of prybar to prevent rotor rotation. Rest its handle on the ground, hold it in place until the wheel lugs take up the slack.

This rotor was frozen to its hub; no need to secure it with lug nuts.

arrow points to front of car

At very low speeds (idling in gear), a warped brake rotor won't spin at a continuous rate. It will alternately speed up, then slow down with every revolution. The point at which it slows is the section that creates the pulsation you feel. If you are having trouble seeing this, ask that an assistant apply a very slight amount of pressure to the brake pedal to slow the rotor further. Be sure to compare both sides.

If you find that only one side is turning, don't conclude that the stationary one is binding. The more likely explanation is that the differential is delivering power unevenly. Inspect that free-turning rotor first as described above. Then have an assistant bring both rotors to a full stop by gently applying the brakes. Have him put the transmission in Neutral–**brakes still engaged**–while you lock up the first rotor using a long, sturdy pry bar positioned as shown in Fig. 12–17.

Now have him re-engage first gear, then *slowly* release the brake as you verify that your pry bar stays put. **If not properly secured, the pry bar could kick out; if placed on the wrong side of the hub, it will quite literally fly off into the fender. . . Using a round shaft, while feasable, can be dangerous.**

An alternate approach to rotor evaluation is to use a micrometer (Fig. 12–18, p.286) to measure the relative thickness of the rotor at random points around its circumference. Variations in these measurements indicate nonparallel surfaces. This approach becomes necessary in more subtle cases where the "quick and dirty" procedure above provides no obvious conclusion. An even more precise test involves the use of a dial indicator (Fig. 12–19, p.286) on each rotor to measure its "run out" (the extent of variation from its central vertical plane). Dial indicators are delicate and can easily be knocked out of alignment. Using them can be frustrating, but their accuracy is undeniable.

Fig. 12 – 18. Micrometer

Fig. 12 – 19. Dial indicator

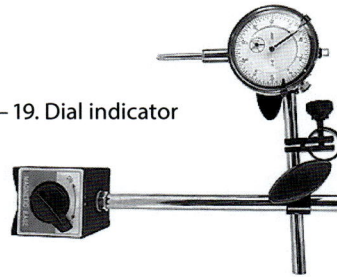

10C2. Front Brake Rotors, Rear Wheel Drive (RWD)

To prevent your fingers from getting pinched between caliper and hub, use the wheel lugs to turn the rotor. Rotate in the direction shown in Fig. 12–17, p.287. This may require both hands–even with perfectly normal brakes–but the amount of effort required should be uniform throughout each revolution. If not, mark the sector that's binding–just scratching two lines through the rust buildup that resides on the rotor's outside circumference will suffice–then see if the binding repeats every time the rotor returns to that spot. If so, the rotor is warped. Repeat for the opposing wheel. The amount of effort required to spin each side should be equal or close to it. If not, the "reluctant" rotor indicates some binding to that caliper, pad set, or both.

A significantly warped rotor will be obvious even to the novice, but slight pulsations may require the use of a micrometer or dial indicator.

10C3. Rear Brake Rotors (FWD)

This inspection is virtually identical to §6B, p.268. The only real difference is that there's a parking brake involved here. Obviously, it needs to be released prior to the inspection.

10C4. Rear Brake Rotors (RWD)

This inspection is virtually identical to §6B, p.268. The only real difference is that there's a parking brake involved here. Obviously, it needs to be released prior to the inspection.

10C5. Rear Brake Drums (RWD & FWD)

On all rear-wheel-drive cars, the brake drum simply slides over the lug bolts onto the rear axle shaft flange, Fig. 12–20A. In most case, the drum will be frozen to the hub. If not, secure it with a pair of lug nuts prior to inspection. On front-wheel drive cars, the rear drum is suspended on a spindle, Fig. 12–20B. A pair of bearings secured by a locknut are hidden beneath the central grease cap; it's ready to go as is.

To assure that the shoes are centered, pump the brake pedal once or twice (handbrake off). Then simply rotate the drum with both hands. Virtually all drum brakes will produce a "shushing" sound as they are rotated. That's not a problem; it's due primarily to a buildup of brake dust. What *could* be a problem is a drum that binds with every revolution in the same quadrant.

Fig. 12 –20A. Drum brake, RWD

Fig. 12 –20B. Drum brake, FWD

12

Brake drums seldom produce brake vibrations. It takes a lot of heat to produce hot spots, and a whole lot more to actually warp the drum. Naturally, if you're in the habit of driving with the parking brake engaged, you might just be that lucky guy. An overtightened parking brake adjuster can accomplish the same thing. Drums, like rotors, can develop hot spots, but only if the parking brake cable is binding or if a wheel cylinder is frozen. In both cases, your gas mileage will be terrible and hard braking should be a dead giveaway.

Drums, like rotors, can be refaced on a lathe. Drums are stamped with a maximum allowable interior diameter.

11. Disc Brake Inspection

11A. Single-Piston Calipers

Viewing the front brakes is facilitated by turning the wheels outward. If necessary, turn the ignition key to the accessory position to disengage the steering lock.

The only way to accurately check pad depth is to separate each caliper from its pad bracket. Even though virtually all calipers offer a "window" to peer through, that view shows only the center of the pad. Most brake pads wear out unevenly due to the way that a car's weight is thrown forward and down when braking; one end of each pad usually wears out faster than the easily visible middle, and one of the four pads will often wear out faster than the other three, Fig. 12–21.

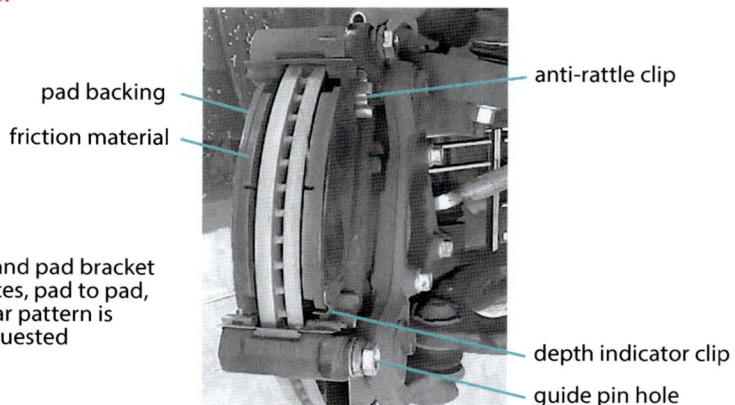

pad backing

friction material

anti-rattle clip

depth indicator clip

guide pin hole

Fig. 12 – 21. Brake pads and pad bracket Note the wear differences, pad to pad, top to bottom. This wear pattern is Normal. Attribution requested

■

Naturally, if the car is not exhibiting any problems and you find that your "window" view indicates that all four pads are (a) close to the same thickness and (b) relatively thick (maybe 3/4 inch or better), then you can probably get away with slapping the wheels back on and getting on with your day. But if you do, you'll miss out on all the fun of lubing the guide pins, verifying that the pistons are moving freely, checking the piston boots for dryness and cracking—not to mention getting really filthy in the process . . . And by removing those calipers, you'll get to see the entire edge of each pad!

Brake pads should be replaced whenever you find:

- that a depth indicator clip has started cutting a groove into its rotor, it's getting close to that point, or

- the friction material is less than 3/16 of an inch (3 mm) thick on any one of them, measured at its thinnest point. The pad backing is not included in this measurement, just the pad itself.

Single-piston calipers are secured by either two caliper guide bolts or by one guide bolt and one guide pin. Like virtually any other bolt, these loosen counterclockwise, which means that if you are facing the car, it will appear that you are backing them out clockwise. These bolts are *tight*; start with a six-point wrench. To save your hand, use a hammer to smack the wrench's free end (Fig. 8–7, p. 142), then finish with a ratchet. Clean and inspect them. Any that have rust or extreme wear will have to be replaced. Wrap them in a rag and put them aside.

❑ Two Bolts: Remove them both. Pull the caliper straight off the rotor. Secure it. Many of you will find that the caliper won't budge (due to rust, crud, or binding). Grab a flat-bladed screwdriver, insert its tip either above or below the caliper between it and the pad bracket, then pry it out.

❑ One Bolt, One Pin: Remove the bolt. The caliper will rotate outward on the pin. If the caliper won't budge, read the paragraph directly above. Once clear of the pads, the caliper will slide off the pin. Secure it.

> **Never let a caliper hang free, dangling solely by its hydraulic flex line.** Rig some wire or a heavy-duty wire tie either from the coil spring or another suspension component above.

The brake pads should pop in and out of their pad bracket fairly easily. The amount of effort for any one of them needs to be within the same ballpark as the other three. If you need to apply force or leverage to make any one of them move, then that brake is binding—assuming that you haven't mis-aligned the pad within its bracket while trying to get it out. Inspect the pad bracket for a buildup of rust where the pad ordinarily sits. Clean up any shims or spring clips that lie between bracket and pad. Use brake cleaner, a wire brush, or a pick to remove any crud hiding beneath them. If they're bent they'll need to be replaced.

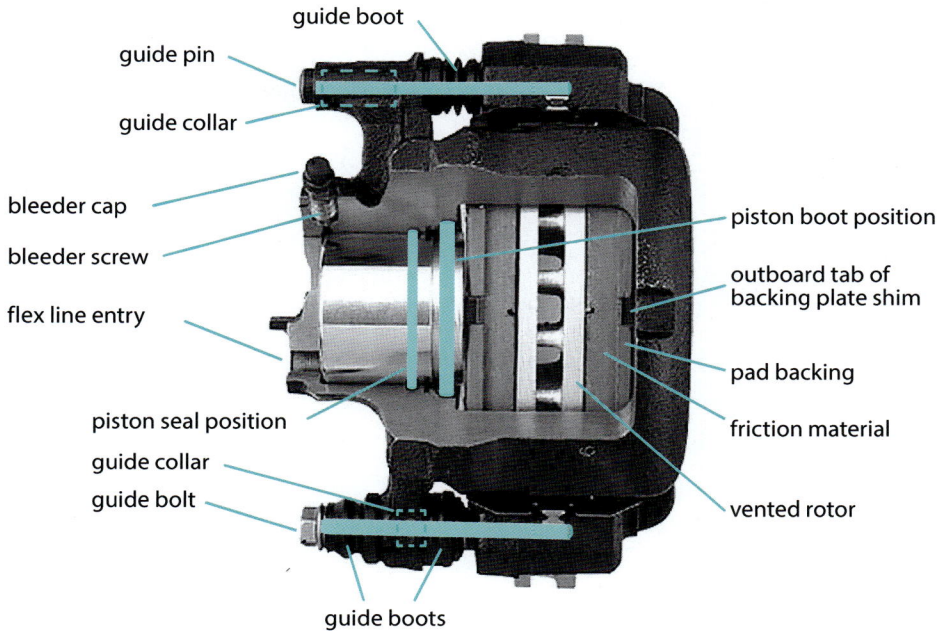

Fig. 12 – 22. Cutaway view of R/F single-piston caliper
Courtesy of Subaru Motors

Inspect the shims that cling to the pad backing–those separating the pads from the caliper's claw and its piston. (These shims are not shown in Fig. 12-21; in Fig. 12–22, just the outboard tabs are visible.) If they're rusted or distorted they'll need to be replaced, discarded, or treated with goopie, § 4B, p.264. Simply removing them without *some* treatment can produce screeching or squealing.

Each guide pin passes through the caliper, then a boot, Fig. 12–22. Guide bolts can pass through two boots, but not necessarily. The holes through the caliper contain a collar which is typically removable/replaceable. To get them out for inspection, first remove the boot(s), either by squeezing and pulling them out, or by using a small flat-bladed screwdriver to pop them off. Some collars just slide right out, others need to be persuaded by tapping them out using a blunt-ended punch. Replace any that are rusted or badly worn. Use a battery terminal brush (or comparable) to clean out any rust within both collar's nesting places. The boots should be supple and flexible. Replace them all if any one of them is cracked or brittle; they come as a set with every caliper overhaul kit.

Assuming nice guide boots and happy collars, clean them all, then reassemble with plenty of caliper grease. Wipe up any excess.

> Caliper guide lubrication requires specially formulated high temperature grease that won't produce swelling or deterioration of the guide's rubber boots.

12

If the piston boots are cracked or dry-rotted, then caliper overhaul or replacement is not far off. To determine which–and how soon–inspect each piston's freedom of movement within it's caliper bore, §11C, p.292. If both pistons are moving freely *and* everything else so far passes muster *and* you're not yet in need of pads, then hold off till you need the pads. Otherwise, there's no time like the present!

In *all* cases involving a torn piston boot, you'll need to disassemble that caliper to inspect for rust, §12A, p.294.

Inspect the rotors for hot spots or glazing, §4B, p.264. Replace the rotor(s) if you find hot spots (Fig. 12–10, p.265) or if they're thinner than their minimum spec. Sand them if they're glazed.

To reassemble, reverse your steps. For information on:

- pad replacement, see §13, p.297
- caliper overhaul, §12A, p.294
- rotor replacement, §14, p.298
- bleeding the brakes, §20, p.309
- parking brake adjustment, §15, p. 300 and §16D, p.304

> Pistons that have been pushed back into their bores will usually not be in contact with their brake pads. The result? NO BRAKES!
>
> Following caliper reassembly, pump the pedal sufficiently to reposition the pistons. You'll feel the brake pedal rising to a firm stopping point with successive pumps.

11B. Disc Brake Inspection, Dual-Piston Calipers

Viewing the front brakes is facilitated by turning the wheels outward. If necessary, turn the ignition key to the accessory position to disengage the steering lock.

Most brake pads wear out unevenly due to the way that a car's weight is thrown forward and down when braking; one end of each pad usually wears out faster than the other, and one of the four pads will often wear out faster than the other three, Fig. 12–21.

These calipers typically have an anti-rattle plate over top of the pads; many of them have to come off to get a decent view. Some just pop off with a screwdriver, others require that one or both of the pins that secure the pads (Fig. 12–23) be removed first. Some pins are threaded, some are held in place by clips, some are a press fit. Once freed, most require a small blunt-ended punch to drive them out. If they are rusted or worn, replace them.

Brake pads should be replaced whenever you find:

- that a depth indicator clip has started cutting a groove into its rotor, it's getting close to that point, or

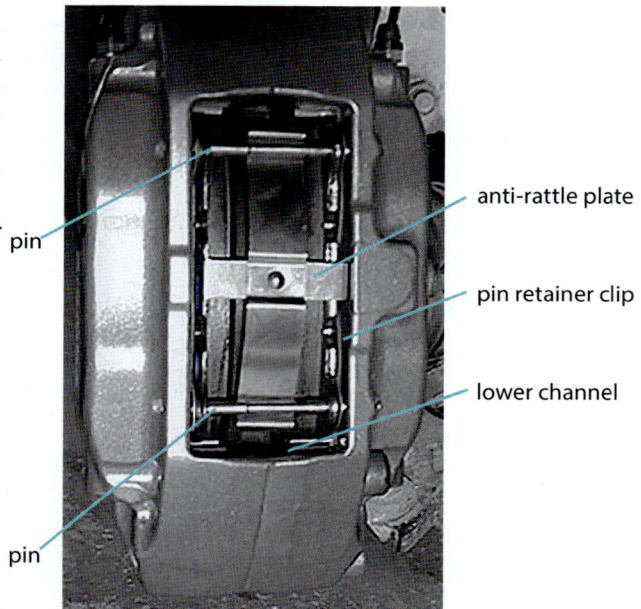

● the friction material is less than 3/16 of an inch (3 mm) thick on any one of them, measured at its thinnest point. The pad backing is not included in this measurement, just the pad itself.

Ideally, all four pads will be close to the same thickness. If not, one or more of the pistons is sticking. Test all of them them individually as decribed in *§* 11C2, p.293.

The pads pull straight out; you'll need a pair of slip-ring pliers or a screwdriver to persuade them. Once they're out, soak the channels that bracket the top and bottom of each pad with brake cleaner. Then use a small wire brush or screwdriver blade to scrape out the brake dust and road crud. Keep away from the piston boots!

anti-rattle plate

pin retainer clip

lower channel

pin

pin

Fig. 12 – 23. Four-piston caliper
Attribution requested

Inspect the shims that cling to the back of each pad. If they're rusted or distorted they'll need to be replaced, discarded, or treated with goopie, *§* 4B, p.265. Simply removing them without *some* treatment can produce screeching or squealing.

Inspect the piston boots as best you can (the hub side of each is impossible to access without removing the caliper). If the rubber is supple and you see no evidence of cracking, then you might decide to move on with your day. If you need to take a closer look, the caliper will have to come off.

In most cases, caliper removal is accomplished by removing two bolts, accessible from the inboard side. Like virtually any other bolt, these loosen counterclockwise, which means that if you are facing the car, it will appear that you are backing them out clockwise. These bolts are *tight*; start with a six-point wrench. To save your hand, use a hammer to smack the wrench's free end (Fig. 8–7, p.142), then finish with a ratchet.

> **Never let a caliper hang free, dangling solely by its hydraulic flex line.** Rig some wire or a heavy-duty wire tie either from the coil spring or another suspension component above.

If the piston boots are cracked or dry-rotted, then caliper overhaul or replacement is not far off. To determine which–and how soon–inspect the piston's freedom of movement within it's caliper bore, *§* 11C, p.292. If both/all pistons are moving freely *and* everything else so far passes muster *and* you are not yet in need of pads, then hold off till you need the pads. In *all* cases involving a torn piston boot, you'll need to disassemble that caliper to inspect for rust.

Inspect the rotors for hot spots or glazing, *§*4B, p.264. Replace the rotor(s) if you find hot spots (Fig. 12–10, p.265) or if they're thinner than their minimum spec. Sand them if they're glazed.

To reassemble, reverse your steps. For information on:

- pad replacement, see *§*13, p.297
- caliper overhaul, *§*12B, p. 296
- rotor replacement, *§*14, p. 298
- bleeding the brakes, *§*20, p.309
- parking brake adjustment, *§*15, p.300 (rear disc brakes) or 16D, p.304.

> Pistons that have been pushed back into their bores will usually not be in contact with their brake pads. The result? NO BRAKES!
>
> Following caliper reassembly, pump the pedal sufficiently to reposition the pistons. You'll feel the brake pedal rising to a firm stopping point with successive pumps.

11C. Testing Whether Caliper Pistons Move Freely In Their Bores

This test requires an assistant. This section is applicable to all front calipers and all Type 1 rear calipers, Fig. 12–6A, p. 262. There is no equivalent test for Type 2 rear calipers (those that have a lever attached to them). See *§*15, p.300 for information on moving Type 2 pistons back into the caliper.

11C1. Single-Piston Calipers

Begin by reading *§*11A, p.287.

Unbolt one of the calipers from its pad bracket, pull it free of its pads. All other calipers must remain in place. For those of you with drum brakes, verify that the rear brake drums are in place. Driving the pedal toward the floor just once is usually sufficient to pop the guts right out of a wheel cylinder. The same warning holds true for the other caliper(s), although with these it's unlikely that one pedal pump would pop out the piston.

If your original brake pads are wafer thin, the caliper piston will already be protruding a pretty fair distance out of its bore. To prevent it from popping free in a hemorrhage of brake fluid, it's a wise idea to first drive the piston at least partway back into its chamber using either a C-clamp applied directly to its center or a large set of channel locks applied across its upper face. Be careful not to pinch or tear the caliper boot.

> If you intend to open the bleeder screw, crack it loose before you unbolt the caliper to simplify opening it later.

Pistons typically feel pretty tight going into the caliper unless the bleeder screw is cracked open partway, but they do move with persuasion as long as the force applied is directed straight down through the center of the piston. Applying force to only one edge can cock the thing and jam it. If yours won't budge, crack open the bleeder on that caliper. If the piston still won't budge, the caliper and/or piston are either rusted or gummed up with crud.

Brake fluid should never be pushed back into the brake lines on higher mileage cars, particularly if the system has never been flushed. Doing so can ruin ABS modulator valves. To avoid that, crack open the bleeder screw prior to squeezing the caliper piston back into its bore. But be advised: As soon as you open that screw, brake fluid will begin to dribble out onto whatever is beneath it. Furthermore, pushing the piston back into its bore will shoot a stream of fluid at whatever is in its path. Close the bleeder as soon as the piston is fully retracted.

Keep an eye on the brake master cylinder's fluid level: Allowing it to run dry requires that the entire system be bled. That's a whole lot of consequence for what should be a simple inspection. See *§* 20, p.309 for more on this.

Have your assistant depress the brake pedal once, then once or twice more. With each pump, the piston should move outward, smoothly, a small distance at a time. **Do not drive the piston out of it's bore unless you intend to overhaul the caliper,** *§* 12A, p.294.

Now drive the piston all the way back into the caliper using either tool cited above. Repeat the process; one or two full cycles should be sufficient unless you've encountered binding. If your assistant has to "stand" on the pedal to make the piston move, you're probably looking at overhaul or replacement. But try one or two more cycles, opening the bleeder every time you squeeze the piston back in, just to see if the situation improves.

Once the piston has been pressed back in, resecure the caliper to its mounting bracket. Reinstall the original pads or a new set before moving on the opposite side. If you opened the bleeder screw during your inspection, bleed that caliper before moving on, *§* 20, p.309.

11C2. Dual-Piston Calipers

Begin by reading *§* 11B, p.290. Follow the procedures of *§* 11C1 (p.292) using the specifics of what follows:

The simplest way to do this inspection is with the caliper still installed. Remove one pad at a time, leaving the other three in place to prevent other pistons from popping out. With four-piston calipers, you'll need to block the other free piston while you test its mate. Use a block of wood, somewhat larger than the piston's face to prevent damage to its boot.

Remove the ignition key to ensure that the steering will lock during what follows. Use a large flat-bladed screwdriver with a square shank to move each piston back into its bore. Slide it through the caliper, rest its tip against the rotor where it meets the hub. Press its shank against the piston's lip and push (or pull, as the case my be). Three problems:

- The force applied is uneven, only one side of the piston's lip will be in contact with the screwdriver.

- The screwdriver can slip off the piston and damage the piston boot.
- It's impossible to closely examine the condition of those boots.

Removing the caliper presents its own set of problems:

- These calipers are heavy: Don't leave them dangling at any time.
- All pistons not being tested must be secured so that they won't pop out. Other calipers are fine as long as they're still bolted up with their pads in place. On the caliper being tested, you'll need one C-clamp for each piston—one to secure the piston that's not being tested, the other to slide the tested piston back into its bore.
- Should you decide to go ahead with an overhaul, you'll find that the bolts that join the two halves together are *tight*. If you have an air gun, that's no big deal. If you'll be disassembling them by hand, you'll probably have to remount the caliper to gain sufficient leverage to break them free.

12. Caliper Overhaul

Before you begin, read all relevant sections in *§*11, pp.287-293.

12A. Single-Piston Calipers

The minimum parts list required is a seal kit and caliper grease. Most of you will be replacing brake pads as well. All OEM seal kits contain a pair of piston seals, piston boots and clips (Fig. 12–13, p.275), plus caliper guide boots and bleeder caps (Fig. 12–22, p.289). Many kits include a packet of caliper grease for lubricating the guides, and many pad kits include grease for their backing plate shims. If not, you'll need some. Get a small tube; it can be used for both applications.

> Caliper guide lubrication requires specially formulated high temperature grease that won't produce swelling or deterioration of the guide's rubber boots. Two good products are BG's 608BK and Permatex's Ultra Disc Brake Caliper Lube.

As for special tools, you'll need a caliper hone, line wrenches (one for the flex line, one for the bleeder screw), a C-clamp (more for multiple-piston calipers), and a few picks, Fig. 12–24. Supplies include rags, a sheet of emery cloth, several cans of aerosol brake cleaner, and the appropriate brake fluid (consult your owners' manual). Brake assembly fluid is a nice touch as well; it's used to lubricate the caliper bore, piston seal, and piston prior to reassembly. Brake fluid can be used, but it's a whole lot less slippery. Compressed air is *essential* for cleaning the calipers. You'll need at least one can per caliper (depends on how many pistons you'll be dealing with). If you own an air compressor, so much the better.

Doing this job requires solutions to three problems:

- ❑ As soon as the caliper is disconnected from its flex line, brake fluid will start dripping from that line. Initially, that's no big deal if there's a drain pan beneath it; but it won't take long before the brake master cylinder runs dry. Overcoming the presence of air within the brake master can be problematic, so you need to be prepared to shut down fluid loss as soon as the caliper comes off.

12

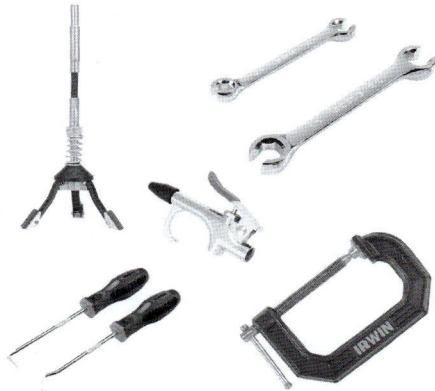

Fig.12 – 24. Required tools for caliper overhaul
(air valve optional)

Fig. 12 – 25. Banjo fitting
Courtesy of Motorcycle.com

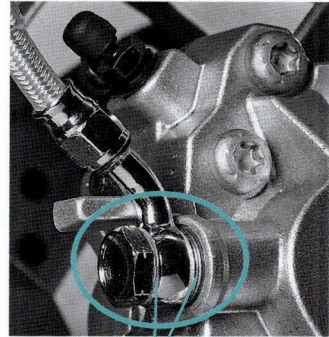

brass washers

- Banjo fittings (Fig. 12–25) – Remove its bolt while the caliper is still secured to its mounting bracket. Immediately slap solid rubber washers over top of the banjo's two holes, then clamp them in place. If the brass washers are stuck to the banjo, just leave them where they sit.

- Threaded fittings – Crack the flex line loose just before removing the caliper, then spin off the caliper as if you were removing a nut from its bolt. Use a rubber cap or a golf tee shoved into the line; if necessary, hold the tee in place with a piece of flexible rubber secured with a wire tie.

  Many mechanics just use a pair of vice-grips on the flex line. Certainly effective, but it can destroy an older line–perhaps just weakening the interior wall right now, but ultimately producing a hard to diagnose brake pull as the line deteriorates, $\mathcal{S}$ 19, p.308.

  It can be hard to stop fluid loss completely. An occasional drip is OK as long as you keep ahead of it, topping off the master cylinder from time to time.

- ❑ Piston removal is a simple thing if you have an air compressor, particularly with the nozzle shown in Fig. 12–24. Just remove a caliper, plug the flex line, then pop out its piston by blowing air through the caliper's flex line inlet port. (Press the air nozzle's rubber tip into the port.) Three thoughts:

- **Keep your fingers out of the way!**

- Do this over a pan; put a few rags in the pan to soften the piston's landing.

- Drain the caliper first! If you don't, you'll be awash in a spray of brake fluid.

The rest of you will need an assistant to pump the brake pedal while you attend to the pistons. Begin by re-reading $\mathcal{S}$ 11C1, p.292. The trick is to move the first piston out about three quarters of the way, stopping shortly before it pops free. Next, set up a C-clamp to prevent that piston from moving any further while you deal with the other caliper.

12

Start by cracking its flex line loose–this to enable disengaging the caliper following piston removal–but snug enough that it won't squirt fluid all over you. Unbolt the caliper, then have your assistant drive out the piston using the pedal. Use your hand to catch the piston, a pan to catch the fluid.

The piston that remains in the first caliper can be removed in several ways. Begin by draining it . . .

- Grab a block of wood, a heavy glove and some rags. Unless the piston is frozen in its bore, most calipers will surrender their piston by smacking the caliper's claw hard into the block. Protect your hand! Use the rags as a shock-absorber between caliper and glove. Typically, one or two smacks will the be sufficient.

  Concerned about the consequences of using your caliper as a hammer? Don't be. Calipers don't crack, they don't dent, they're damned near indestructible–unless, of course, rust finds a way in.

- Insert a can of compressed air, pinpoint sprayer attached, into the flex line port. Shove a rag down tight around the caliper's opening and blow. Two problems with this method: it's expensive, and it's hard to secure a good seal at the caliper.

- Head to your go-to gas station and ask to use their air.

❑ Speaking of heading out . . . Obviously, this isn't a problem if you have a second car, but you may need to get more parts. Rusted guides must be replaced if they won't clean up with 120 grit emery cloth and brake cleaner. The same goes for guide collars and pistons. Light rust can be totally superficial; it will clean up easily and leave no trace of pitting. But if you encounter pitting on any of these pieces, they need to be replaced, period. Then the question becomes whether it would be cheaper to just replace the calipers; it would certainly be simpler.

On to the overhaul itself! Read the following sections for information on:

- caliper inspection, see *§* 11A, p.290
- caliper overhaul, *§* 12A, p.294
- brake bleeding, *§* 20, p.309

> Following caliper reassembly, pump the pedal sufficiently to move the pistons out into contact with the brake pads. You'll feel the brake pedal rising to a firm stopping point with successive pumps.

12B. Dual-Piston Calipers

Overhauling dual-piston calipers successfully requires that all pistons respond equally to fluid pressure; otherwise the car will pull–sometimes severely–when the brakes are applied. I never found that to be particularly problematic as long as there was no rust in any one of the bores, but I can't say the same about the four-piston calipers on high-performance cars. To me, they were always a nightmare. Caliper replacement is expensive on these cars, but what isn't?

Everything in *§*12A applies here as well, with the following exceptions:

- The caliper kit will not have caliper guide boots. Instead, it will have two o-rings, one for each end of the caliper. These seal the ports that transmit fluid between the halves.

- No pesky collars or guide pins to clean, just those extra pistons and bores.

13. Disc Brake Pad Replacement

Most brake pads look like those in Fig. 12–21, p.287. Basically, these are interchangeable–their placement is irrelevant. But there is one caveat: With single-piston calipers, backing plate shims can differ–they are site specific, one type for the piston interface, another for the claw. Another style of pad has spring clips riveted to its backing plate. These too are site-specific. So keep your wits about you; perhaps take some pictures prior to disassembly to assist you with reassembly.

Reread *§*11A, p. 287 or *§*11B, pp. 290; it contains most everything you need to know. Usually, the pad bracket can be left in place throughout this procedure unless the rotor is being sanded or replaced.

Because of the countless variations in brake pad shims and anti-rattle clips, all I can offer are generalizations applicable to most. First, their presence and proper placement are essential to having quiet brakes. Remove the original anti-rattle clips one at a time, and examine the surfaces on the pad bracket with which they were in contact. Scrape off any crud with a screwdriver blade, use a pick to dig out the corners. If these surfaces are rusty, scrape them down as best you can, then apply a *thin* coat of caliper grease to the area that will contact the clip. A thin coat of grease applied overtop can facilitate pad installation, but in most cases it's unnecessary and is likely to attract a new layer of crud.

Many (most?) OEM pad kits provide a small packet of grease and instructions on where to place it. It's used on backing plate shims to help them shift incrementally during braking. In *§*4B, p.265, I referenced several products that can be useful in combating brake squeal. New OEM pads, properly installed, do not need their help.

For information on:

- caliper inspection, see *§*11A, p. 287 or *§*11B, pp. 290

- caliper overhaul, *§*12, p.294

- rotor replacement, *§*14, p.298

- brake bleeding, *§*20, p.309

- parking brake adjustment, *§*15, p.300 (rear disc brakes) or 16D, p.304.

- disc brake squeal, *§*4, p.263

❑ Retracting Pistons on Type 2 Rear Calipers

To accommodate a new set of pads, all rear calipers with integral parking brakes (Fig. 12–6B, p.121) have pistons that must be screwed back into place, Fig. 12–26. Many of them requires a special service tool to do so; some will accept heavy-duty needle-nose pliers. Fortunately, most parts jobbers carry some variations. If you find any piston that's unwilling to rotate freely when the proper tool is applied, then that caliper should be overhauled or replaced.

Fig. 12–26A. Type 2 innards
Courtesy of quattro123.com

Fig. 12–26B. Type 2 rear caliper
piston tool for a VW Eos

14. Rotor Removal and Replacement

Brake rotors come in two flavors: those positioned on the outboard side of the hub (Fig. 12–27) and those that bolt to the inboard side (Fig. 12–15, p. 278).

14A. Outboard Rotors

Outboard rotors are removed by simply removing the wheel, caliper and pad bracket (and where present, the rotor's retaining bolts). Because the hub stays home, its wheel bearing(s) require no attention. If the caliper requires inspection or if the pads are to be replaced, the caliper should be removed first, separating it from the pad bracket.

14B. Inboard Rotors

14B1. Front brake rotors (FWD)

Most of the modern-era front-wheel-drive cars that carry inboard rotors look identical to Fig. 12–27, except for the position of the rotor. Access to it typically requires both a 1/2 inch drive air gun to break the driveshaft's lock nut free and a 1/2 inch drive torque wrench for reassembly (these nuts are torqued north of 120 ft-lbs–the factory spec is critical). A hub puller is usually required to free the hub/rotor assembly from the steering knuckle's bearing.

There is no reason to mess with a single, sealed bearing (Fig. 12–27:C) if it's not causing a problem. Front wheel bearings are seldom an issue below 120,000 miles; some can go twice that far. Those that do need attention typically make a growling or grinding noise, evident primarily when cornering. If you're questioning whether the bearing is involved, just look at the rotor–if it's torn up, most of you already have your answer. For bearing inspection procedures, see §11, p. 342.

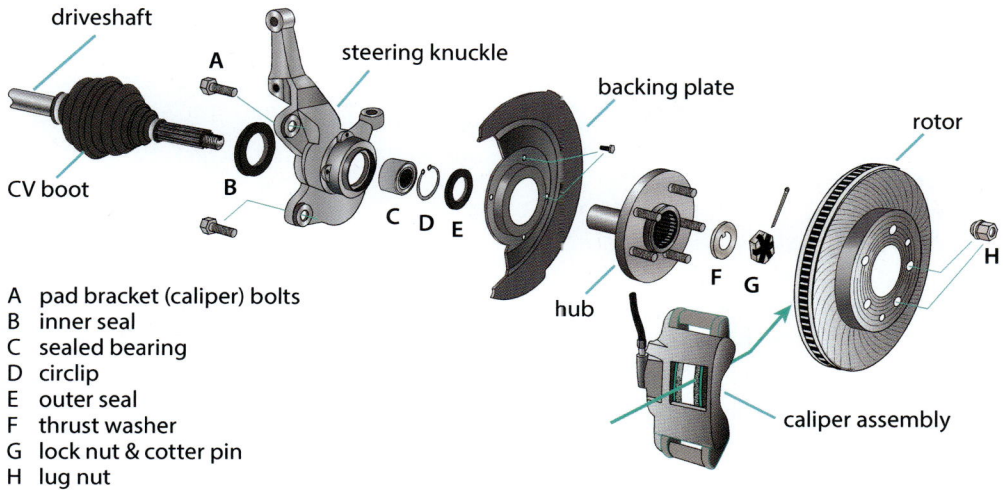

Fig. 12–27. A typical FWD steering knuckle assembly with outboard rotor © PT

A pad bracket (caliper) bolts
B inner seal
C sealed bearing
D circlip
E outer seal
F thrust washer
G lock nut & cotter pin
H lug nut

Use light oil on the facing surfaces between bearing and hub prior to reassembly; grease the splines on driveshaft and hub as well. A 4 lb hammer, judiciously applied around the circumference of the hub's center, will start the hub inward. Tightening the axle nut will gradually pull the hub home.

Now comes the torque wrench: **Tightening this nut to spec is critical!** Lock it using a new cotter key or stake nut (Fig. 12–28) to prevent the assembly from loosening over time.

14B2. Front brake rotors (RWD)

Refer to Fig. 12–15, p. 278. A small steel chisel (the sharper the better) and a pair of slip-ring pliers are needed to remove the grease cap. Use wire cutters to pull the cotter pin and slip-ring pliers to remove the lock nut. Note the amount of torque applied to the nut: Basically, it is just snug, sufficient to remove the wobble, but not so tight that it produces strain on the bearings.

The outer bearing typically falls out as the hub is pulled off its spindle, so have a clean rag handy to catch and store it. To prevent rust from entering the hub, shove a rag into the inner seal. Remove its bolts, then smack the rotor with a hammer to loosen it. Pop out the hub seal with a large, flat-bladed screwdriver (doing so will destroy the seal), then lift out the inner bearing.

Both bearings should be thoroughly cleaned (varsol or brake cleaner), then repacked with fresh grease prior to reinstallation. Ditto for the hub. The hub seal should always be replaced, its inner rubber seal coated with grease just prior to reinstalling the hub. Just tap it into place. Spin the hub as you tighten the lock nut; back it off a half-turn, then tighten again. Remember, just snug. A new, stainless-steel cotter pin is always simpler to deal with than the original.

14B3. Rear Rotors

Outboard rotors are removed by simply removing the wheel, caliper and pad bracket, and where present, the rotor's retaining bolts. If the caliper requires inspection or if the pads are to be replaced, the caliper should be removed first, separating it from the pad bracket.

As for the parking brake:

- Type 1 (rear shoes housed within the rotor's hub): There'll be some brake dust to contend with; use an aerosol brake cleaner on the shoes and backing plate. See *§*16D, p.304 for parking brake adjustment.

- Type 2 (parking brake internal to the caliper): The parking brake cable could be a nuisance; disconnect it only if necessary. If you must, mark the placement of the cable sheath relative to its mounting bracket prior to disassembly. (It's configured just like the brake cable on a bicycle.) If you're satisfied with your parking brake's height, then restore it to its original position. If you're not, shift the sheath away from the caliper–gradually, one full turn at a time–until you are.

Fig. 12–28. Front driveshaft locknut, staked
Courtesy of clublexus.com

15. Rear Disc Brake Inspection and Repair

Aside from being smaller, Type 1 rear calipers (those with a separate drum brake to handle parking brake duties) are essentially the same as those on the front. See *§*11A, p. 307 and *§*12A, p.294.. Type 2 rear calipers (those with an integral parking brake) are fundamentally different: their pistons thread onto a central pin, Fig. 12–26B, p.298. Many require a special service tool to do so. For more information on these calipers, see *§*13, p.297.

❑ Parking Brake Adjustment, Rear Calipers Only

Type 1 assemblies are fully adjustable, usually through a peephole in the face of the drum (hub). See *§*16D, p.304 for adjustment procedures. These brake shoes are teeny, the linings are thin to begin with. Driving with the parking brake on will burn them up in a hurry.

With Type 2 calipers, the parking brake cables can be tightened to compensate for any stretching that might occur over time, but they cannot be tightened to provide better bite because the brake pads are already in contact with their rotor. See *§*14B3 (above) for adjustment procedures.

Since calipers have no real leverage, Type 2 parking brakes will not lock up; so when setting the parking brake on a steep hill, always turn your front wheels toward the curb. Be kind to your tires: Don't leave them resting against the curb.

16. Drum Brake Inspection and Repair

16A1. Drum Removal (on RWD cars)

Release the parking brake. Do not continue unless you have at least one can of Brakleen.

Put the transmission in neutral. Verify that the drums turn freely. Typically, you'll find that they spin significantly longer and with far less effort than front disc brakes: The return springs that retract brake shoes are far more effective than those dinky piston seals that calipers rely on, Fig. 12–29 and Fig. 12–30.

12

Fig. 12–29. Drum brake, FWD

- wheel cylinder boot
- return spring (shoes)
- parking brake lever
- portion of parking brake adjuster
- parking brake cable

Brake drums are often frozen to the rear axle flange. If so, use a 4 lb hammer to smack the drum hard across its front face between each pair of lug bolts. You should see the drum pop away from the hub. If not, inspect the seam between the hub's protruding center and the drum; a buildup of rust or crud could be interfering. If so, shoot some *PB Blaster* around its circumference (Fig. 12–30), follow that with a wire brush or emery cloth, then smack it again.

> Some drums have a retaining screw in the front face that must be removed before you start smacking the drum.

Some drums provide another option: A pair of small holes in the drum's front face, placed opposite each other on either side of the axle flange hub, Fig. 12-5, p.258. These are threaded to accommodate a pair of bolts–deducing their size and pitch I leave to you. Thread them in, then use a ratchet to drive off the drum. This procedure works unless the drum is badly frozen; then the threads will just strip out.

If the brakes have been grinding or the car has a lot of miles on it, then it's possible that the shoes have worn down the drum's inner face, producing a lip around the drum's outer edge. In these cases, the only solution short of destroying everything inside is to back off the adjusting mechanism, §16E, p.305.

16A2. Drum Removal (on FWD cars)

Release the parking brake. Do not continue unless you have at least one can of Brakleen.

Refer to Fig. 12–31, p.303. Use a small steel chisel (the sharper the better) and a pair of slip-ring pliers to remove the grease cap. Use wire cutters to pull the cotter pin and slip-ring pliers to remove the lock nut. Note the amount of torque applied to the nut: Basically, it's just snug, sufficient to remove the wobble but not so tight that it produces strain on the bearings.

Fig. 12–30. Drum brake, RWD

The inner and outer wheel bearings are packed with grease, as is the hub (though not to the same extent). The outer bearing typically falls out as the hub is pulled off its spindle, so have a clean rag handy to catch and store it.

If the brakes have been grinding or the car has a lot of miles on it, then it's possible that the shoes have worn down the drum's inner face, producing a lip around the drum's outer edge. In these cases, getting the drum off might require you to back off the adjusting mechanism, §16E, p.305.

Pop out the hub seal with a large, flat-bladed screwdriver (use it as a lever against the hub's rear lip). Lift out the inner bearing.The bearings should be thoroughly cleaned, then repacked with fresh grease prior to reinstallation. Ditto for the hub. The hub seal should always be replaced, its inner rubber seal coated with grease just prior to reinstalling the hub. (Don't forget to place the inner bearing in its race before installing the seal!) Tap the seal into place with a small hammer. Once the drum is on its spindle, slide in the outer bearing, then its thrust washer, then the lock nut. Spin the hub as the nut starts to tighten; back it off a half-turn, then tighten again (channel locks). Remember, just snug. A new, stainless-steel cotter pin is always simpler to deal with than the original.

16B. Inspection

All brakes produce dust as they wear; brake drums keep it contained. So when the drum comes off, that dust will greet you. Time was that all brake linings contained asbestos; today, none of them do. But the stuff is still nasty; you. might want to consider a mask prior to pulling the drums. As you pull it off, walk it away a few steps–downwind if relevant–then drop it face down, preferably on a hard, flat surface. Most of the loose dust will end up there. Don't bother cleaning the drum, it's a waste of brake cleaner.

Before cleaning the brakes, look for signs of fluid leakage, either from the wheel cylinders or from the rear axle seal. Seal leakage shows up as a darkened area, dampening the area below the axle shaft in an ever-increasing arc. Because brake fluid is so much thinner than gear oil, a *bad* wheel cylinder leak won't really darken the area below the cylinder but it will get things wet. You'll see the fluid exiting from the bottom of the cylinder at either end, right where its rubber dust boot closes over top. Most wheel cylinders never get that bad; most drivers notice the loss in braking effectiveness long before.

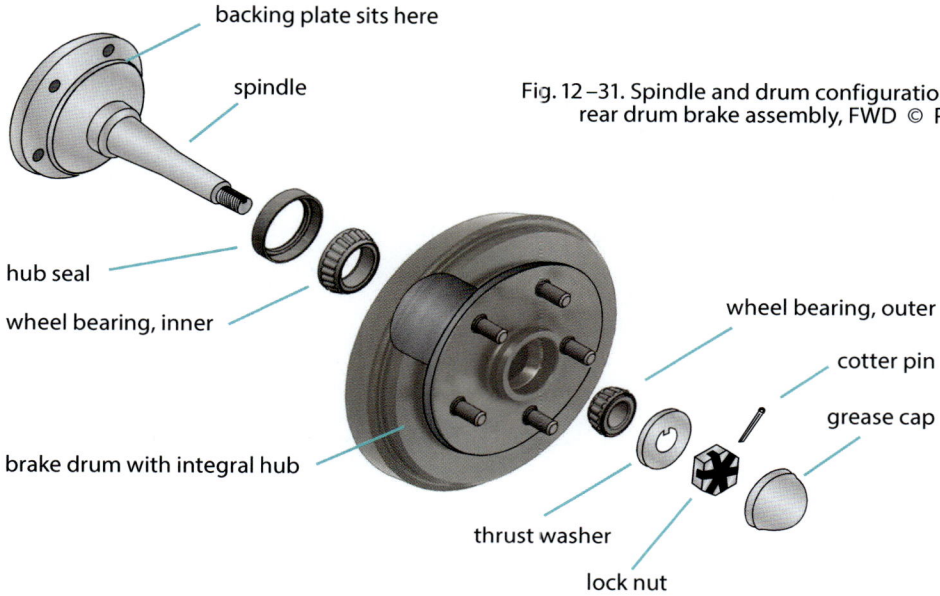

backing plate sits here

spindle

Fig. 12 –31. Spindle and drum configuration, rear drum brake assembly, FWD © PT

hub seal

wheel bearing, inner

wheel bearing, outer

cotter pin

grease cap

brake drum with integral hub

thrust washer

lock nut

Clean the brakes with an aerosol brake cleaner, spraying from the top down. Soaking the dust allows it to run off in liquid form. Paper towels spread beneath help with cleanup.

Back to the wheel cylinders: Use a screwdriver or blunt hook to pry back the most accessible portion of each dust boot. Leave the half that faces the backing plate alone; doing so will simplify reinstalling the boot. The inside surface of many older boots can be a bit moist. That's not a problem, but a cascade of muddy fluid pouring out as soon as the boot is pulled back is. Replace them.

Assuming that the cylinders are dry, alternately shove the trailing shoe forward, the leading shoe rearward. Pushing on the one should move the other; they should each move a half-inch or so in both directions. This ensures that neither of the cylinder's pistons is frozen in its bore. A frozen piston prevents its shoe from working; a frozen wheel cylinder is the predominant cause of cars that spin out of control when braking, particularly while slowing through a corner in the rain.

On to the linings: You'll find that the leading (front) shoe is significantly thinner than the rear one. Replace the shoes if the friction surface on any one of them is less than 1/8th of an inch (3 mm), measured at its thinnest point.

driver's rear

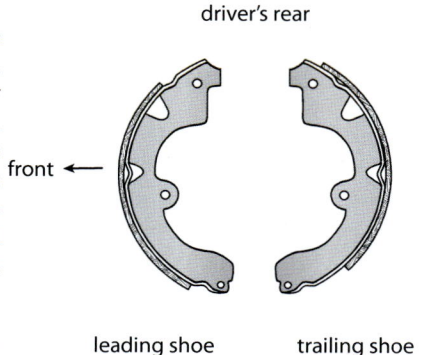

front ←

leading shoe   trailing shoe

Fig. 12 – 32. Placement of rear brake shoes on drum brake systems © PT

The leading shoe and trailing shoe differ one from the other in the positioning of their lining material (relative to the shoe's metal backing), Fig. 12–32.

12

To inspect the drum, start by dropping it again on the floor, away from your original dust pile. Soak that pile with brake cleaner, then mop it up with paper towels. Repeat on the second pile. Now inspect the drum's interior face. If it's smooth and free of hot spots, you're done. If it's been ground up by metal-on-metal contact, if it has hot spots (Fig. 12–10, p. 265), or if its outer edge has a significant lip (>3/16th inch), replace the drum or get it refaced.

For spindle and drum bearing adjustment, see the end of *§* 16A2, p. 301.

16C. Parking Brake Inspection

This section relates only to cable-operated parking brakes. There are two cables, one for each rear wheel, entering through the back side of each backing plate. The cable attaches to the bottom (free) end of the parking brake lever, located on the trailing shoe. With the parking brake completely off, this lever should be at rest against the shoe's backing, Fig. 12–29, p. 301. If it's positioned ahead of that surface, either the parking brake cables are adjusted too tightly, a between-the-seats handbrake mechanism is jammed with a coin or other junk, the ratchet assembly needs some lubrication (aerosol spray), or one of the cables is binding within its protective sheath (frayed cable or rust–only on high-mileage cars).

Have an assistant apply the parking brake and release it several times while you observe what's occurring at each wheel. If the handbrake lever consistently fails to completely return to the inside of the rear shoe, start by loosening the cabling. A turnbuckle can usually be found somewhere along its path beneath the car. If not, look at the parking brake mechanism inside the car.

16D. Rear Brake Adjustment

Most adjust through a slot in the backing plate once the drum is back in place and the wheel has been bolted up. The rest are adjusted through a hole in the drum while the wheel is still off. Both methods employ a rubber plug to seal the opening; remove it. Before you install the drum, find a good tool for rotating the toothed wheel through its access slot/hole. In most cases, a wide-bladed screwdriver is all you'll need, but some require a bent shaft to facilitate access. (There are special service brake tools for this available at any parts jobber, Fig. 12–33, p. 306.) Determine the direction of rotation: Typically, you'll find a spring-loaded plate that prevents rotation in one direction.

Reinstall the drum, centering the shoes if necessary. If you'll be adjusting the brakes through an access hole in the drum, secure it with a pair of lug nuts, hand tight. Otherwise, install the wheels. To assure a good adjustment, pump the brake pedal a few times, engage the parking brake once, then pump the pedal once more to recenter the shoes. All rear-wheel drive cars must be in neutral.

Spin the tire/drum by hand while you ratchet the adjuster tighter. Stop tightening at the first sign of resistance–slowing rotation or the sound of rubbing. If you reach ten clicks on the first wheel, move on to the other side, alternating back and forth between the two drums until the adjustment is completed. This will take up the slack evenly between the two handbrake cables.

16E. Backing Off an Adjuster

For those unfortunate few who need to back off an adjuster: You'll need both hands; one to press a small-bladed screwdriver against that spring-loaded plate (see above), the other to ratchet the wheel backwards (looser). To permit that, you'll need to push that plate away from the adjuster so that the wheel can rotate in reverse. The hardest part of all this is finding a spot within the backing plate opening for the small driver that permits sufficient purchase on that plate while still leaving room for the ratcheting driver. Never impossible; always a pain!

17. Overhauling Drum Brakes

Examine Figures 12-29, p.301 and 12-30, p.302 closely. There's a lot going on there–much more than you'll find in any disc brake configuration. No big deal if you've done it twenty times, a whole lot to learn if it's your first. Because of the countless variations in springs, retaining clips, and parking brake hardware, all I can offer are generalizations applicable to most.

❑ Regardless of configuration, take some pictures before you begin. Always complete one side before starting the other. Read § 8B, p.279 and § 16, p.300.

❑ Wheel cylinders *must* be replaced in pairs. In most cases, they can be replaced with the shoes in place. Disconnecting the brake shoe return spring(s) will provide the space necessary to swap them out.

Start by soaking each line fitting with *PBBlaster. Always* use a line wrench on the hydraulic line fittings–they've been exposed to the elements for years. The line will begin leaking almost immediately, so:

- Top off the brake master cylinder before you begin

- Have some means to seal off the line while you swap out the cylinder. A flexible piece of rubber and a wire tie work well.

  Don't try to pull the line out of the cylinder–that line is rigid. Instead, loosen the line 3-4 turns, remove the bolts securing the cylinder to its backing plate, smack the cylinder from above if it's stuck to the backing plate, then finish loosening the line. Cap it once the cylinder has been pulled away.

Bleed the brakes after both cylinders have been replaced and both drums are back on, § 20, p.309. If you've replaced the shoes as well, adjust the brakes at least somewhat prior to bleeding.

❑ Replacing the shoes without hurting yourself or going nuts requires a few special service tools, Fig. 12-33.

- The spring tool is particularly useful on the Fig. 12–29 configuration (p.258). Either a hook or thin-bladed screwdriver (used as a lever) can be substituted on many small cars (small brakes).

12

- The cup and spring retainer tool makes short work of these. Fingers and a pair of needle nose pliers can be used in a pinch. In either case, put a finger on the head of the pin (behind the backing plate) to prevent the pin from shifting while you twist the tool.

Free the return spring(s); remove those that want to fall away, leave the rest for now. Free the retaining clip securing the front shoe to the backing plate. Unless you're holding it, the shoe will drop away and end up dangling from its lower spring. With most configurations, all kinds of pieces will start falling out–nothing like photos to help you get it all back together again! For the rest of you,

adjuster

spring tool

cup and spring retainer tool

Fig. 12–33. Tools for rear drum brakes
Courtesy of Harbor Freight Tools

releasing the front shoe is a non-event. When you remove the rear shoe's retaining clip, more stuff will start falling out. To disconnect the handbrake lever from its cable, use a pair of wire cutters to slide the cable's tensioning spring forward, bite the cable while you extract the lever. The cable stays where it is unless you're replacing it. (No need unless it's binding.) Remove the parking brake lever from the rear shoe–a thin-bladed screwdriver or small punch to initiate spreading the clip, a pair of needle-nose pliers to finish the job. event. When you remove the rear shoe's retaining clip, more stuff will start falling out. To disconnect the handbrake lever from its cable, use a pair of wire cutters to slide the cable's tensioning spring forward, bite the cable while you extract the lever. The cable stays where it is unless you're replacing it. (No need unless it's binding.) Remove the parking brake lever from the rear shoe–a thin-bladed screwdriver or small punch to initiate spreading the clip, a pair of needle-nose pliers to finish the job. Clean up the backing plate and all your pieces. The brake adjuster's turnbuckle should be disassembled, cleaned, greased, then reassembled. Collapsing it to its shortest length helps with reassembly (mounting the springs). Now's the time to replace wheel cylinders.

Pick out a front and rear shoe (Fig. 12–32, p.303) and show yourself where they'll go. On the backing plate you'll see three elevated spots on which each shoe rests. Put a dab of grease on each. Install the handbrake lever, put a thin film of grease on the areas that contact the shoe. Connect the handbrake cable, then mount the rear shoe.

From here you're pretty much on your own. Once it's all back together again, verify that the free end of the parking brake lever is at rest against the shoe's backing. If not, loosen the parking brake cables.

Adjust the brakes by spinning the turnbuckle. Complete the bulk of this before you install the drum; it's a whole lot easier to access the turnbuckle while you're staring at it than it is to do it all through that little slot on the backside of the backing plate.

For information on:

- drum brake inspection, *§* 16B, p.302

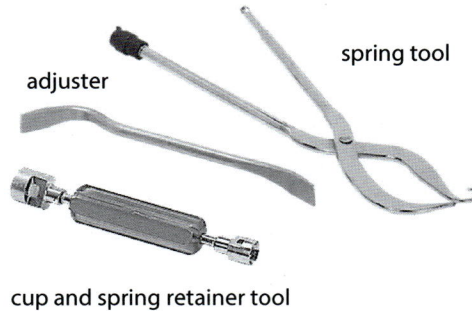

- drum brake adjustment, *S* 16C&D, p.304
- brake bleeding, *S* 20, p.309

18. Low Brake Pedal (Excessive Pedal Travel)

12

18A. Rear Drum Brakes

Three possibilities: The first two are the most likely, the third is the most dangerous.

1. Your rear brakes are out of adjustment: Drum brakes require routine adjustment, once a year or every 12-15 thousand miles. As their linings wear, the brake shoes have to travel farther before they start to contact the brake drum. You will usually find that pumping the brake pedal improves brake pedal height: The first application moves the shoes outward while the second puts them hard against the drum. This has to be repeated at every stop because the shoes consistently retract under the unceasing pull of their return springs, Fig. 12–29, p.301 and Fig. 12-30, p.302. Since you can't evaluate the condition of your brake linings without taking a look, simply adjusting them is not good enough. Do a complete brake inspection to cover your bases.

2. Your rear brake linings are worn out: Everything said in (1) above is true here as well, but with one variation: The cost of repair goes up exponentially if you wait until the drum is destroyed.

3. Air in the Lines: Air compresses easily under pressure, brake fluid doesn't compress at all. Its presence results in a spongy or bottomless feeling to the pedal. Air can get sucked into the brake system through a number of pathways and can end up either behind a caliper or wheel cylinder piston, in the hydraulic lines, behind the brake master cylinder where it bolts up against the vacuum booster, or within the ABS modulator valve.

    Cars with a significant amount of air in the lines respond somewhat differently than those that are just out of adjustment. When pumped up, the latter provide a hard pedal until they're released, whereas the former continue to offer that mushy, bottomless feel. Even though the brake pedal is higher following the act of pumping, it will gradually creep toward the floor.

    How did the air get in there? There are at least five possibilities:

    - Recent brake bleeding left some air in the line.
    - A leaking wheel cylinder piston seal. Whenever the brake pedal is released, air can be sucked into the cylinder by the loss of brake fluid pressure.
    - A worn seal at the rear of the brake master cylinder (where it bolts up against the vacuum booster) can do the same. A fair amount of fluid can seep into the space between the cylinder and booster before you'll see any sign of it–a darkened trail of wetness creeping down the forward face of the booster.
    - A worn seal within the brake master cylinder or ABS modulator,
    - Or crud lodged adjacent to that seal. A thorough brake system flush (bleeding) can cure this, or not: The pedal will either come up or it won't. If it does improve, you might try out the brakes for a week or two, closely

12

monitoring pedal height and brake fluid level within the master cylinder reservoir, *§* 3D, p.105 and Fig. 6-12, p.106. Check the level in the morning before the car is driven in order to maintain a consistent baseline.

18B. Disc Brakes

The only possibility on cars with disc brakes at all four wheels is air in the brake lines. Air compresses easily under pressure, brake fluid doesn't compress at all. Its presence results in a spongy or bottomless feeling to the pedal. Air can get sucked into the brake system through a number of pathways and can end up behind a caliper piston, in the hydraulic lines, behind the brake master cylinder where it bolts up against the vacuum booster, or within the ABS modulator valve. The bullet points directly above list possible points of entry.

18C. Brake Pedal Goes Straight to the Floor

Two likely causes and one other possibility. The first two can occur rather suddenly; the last one has been evolving over a very long period of time:

- Defective brake master cylinder
- Low or empty master cylinder due to significant fluid loss
- Rear drum brakes so badly out of adjustment that the shoes no longer contact the drums when first applied.

Check your brake fluid level, Fig. 6-12, p.106. If the system is empty, you have a significant leak. Call a tow truck.

If the reservoir is a long way from empty, suspect a faulty master cylinder. See if the brake pedal will pump up in two or three cycles, and then hold that height. If it does, how long does it takes before the pedal starts moving toward the floor? A decent result might give the bold and the broke among you some leeway to limp your way to the shop.

19. Flex Lines (Brake Hoses)

There are only two reasons to replace flex lines:

- Significant cracks in their exterior casing: Most flex lines have a useful life of 120,000 miles or more. Ask to see the evidence if you've been told you need them sooner. In my opinion, a thousand shallow cracks are insufficient cause, but one deep crack is.

- If your brakes are still pulling after you've dumped vast amounts of money into pad replacement, caliper overhaul, rotors or tires, then odds are that either a flex line's interior wall is deteriorating–blocking the free flow of brake fluid–or the ABS modulator has a problem. Obviously, flex lines are a whole lot cheaper. How old are they? Has the brake system been routinely flushed?

20. Bleeding the Brakes

The brake system's hydraulic fluid should be replaced in accordance with factory service intervals, commonly once every five years or so. Additionally, any time a bleeder screw is opened, at least that one quadrant must be bled to remove any air that may have entered. Typically, brake bleeding is performed symmetrically: If you open up the left front, bleed both front calipers. Ditto for the rear.

You'll need the following:

- hydraulic jack and four jack stands; socket and breaker bar to remove the wheels
- an assistant
- one or two bottles of fresh brake fluid, check your owners' manual for type[9]
- *six-point* box-end wrenches (Bleeder screws are particularly susceptible rounding off.) Bleeder screws on calipers are usually bigger than those on wheel cylinders)
- one or two cans of brake cleaner
- drain pan
- a ton of rags
- *Spray Nine* or comparable to clean up your paint if necessary.

See *§* 7A, p. 139 and *§* 7D, p.140 for instructions on lifting and supporting the car. Bleeding the brakes is usually a lot cleaner and more straightforward with the wheels removed. You'll want to crack the lugnuts loose before the wheels get off the ground.

20A. Old School I

1. Remove the master cylinder's cover. Top it off.

> Never try to drive a car anywhere relying solely on the handbrake; they are not designed for stopping the car, and they do a very poor job of it.

2. Remove the rubber cap from the bleeder closest to the master cylinder (driver's front), then open that bleeder roughly one full turn. Put the tip of your index finger over top of the bleeder's opening and have your assistant pump the brake pedal, pushing it to the floor repeatedly until the bulk of the fluid within the reservoir has found its way to your drainpan. **DO NOT let the master cylinder run dry**.

    The pedal should move toward the floor with little to no resistance; if your assistant suggests otherwise, open the bleeder another half-turn or so. Using your index finger across the bleeder's opening is to prevent air from being sucked back into the system while the brake pedal is rising from the floor back to its starting position. Depending upon how you position your fingertip, the fluid that exits the bleeder could either end up in your eye, on your car's paint, or in your pan. I'd choose the latter. The purpose of this initial bleed is to remove the bulk of fluid from the reservoir, *not* to empty it.

---

9   Brake fluid is hygroscopic –it snatchs water right out of the atmosphere–so don't use that can of fluid that's been sitting on your shelf for the last four years. Invest in two brand new bottles that still have foil seals protecting them from invasion.

12

Should you run the reservoir dry, bleeding the system becomes dramatically more difficult, because you'll be attempting to compress air rather than fluid. So watch the level like a hawk! Most master cylinders have a little screen in there that can block your view. If so, pull it out for inspection and cleaning (brake cleaner).

If the brake pedal refuses to return from the floor, your assistant may need to position her/his other foot behind the pedal to assist its return.

Complete this step by closing the bleeder *finger tight* **while your assistant holds the pedal firmly to the floor.** In *every* ensuing step, **the brake pedal must remain on the floor until the appropriate bleeder has been closed.**

Bleeder screws are not as strong as bolts –they have a channel drilled right through their central axis. They will snap off if overtightened. Just past snug is plenty tight enough.

3. If there's a lot of crud in the bottom of the reservoir, use a clean rag to soak it up. Make sure that you don't just push the stuff into the system's inlet(s). Don't use any chemicals. When you bring that rag back across your car's fender, be careful not to leave drops of fluid resting on its surface. Brake fluid will discolor painted surfaces.

   Reinsert the screen and top off the master with fresh fluid. You will need to replenish the master throughout the following steps. **Do not let it run dry!**

4. Return to the left front wheel and repeat Step 1, having your assistant depress the pedal completely to the floor 8–12 times. **Always close the bleeder while the pedal is still fully depressed.** That caliper and the line leading to it will now be flushed and filled with clean fluid.

5. Top off the brake master, move to the right front of the car, and repeat Step 4.

6. The right rear comes next, followed by the left rear. Because the lines leading to the rear of the car are longer than those to the front, you'll want to pump each of them through perhaps 15 strokes. Have your assistant watch that fluid level!

7. The final phase of brake bleeding involves a revisit to each wheel, but this time your assistant will pump the pedal 3–5 times while you have the bleeder closed, then as he holds the brake pedal down firmly, crack open each bleeder perhaps a quarter to a half turn, then immediately slam it shut. You won't be using your fingertip over the opening this time; the fluid stream that escapes will shoot straight out at whatever is in its path. The point of this is not to make a mess; rather, it's to dislodge any air bubbles that might remain in that particular quadrant, so you'll need to get the bleeder closed before the brake pedal reaches the floor. Repeat this process 2–3 times per wheel, starting at the left rear and moving counterclockwise around the car to the left front.

8. Add just enough brake fluid to the master cylinder to bring its level to the full line. Reinstall the cover.

9. Clean up the mess you've made, starting with the area surrounding each bleeder, perhaps with a rag first, followed by brake cleaner. (You'll want to recheck each bleeder following road testing to assure that nothing's leaking.) **Don't forget to reinstall your bleeder caps.** Fluid on the pavement should be mopped up with your grimiest rags. To eliminate potential staining, grind kitty litter into the pavement once you've got the car back on its feet.

10. Road test, then recheck for leakage.

20B. Old School II – The "Tube and Jar" Approach

12

If you are working alone, you can bleed the system using a large mayonnaise jar, a length of clear plastic tubing of appropriate diameter and length, and at least one extra can of brake fluid (totalling three).

The procedure is essentially the same as described above, except that:

- The bleeder won't be sealed with your fingertip during the pumping process, and
- You'll only bleed each wheel once.

Put about two inches of clean brake fluid into the bottom of the jar and insert the free end of your tubing beneath its surface. This permits air and brake fluid to be displaced from the brake lines as the pedal is pressed to the floor, but prevents air from re-entering the system as the pedal comes back up. (Any air contained within the lines bubbles up through the fluid residing in the jar.) Since the fluid in the jar is clean, the amount that's sucked into your tubing as the pedal returns from the floor can't contaminate the system. Assuming a clear glass jar, you can discern whether the brake fluid that's exiting is clear–indicating fresh fluid–or murky, indicating that you still have a ways to go.

Naturally, to watch the jar while pumping the brake pedal requires that you stand on your head–my first objection to this procedure. Other problems follow:

- Any crud that you pump out settles to the bottom of the jar. If the tubing is positioned with its free end resting right on the bottom, you'll suck this sediment right back into the tube as the pedal comes off the floor. You can get around this by punching a hole of appropriate diameter through the lid of the mayo jar, inserting your tube through the hole, and positioning the free end a full inch above the jar's bottom.

- You'll waste a lot of fluid, particularly if you conscientiously dump the crud and refill with fresh fluid at each wheel.

- Opening and closing each bleeder can be a major source of frustration, especially if the tubing you're using has been looped around some hardware store display spool for the past several years. The stuff will want to twirl and jump, spraying brake fluid and–God help your composure–occasionally knocking over the jar.

- And then there's the problem of opening and closing the bleeder with its outermost portion shrouded in plastic. Put your wrench on first, then mount the tube and pray that you can rotate the damned thing far enough to sufficiently open the bleeder without having to reposition your wrench. (Remember that these bleeders love to strip when approached with a twelve-point wrench–imagine what that three-sided open-end wrench could do to your weekend!)

On the plus side, it can be done alone. Furthermore, if you don't knock over the jar, your pavement stays clean.

20C. Vacuum Assisted Bleeding

Inexpensive hand-operated vacuum bleeders are available that eliminate the need for an assistant. Assuming you do it right, you'll be less likely to spill any fluid. You still have the issues surrounding opening and closing the bleeders. That can lead to air bubbles, but again, if you do it right . . .

12

20D. Bleeding an Empty Brake Master Cylinder

If you've let the master cylinder run dry, then bleeding the system without a vacuum bleeder can be especially difficult, but doable. Be prepared for a whole lotta pumping, probably with your assistant's left foot between the pedal and the floor to enable returning the pedal after each "pump".

If you're installing a new master cylinder, invest in a vacuum bleeder. Otherwise, you'll need to bleed it initially on the bench in a vise, as follows:

You'll find that the master cylinder comes with two rubber plugs protecting the cylinder's "innards" from dirt, etc.

- Fill the cylinder with fluid.
- Remove one of the plugs, substituting your fingertip for the bleeding. (See §20A Step 2, p.309)
- You'll need to hold the second plug in place either with a clamp or another finger.
- To "pump" the cylinder, you'll use a phillips-head screwdriver tip inserted into the end of the unit (it's recessed).
- Once you have the first port squirting nicely, move on to the second.
- Install the master cylinder once it's full of fluid, then bleed the entire system to eliminate all air from the lines.

Chapter 13. The Wheels: Tires and Alignment;
Steering, and Suspension

13

Wheels, Tires, and Suspension

13

Notes

Notes

Wheels: Tires and Alignment; Steering, and Suspension

1. Tire Wear

Did you ever hear the one about the guy who drops his car off for a front-end alignment? He's been experiencing vibrations in his steering at 55 mph and up! . . . You don't get it? You're not alone–we must see ten people a month with that very request. And it's no laughing matter, because the punch line is that an alignment will never cure the vibration; for that you need a tire balance.

The purpose of an alignment is to correct the tires' positioning relative to the road, so that the tires roll over the pavement rather than "scrub" across it. Improper alignment will definitely cause a tire to go out of balance, but that takes time–first the tire must lose some rubber by being dragged down the road. The vibrations evolve from the uneven wear patterns being produced on the tread. So, while it's true that improper alignment causes tire wear, subsequent tire vibrations and eventual tire noise, you won't get rid of those vibrations without either balancing or replacing the tire.

Tires go out of balance in two other ways:

- Owner neglect (leaving the tires low on air for extended periods of time)
- Smacking a pothole or curb with sufficient force to either bend a wheel, knock off a wheel weight, or produce the start of a ply separation within the tire, §3A, p.320. These very same "accidents" can knock out the car's alignment as well.

Whenever a "close encounter" with the roadway produces a new vibration in motion, you need to check your tires and wheels for damage. This inspection should be performed as soon as possible because of the possibility of a blowout. An alignment check should always follow an impact of this magnitude.

The odds of a vibration suddenly developing are fairly rare. In most cases of impact with pothole or curb, tires don't suffer measurable damage. But all of those lesser shots–the simple, repetitive stuff that we're all guilty of, like using the "braille method" to back into a parking space–will increase the odds of producing a ply separation in that tire, possibly even knocking out your alignment, all without leaving a single memorable event that you could blame on the city's road crews. Do what you can to prevent the right front wheel from smacking into the curb with the steering wheel turned hard over.

2. Tire Inspection

This is a four-step process that involves a visual exam, physically running your hand over the tread, checking tire pressures, and road testing.

❑ Tire Pressure

Tire pressures should be checked on a cold tire. As you drive, the frictional forces applied to your tires heat the air that inflates them. As their temperature climbs, so too do their pressures, rendering hot tire pressure checks grossly inaccurate. Having your own $5 tire

13

pressure gauge enables readings that are accurate to within a few pounds, out in front of your house while your tires are truly cold. The gauges found in gas stations are often inaccurate; they've been abused and are seldom calibrated. The hand crank style is the worst offender.

The driver's door jamb should have a sticker showing the manufacturer's recommended tire inflations. Otherwise, consult your owner's manual. Because of the difference in a car's relative weight front to rear, the inflations recommended for the front are usually different than those for the rear. If you want to maximize mileage and resistance to wear, inflate all four tires 2 psi (pounds per square inch) above the manufacturer's recommendation. On most passenger car tires, that translates to something in the vicinity of 32-34 psi.

Low tire pressures in three or four tires usually indicate owner neglect. One low tire out of four usually indicates that you've picked up a nail that punctured the tire. The cheapest solution involves removing the nail, then plugging the tire from the outside. This requires a tire repair kit available from parts jobbers. The tire has to be off the ground. In most cases, it's easier to head to a tire store or gas station.

Punctures in the edge of the tread can be repaired by installing a patch from the inside of the tire. This requires that the tire be "broken down" (removed from its wheel), then rebalanced upon reassembly. Use a tire store for this latter type of work. Don't choose this remedy unless the tire is very new or was very expensive.

> Using a plug to seal a nail hole along the inner or outer edge of the tread can be fatal. The forces involved in cornering often work the plug free, causing it to pop right out.

Tires that squeal when cornering are usually just low on air, but squealing can also indicate a worn tire long past its useful life. !

❑ Front Tire Inspection

Because front tires are easier to access, start there by turning your steering wheel to the limit of its travel, either full left or full right. The forward half of one tire and the back half of the other will now be in view.

Figure 13-1 should give you an idea of what you don't want to see. What you do want to find is even wear across the surface of the tire, with plenty of tread depth remaining above the wear bars, Fig. 13-2. Any tire that has worn down to its wear bars should be replaced; otherwise, traction on wet roads will be virtually nonexistent and braking power will be greatly reduced—even on dry roads

Run your hand across the tire in the eight directions shown in Fig. 13–2. Ideally, you'll find the surface uniform in all directions. Differences in tread height are usually found along the inside or outside edges, and feel much like ripples on a pond. This type of wear indicates the need for tire balancing (to reduce the amount of vibration delivered to the steering wheel), rotation (if there is too much wear to justify the expense of balancing but too much rubber to throw away), or both.

Depending upon the condition of the front tires, a front-end alignment may be necessary to eliminate the cause of the abnormal tire wear. It's hard to justify an alignment on used tires that are already chopped up (rippled), particularly when mounted on a car used strictly for low-speed commutes over city streets. On the other hand, if the wear pattern is slight or if you are investing in a new set of tires, it makes sense to bring them into the world with as much opportunity as possible for a long and healthy life.

Fig. 13-1. Three tires that rolled through our doors: (L) blown strut or shock; (M) severe misalignment (camber); (R) toe-in problem that's been going on for way too long.

Fig. 13-2. The eight inspection planes on a relatively new tire

13

Whenever you buy tires, don't spring for department store brands or generic cheapies–this is your life we're talking about here! Buy a high-quality tire from an independent tire store or alignment shop that sells several different brands, §8, p.331.

A more expensive condition (cupping) looks much like the surface of ice cream after several scoops have been removed, Fig. 13–1(L). These tires have to be thrown away to eliminate their noise and vibration, and repairs must be made to correct the cause of their demise. The most likely causes of cupping include a blown strut (or shock absorber) or bad ball joint, but other loose suspension/steering components can also be to blame. It's an absolute waste of money–and possibly hazardous to your health–to replace the tires and align the car without first addressing the reason for the cupping. If your budget can't absorb the outlay, find some credit! (Consider the cost of repair against the costs of losing your car to a body shop for two to three weeks.)

Two final thoughts on badly chopped or cupped tires:

- Seek out a shop that specializes in alignment work before you replace your tires. Their wear patterns display valuable information to the skilled observer.

- Badly worn tires make a distinct and often loud noise that is regularly mistaken for a bad wheel bearing. Unless the wheel is actually loose when lifted off the ground, or noisy when spun, I would opt for another road test following tire replacement before authorizing replacement of wheel bearings. See §3A, p.320 and §10, p.338.

Meanwhile, back at the ranch . . . Inspect both the inner and outer sidewalls of each tire, looking for any cuts or "eggs" (a lump or balloon jutting out from the side of the tire that results from a weakness in the sidewall).

- Cuts are most prevalent in the outside walls of passenger-side tires, and are caused mainly by curbing. The seriousness of a sidewall slash must be evaluated by inspecting its depth; if you can expose the cords in the tire by pulling back the rubber flap, that tire must be replaced, Fig. 13–3A.

- You might find an egg in either wall of any tire, Fig. 13-3B. If so, that tire must be replaced at some point. Depending upon its size, that could be sooner or later, because little eggs eventually grow into bigger ones.

Assuming that your front tires pass muster on half their circumference, rotate the steering wheel to its opposite extreme and repeat the visual and physical inspections above.

Fig. 13-3A. Torn sidewall with cords visible

Attribution requested

Fig. 13-3B. Tire "egg"

❑ Rear Tire Inspection

The rear tires are no fun to inspect. Instead of simply turning the steering wheel to expose the front tires' tread, on the rear you must literally get down on all fours or lie down on your side to see much of anything. Even then you can only see those sections of the tire positioned between 2 and 5 o'clock. To see the rest, roll the car two to three feet at a time. (First chalk the tires to identify what you've already seen.) Clearly, this inspection is a lot easier when your car is on a lift, perhaps while it's getting its next oil change.

3. The Road Test

3A. At Low Speed – Ply Separations

Road testing is the final step in tire inspection; searching for a ply separation is its first component. Ply separations usually occur when a tire is smacked hard against a curb or pothole; one layer of the tire literally loses its bonding to the next. These problems typically start in one small area of the tire and spread over time.

Ply separations are most evident in tires that are mounted on the front. Early in their deterioration, you might only experience the vibration while the tire is still cold (during the first two to three blocks of driving). At very low speeds (0 to 10 mph) the steering wheel wobbles back and forth through an arc of 10 to 30 degrees. Naturally, having your hand on the wheel dampens this out, so make your observations with your hand just off the wheel on a smooth, straight, dry road. Try to accomplish the following three observations before the guy behind you absolutely loses his mind. That's actually a little joke: The cardinal rule of any road test is choosing a safe location–an area with good visibility and no traffic.

- Accelerating gently from a stop (0 to 10 mph)
- Maintaining a constant speed (0 to 10 mph)
- Gradually braking to a stop

Because a warped front brake rotor creates vibrations in the steering wheel when slowing to a stop, you need to be very precise in your observations. If you have absolutely no oscillations except when braking, suspect a warped rotor, *5, p.266. Should you experience oscillations under the first two or all three conditions, suspect a ply separation.

Severe ply separations are often accompanied by noise. At speeds below 5 mph, a "galoomph–galoomph–galoomph" sound is common. At higher speeds, a "b-d-d-d-d-d-" sound is likely, that increases in volume with increasing speed. (This latter sound is characteristic of a lumpy tire, whatever the cause. See *4C, p.325.)

13

Ply separations in rear tires are much more difficult to detect. Because you won't feel it in the steering, most of them aren't noticeable until the tires are rotated or until the deterioration has evolved into a vibration felt at all speeds throughout the entire car. This "seat-of-your-pants" vibration is usually accompanied by the noises described above. See *4A, p.322 for directions on visual inspection for ply separation.

> Any tire with a ply separation should be replaced or moved to spare. Severe ply separations can be life threatening.

3B. Testing for Pull

The next portion of this road test requires a smooth, straight, level road. Since most two-laned roads are "crowned" to promote water drainage, even a perfectly conditioned car will wander toward the right shoulder as soon as you let go of the wheel. So, a four-lane road that permits speeds in the vicinity of 30-40 mph or a parking lot in your nearby mega-mall can be ideal.

The road should be dry. Travel it at constant speed, initially with your hands placed firmly on the wheel. On a straight stretch where your car is nicely centered in its lane, move your hands a hair's breadth off the steering wheel. Be prepared to take control if the car starts to change lanes, because once the car starts to wander, it will continue to move with increasing pace. In most cases, your fingertips will actually be off the wheel for periods of no more than a few seconds at a time–very few cars travel in a perfectly straight line. Repeat the test several times until you are certain that you are seeing consistent behavior. This portion of the drive will identify any tendencies that the car might have independent of braking, such as a misalignment of the front end or a problem with one of your front tires. Section 3, p.260 is a continuation of this road test that explores issues related to braking.

A pull can be caused by any of the following:

- Uneven tire pressures
- A mismatched set of treads across the front
- Improper front-end alignment
- A developing ply separation in either one of the front tires
- Binding in one of the front brake calipers

While other possibilities do exist, these five are primary.

13

3C. High-Speed Vibrations

The final aspect of this road test is performed at speeds between 45 and 70 mph. Again you need a smooth, straight road with nobody around. See if the steering wheel vibrates as you lift your hand just off the wheel. You'll need to do this at several different speeds within that range: Vibrations felt at 55 mph may well disappear by 62 mph.[1] A vibration at high speed is usually indicative of one or more tires being out of proper balance.

If the vibration comes primarily through the steering wheel, the front tires should be balanced. Vibrations coming through the seat of your pants, particularly when coupled with a rock-steady wheel, are clearly arising from the rear. Both front tires are balanced as a pair, same for the rears.

4. Interpreting the Road Test, and What To Do About It

4A. Low-Speed Vibrations; Inspecting for Ply Separation

To isolate the offending tire, you need to have the car on a lift so that the tires can be viewed as they spin. This can be accomplished using jackstands, although it's harder to acquire the necessary end-on view. Then, standing (or laying) two to three feet back, in line with the tire's rotation (not off to one side), watch the tread as it comes around at low speed. Focusing on the very bottom of the tire provides the best vantage–you'll see a well-defined edge with air as a backdrop.

> Never attempt this without at least two jackstands, properly placed, §7D, p. 140. You'll need glasses or goggles.
> Be sure that all loose rock and dirt has been cleared from the tread surface to prevent it from being thrown in your face as the tire spins.
>
> Don't lose sight of the fact that these tires will be spinning: Keep clear of them to avoid injury.

❑ Front Wheel Drive

Because each front tire is linked to the transmission through a driveshaft, spinning the tire by hand is a thankless task, even if you've placed the transmission in Neutral (engine off). As long as the front end is on jackstands, you can start the engine, and with your foot firmly on the brake pedal, put the transmission either into Neutral or Low–whichever best accomplishes a slow tire rotation. If the tires won't turn willingly with the transmission in Neutral, and turn too fast in Low gear, have an assistant apply light pressure to the brake pedal while you observe the rotation.

A well-developed ply separation is clearly visible: Either the entire tread or a portion of it will appear to hop up and down as the tire spins; one portion of the tread will appear flatter, the rest will seem to bulge. It's as if you're looking at an egg coming round, rather than a tire that's meant to be round. Don't get hung up on little variations: Even a brand new tire in perfect condition will not rotate in pure concentric fashion–they often display

---

1 Harmonic frequencies are just as applicable to tires as they are in music. Distinct vibrations, when in sync, amplify each other; go off that speed just a bit and they begin to cancel each other out.

a variance (hop) of as much as an 1/8 inch. A true ply separation produces a strong, lumpy hop, often with as much as 1/2 inch of distortion. Note that you can have a ply separation in a tire that in all other respects appears perfectly normal–plenty of tread, a uniform wear pattern, etc.

4B. Does the Car Wander? Does the Steering Consistently Tug at You?

Assuming proper inflation, proper alignment, a matched set of tires, and a flat, dry road surface, your car should go straight when you take your hands just off the steering wheel. This should be true at any constant speed. Cars that drift a little constitute the norm–this is not a perfect world.

13

If the car tracks straight consistently, then at least for this one road test you don't have a demonstrable problem. Still, it wouldn't hurt to check your tire pressures and inspect the surface of your tires for wear, $\mathcal{S}$2, p.317.

If you find that your car has an intermittent problem with pull, you can usually demonstrate that it's related to temperature: Both your brakes and your tires take a mile or two to heat up; does the problem only manifest itself within that time frame? That could easily suggest:

- An evolving ply separation in one of your front tires
- A problem with binding in a front brake caliper, $\mathcal{S}$6, p.268.
- Binding in the bearing at the top of a McPherson strut, Fig.13-13, p.340. For those that aren't sealed, a dollop of grease might just solve the problem, $\mathcal{S}$13D, p.350.

If your car wanders consistently in one direction while driving but seems to stop in a straight line, you are experiencing a problem with one of your front tires, a problem with the car's front-end alignment, or both. Even though it's conceivable that a rear tire or rear-end misalignment could cause your wander, it's highly unlikely. This is not to say that rear-end problems don't exist, but remember this: A little problem in the front end will produce a pull that's equivalent to a major problem in the rear.

Check your tire pressures first, then inspect for tire wear, $\mathcal{S}$2, p.317. Tire pressures must be equal, left and right. Both front tires should be identical: same manufacturer, same tread pattern, same depth of tread. If they're not, you might have to accept a certain amount of wanderlust, even on a perfectly aligned car.

If the front tires are not wearing well, you'll need an alignment check to uncover the reason. Should you notice the problem soon enough, you'll need nothing more than a standard alignment in conjunction with tire balance and rotation. But if you have badly worn tires, don't bother with an alignment until they've been replaced. It makes sense to have a good alignment man look over the front end before you invest in tires, just in case some ugly news is lurking there. This is particularly true on high-mileage cars (100,000 miles and up) and those that have seen entirely too many potholes. If tire replacement is unnecessary, move your best tires to the front (as long as they're a matched set). Spin balance ($\mathcal{S}$4C, p.325) any tire that exhibits light-to-moderate wear, particularly those that are going to the front.

When tires have some miles on them, either the wear and tear they've encountered or a ply separation is most likely to blame, *3A, p.320* and *4A, p.322*. With a minor ply separation, swapping right to left often cures the pull—at least in the short term. But a more severe separation necessitates replacement.

It's always a good idea to ask your alignment man if he had to swap tires left to right. After all, it's your car—you have a right to know what it took to make it go straight.

We often see cars with what appear to be decent tires—occasionally even brand-new ones—that drift while driving. This consistent wandering in one direction can be light or even moderate. And it sure feels like an alignment problem! We'll send these cars over for alignment, only to find that they're right on spec. So what's wrong? A small percentage of tires seem to have a natural predisposition toward causing a car to pull. I'm not sure why this occurs with brand-new tires, except that they clearly encountered some glitch in the manufacturing process. Try swapping the front tires left to right and right to left. Quite often, that will cure the problem. *This can only be done with tires that are bi-directional.* (High-performance tires are typically engineered to roll in one direction only. They sport asymmetric tread patterns and are clearly marked with an arrow molded into their sidewall. These tires can never be swapped side-to-side without first being broken down (off their rims), then remounted on the opposite wheel so that their arrows remain facing in the direction of forward rotation. This is both expensive *and* a pain in the ass, worth it only if you can't get your money back.

❑ Potholes and Curbs – If your car pulls strongly in one direction while driving and pulls even more strongly in the same direction when braking: You either have a defective tire—one with such a severe ply separation that it will clearly require replacement—or your kid smacked something hard enough to knock the alignment clean into Kansas!

If you suspect that the front end took a hit, you'll need an alignment check at minimum. Not only is the car potentially dangerous to drive in its current condition, but also it could get quickly worse. Remember that one of those tires was the first part of your car to feel the blow; a severe ply separation is almost a given. In addition, the wheel itself might be bent, creating vibrations at speed. Suspension and steering components could have been shoved rearward.

❑ Brake Calipers – If your car tracks reasonably well at constant speed but pulls when you apply the brakes, you have a problem with one of your front brake calipers, specifically with binding of one sort or another. See *6, p.268*.

❑ Worn Bushings – Every car has rubber bushings across the front end, whether they be on a pair of strut rods that prevent the the lower control arms from moving forward or backward or along a stabilizer bar that ties the left and right steering components to the chassis and to each other, Fig.13-16, p.346. On aging cars, these fat rubber bushings can wear out, allowing one or more components some freedom to shift under acceleration, deceleration, or cornering. As a result, for example, the car can pull in one direction during hard acceleration, then head back the opposite way under deceleration. When traveling at constant speed and during moderate braking, the car tends to track fairly straight.

❑ Rear Alignment Issues – The above road test will not tell you much, if anything, about the rear of your car, simply because the rear must follow wherever the front end leads. Drivers seldom notice that they have a problem back there until it becomes a major issue. The best way to evaluate rear-end alignment is through tire inspection, §2, p.317 or an alignment shop, §10, p.338.

4C. Vibrations While Driving at Higher Speeds; Tire Balance

Because of the number of possible causes for vibrations that occur at higher speeds, it's important from the standpoint of diagnostics to accurately distinguish your particular vibration's signature.

Is the vibration tied directly to road speed (as registered by your speedometer), or is it linked to engine speed (as registered by a tachometer or your ear)? Tire and wheel vibrations are tied solely to road speed; changing gears and the consequent rise and fall in engine RPMs has no effect on them. Vibrations that are related in some way to the engine or transmission follow instead the sound or pitch of the engine; they correspond directly to how fast the engine is turning.

Does the vibration appear to be independent of the brake pedal? If so, you have a tire problem. If the vibration starts up as soon as the brake is applied, you are experiencing a brake pedal pulsation, §5, p.266.

Is the vibration felt primarily in the steering wheel or in the seat of your pants? If it's in the steering wheel, have both front tires balanced. If it's coming through the body of the car, particularly when coupled with a rock-steady steering wheel, you'll need both rear tires balanced.

At what speed does it start, and when does it go away? Vibrations occurring at speeds of 45 mph and up are usually due to a problem with improper tire balance. These vibrations often come and go; the car could be steady as a rock through 55 mph, then shake violently between 56 and 64 mph, only to smooth out again above 65. The causes of tire vibration are as follows:

- Improper tire balance
- Tire wear from consistent underinflation
- Tire wear due to an alignment problem
- Ply separation, §3A, p.320 and §4A, p.322
- Bent rim (wheel)
- Missing wheel weight
- Loose wheel bearing, §11, p.343
- A problem with the suspension–either a blown strut (or shock), a worn bushing, or a bent component–creating wear on the tire.

Inspect your tires and wheels, §2, p.317. You might find that there is just not enough tread there to bother with or that a portion of the tread has actually worn away. If so, you'll be in the market for new tires and an alignment, §8, p.331 and §10, p.338.

The possibility of a ply separation or bent rim become obvious on a tire balancer–assuming that you have an observant mechanic. The problems are visually obvious as the machine spins your tire. In addition, these two conditions make it hard for the mechanic to achieve a good balance. If you have a bent rim, it will either have to be replaced, moved to spare, or at the very least rotated to the rear. Should you need to, it's perfectly legitimate to swap tires around to get the best tread onto the best wheels, and then to get them to the front. Don't waste your money trying to balance distorted rubber, particularly on a bent wheel.

Because all original wheel weights are supposed to be removed prior to rebalancing, a wheel weight that's been missing should not be an issue.

A good way to verify whether your tires have been properly balanced is to look for the telltale line of dirt that should remain, ringing the area where the original weights used to be. This evidence will remain for a day or two–as long as it doesn't rain. Naturally, a set of shiny new wheel weights are a certain sign that the wheel was balanced.

All modern-era tire balancing is performed on dynamic spin balancers–computerized machines that actually spin the tire, usually at fairly low speeds. Each tire is removed from the car and mounted vertically onto the machine. The amount of weight required, as well as its position relative to some fixed point (the tire valve, for instance) is indicated on an LCD screen. Typically, weights are required on each side of the wheel.

Should you find yourself in a shop that's about to balance your tires horizontally, find some means to politely disengage yourself. You are looking at an antique that's known as a static bubble balancer.

Should your vibration persist following tire balancing, ask the service manager or shop foreman to road-test the car with you, and then reinspect your tires and wheels. Perhaps their mechanic missed something obvious the first time around. Or perhaps the vibration is due to a loose wheel bearing, a worn suspension component, or something weirder– perhaps a frozen caliper overtop a warped brake rotor.

Is there any noise associated with the vibration? The most common is a "bd-d-d-d-d" sound that increases in volume with increasing speed. It's a softer sound than a noisy wheel bearing–there's no metallic edge (grinding) to it. The noise can indicate a ply separation or a badly chopped (lumpy) tread. See *§3A, p. 320 and §4A, p. 322* for the means to identify a ply separation. Any ply separation that is actively noisy should be replaced immediately. A badly chopped tire is easy to diagnose: Simply run your fingertips across the tread; if you find that it's lumpy, you have a noisy tire.

A badly chopped tire is long past the point of benefiting from a tire balance. Although balancing might make it last a bit longer, it won't help either the noise or the vibration. Unfortunately, many of these tires have too much tread on them for the budget-conscious owner to simply throw away, so they linger on the car until the owner just can't take it anymore. (Comfort yourself with the fact that they might just make decent snow tires.)

Most lumpy tires get that way because of a weak strut or severe misalignment. It pays to determine the cause. A blown strut can be extremely dangerous–all too often, some jerk will put you and your car to the test, and just when you need it most, the absence of that strut will hand you mush rather than maneuverability. If the struts are O.K. (§12, p.345), wait on the alignment until you replace the tire.

5. Tire Rotation: Shifting Them to Their Best Advantage

Tire manufacturers recommend routine tire rotation. So do the car manufacturers. But as a mechanic working in the real world, I'm not convinced of its efficacy. The number of cars with more than 10,000 miles on them that are running four decent, equivalent tires is surprisingly small. Chronic under-inflation, misalignments, tire wear, and an occasional bent rim (wheel) all take their toll. I'm usually looking for two decent tires out of the four or five provided to put up front! The worn tires or bent rims go to spare or to the rear so that their vibrations will no longer travel up the steering column. (When mounted on the rear, wheel and tire problems are dampened by distance and upholstery.)

13

If you own a front-wheel drive car and intend to rotate only at the major service intervals (when it will be performed as part of the package), you will find that your front tires are significantly more worn than the rears–even at 10,000 miles–and that's assuming proper alignment and inflation at both ends of the car. For all its advantages, front-wheel drive is tough on tires. Consider the frictional forces of steering and powering the car, then add to that the extra weight of an engine and a transmission.

If you like the idea of tire rotation, you'll have to start the process early–and maintain that schedule consistently–once every 6,000 miles. Doing so will probably cost $15 or more, not to mention the cost of all those "critical" repairs that your mechanic will find along the way. Is it worth it? Owner's choice . . .

On a four-tire rotation, move the front tires straight back, but criss-cross the rears coming forward so that the right rear tire takes up residence on the left front, and the left rear goes to the right front. With five tires, put your worst tire to spare (the right front, if all seem equivalent) and rotate the rest as above.

> There is one exception to this rule: Uni-directional tires engineered with asymmetric tread patterns, designed to turn in one direction only. As mentioned previously, they have a directional arrow molded into their sidewall. These tires can only be switched from the left to the right side of the car (and vice-versa) if the tires are broken down and remounted on the opposite wheel such that their arrows remain facing in the direction of forward rotation.

This approach to tire rotation began to supplant old-school notions that radials should not be swapped from one side of the car to the other some twenty years back. At issue is the idea that once mounted, each tire would develop a preference for its particular direction of rotation; if that direction of rotation were reversed by swapping it side-to-side, then a ply separation would likely result. Frankly, I haven't a clue which view is correct, but I lean toward the pragmatic: Position your tires wherever they need to be, considering tire condition, tendency to pull, etc.

Common Treads

- Any tire that's going to the front should be balanced if it shows signs of visible wear, or if the car develops a vibration at higher speeds following the rotation.

- Whenever possible, put a matched set of treads across the front to reduce the tendency toward drift or pull.

- On front-wheel drive cars, putting your best treads up front provides maximum traction and keeps steering vibrations to a minimum. Once the front tires wear to the point that steering vibrations do become a problem, either rotate them to the rear, to be replaced by those still "pristine" rear tires (assuming, of course, that your rear-end alignment is sound), or buy two new tires for the front. You'll need a front-end alignment check in either case *if* that wear has been uneven or otherwise suspect. This approach wears out only one pair of tires at a time rather than all four.

- If both right-side tires have been wearing poorly and both left-side tires are in decent shape, move one of those lefties over to the right rear and put a new, matched set across the front. Both the front end and the rear will require alignment to prevent history from repeating the damage to the right-side tires.

- Imagine a car with four decent tires, the fronts are matched, the rears are not. Following alignment, the car still pulls left. Many alignment men would swap the front tires right↔left and repeat the road test. In most cases, the pull disappears. This is an old mechanics' trick used on cars whose tires are no longer covered under warranty. If yours are covered, I would insist on a new tire. Although a significant proportion of brand-new tires create pulling problems all by themselves, there's no reason why you should have to own one or accept its being swapped to the other side of your car. To ensure that it's not, you might want to mark the interior side wall of selected tires before going back to the shop.

- Another case: This car has four decent tires on it; the fronts are a matched set and so are the rears. Following alignment, the car still pulls to the left. The best solution is to bring the rear tires forward; the "quick and dirty" approach is to swap the front tires right↔left as mentioned above.

- So what do you do if you have a mismatched set of tires, some more worn than others, but all too decent to throw away? If you are not experiencing much of a pull and don't see a particular wear pattern that you clearly need to address, ride with what you have until such time as your treads start to wear out. Then spring for two or four new tires and alignment work. If, however, a wear pattern is setting in, then an alignment is in order, even though the car might still wander to some extent due to the mismatched tires across the front.

- If you have one bent wheel and little desire to spring for a new one, recognize that it is perfectly legitimate to swap tires around to get the best tread on the best wheels, and then to get them to the front. You can swap good rubber off a bent wheel, mount the worst tread you have onto it, and throw it in the trunk unbalanced–just as a safety net should you end up with a flat. Don't waste your money trying to balance distorted rubber, particularly on bent wheels.

- Your rear tires need not have identical treads, with one caveat: *Never mix a uni-directional tire with a bi-direction tire.*

- Because the rears follow wherever the front tires lead them, you needn't be concerned with wander originating from the rear.

- For those of you who must suffer through snowy winters with the spinning and sliding of rear-wheel drive cars, make sure that you have your best tires on the rear, at least during the winter months, or spring for one, preferably two pairs of snow tires.

  To save on the twice yearly breakdown and remount of winter to summer and back again, consider buying an extra set of wheels, either from a junkyard if you want to match what you already have, or online: Old-fashioned steel wheels of the same dimensions may not be pretty, but they're cheap. Doing so eliminates the risk of a torn tire bead developing as part of all those tire changes.

6. Wheel Locks

If you live in a crime belt, there's nothing wrong with a set of wheel locks, particularly if you have alloy wheels or a nice set of tires. At least they'll keep the punks at bay. But if a professional wants your wheels, he'll take 'em within five minutes, locks or no.

The two biggest problems with wheel locks are that:

- Customers put their locks away in "special" spots so they won't get lost, then forget where they put them by their next trip through the shop.

- Mechanics have a fondness for putting locks down haphazardly on their work benches while the car is still in the air (we tighten wheels at shoulder height), then forget to give the lock another thought until long after the car is gone.

If you use wheel locks, be smart about it:

- Choose a spot within the passenger compartment–an unused ashtray, map pocket, or console tray–that's small enough to facilitate locating the thing the next time you need it, then consistently return it to that location.

- When you go to the shop, pull the lock out of that spot yourself and place it conspicuously on the console or passenger's seat with a note–as much to yourself as to your mechanic–to double check that the lock is present and accounted for before you leave the premises.

A set of locking lug nuts that have lost their lock can be driven off quite easily with an air chisel. Unfortunately it destroys the nuts, and if the chisel slips–a frequent occurrence–it can damage (nick) your wheel.

7. Changing a Flat Tire

The secret to not having to change tires is to run good rubber beneath you! Granted, there is always the chance of driving over a spike that's large enough to land a whale, but the odds against are in your favor.

If you must change a tire on the road, try to park on a flat surface. Grades of any sort can be extremely dangerous–particularly considering the size of most hand jacks. If you can't avoid an incline, make it slight, block the wheels with a rock, log, or suitcase, and stay out of its path should the car pitch off its jack.

Before you even begin, make sure that your spare has air in it . . .

Crack the wheel lugs loose *before* you set up the jack. (If you can't make 'em turn while the wheels are rooted to the ground, you sure as hell won't be able to do it once the car is teetering on that jack!) If one of those lugs has a wheel lock on it, start with it. They can be very difficult to remove without an air gun. Not always; be hopeful!

Because mechanics have to be responsible for ensuring that your wheels won't come loose, they are not particularly concerned about the possibility of a roadside emergency. And because virtually all of them use high-torque impact guns, most of you will be in for a bit of a tussle trying to bust these nuts loose. Take a close look at Fig. 13-4; it illustrates the best technique that I know of for success. Don't jump on the lug wrench, the unlucky among you will find that you have the strength to break that lug right off! You also run the risk of injury should the wrench slip. All lug nuts loosen counter-clockwise ↶. Back off each one no more than one revolution.

Fig. 13-4. Loosening lug nuts
ALWAYS counterclockwise

Fig. 13-5. Tightening lug nuts

Now comes the easy part. Get everyone out of–and well clear of–the car. Place your jack, raise the car, swap the wheel, then tighten the nuts as snug as is practical (finger-tight) with the car teetering there. Don't tighten the nuts going around in a circle; this can misalign the wheel against the hub. Tighten them in a star-shaped pattern, Fig. 13–5 . Finish tightening them once the car is safely back on earth.

By the way, you know those aerosol products that profess to repair flats simply by injecting them with space-age sealants? Unless the tire is shot anyway, don't ruin what's left of it–the goo that's injected is virtually impossible to remove later on. Because it bonds to the wheel just as easily as it does to rubber, cleaning up the mess can create significant additional expense when you buy that replacement tire.

❏ Space-Saver Spares

Have you ever considered what would happen if you were in the middle of a trip when a tire went flat? First, you empty your trunk. Next, you'll change over to this absurd excuse for a tire–not much wider than you'd find on a motorcycl e–and if you are attentive to the large yellow CAUTION emblazoned all over the wheel, you will find yourself reduced to traveling at its maximum safe speed, usually about 45 mph. This could easily be down an Interstate with cars flying past at 80. There's a gas station a few exits down the road and fortunately they have your tire size in stock. Unfortunately, they only stock one brand–the expensive kind. Now you might get lucky–they'll plug your flat tire, reinstall it, and you'll be down the road. But if not, you'll end up owning some no-name generic with tread that will never match any other tire you will ever see.

By now you've had to stop twice within twenty miles, and most of you have unloaded your trunk twice.

Why on earth would so many factories engage in this idiotic practice? Most would probably reply that the "space-saver" provides more trunk space. Personally, I don't buy it, because the tire well on most of these cars is big enough to accommodate a full-sized tire. Frankly, I think it has more to do with the few bucks they can save on every car, not to mention the money they make selling full-sized rims to those who swear they'll never subject themselves to that experience again!

Those of you with space-savers have two alternatives to the above scenario: You can either replace your worn rubber before the trip, hopefully with a pair of quality tires at a reasonable price, or you can replace that tinker toy with a full-sized tire on a real rim (wheel). Naturally, you'll need to check with your dealer before you invest in the extra rim, just to make sure that a full-sized tire will actually fit into your wheel well. If so, consider buying a junkyard wheel; quite often you'll be able to find one with a decent tire already mounted.

8. Buying Tires

All tires sold in this country are rated for expected treadwear life, traction, and heat resistance. These ratings are stamped right into the sidewall of every new tire, along with a series of codes, Fig. 13-6. It should be pointed out that these ratings are provided by the tire manufacturers–not by independent testing–based upon guidelines set down by the Feds. Although not a perfect system, it's a whole lot better than no rating system at all.

❏ Treadwear Rating

Treadwear is rated numerically; the higher the number, the longer that tire should last. Supposedly, a tire that's rated at 200 should last twice as long as one that's rated at 100. Naturally, that assumes proper inflation, proper alignment, and reasonable use.

❏ Traction

Traction is graded alphabetically: AA, A, B, then C. Those ratings are based on high-speed braking on wet surfaces–the distance it takes to get from, say, 60 mph to a full stop. Tires with a AA rating should possess the best grip when slamming on your brakes. But those tests are performed on a straight piece of road and say nothing about cornering abilities. C-rated tires bring up the rear and should be priced accordingly.

13

To me, the traction rating is by far the most important of the three. Now, if you're driving an old heap that you intend to keep on the road for no more than another year, there is a certain logic to saving a few bucks on a tire with less promise of longevity. But your treads are the only surface that connects you with the road; the amount of traction you have in corners and when braking–particularly in the rain–can have everything to do with *your* life expectancy. Quality counts here. Whenever possible, buy tires that boast an A rating.

Fig.13-6. Tire nomenclature
Courtesy of Dunn Tire Company

❏ Heat Resistance

Heat resistance is also graded alphabetically, A, B, or C. The same rules apply with respect to quality (A) vs, well, less than that (C), but many cars really don't need A-rated tires. Commuter cars used solely for banging around city streets come to mind; the grandmother who wouldn't be caught dead out on the Interstate is another example. Heat becomes a much more significant factor if extended high speed driving is a big part of your life, particularly in the sunny South and those long, hot stretches to the West.

Tire manufacturers use different rubber compounds to achieve particular characteristics or levels of quality; changes in composition produce a trade-off between traction and durability. Clearly, the harder the rubber, the longer it will last, but at the price of reduced traction. Frankly, I'm far more interested in how long *I'm* going to last than in how many miles I can eke out of my treads. As with all else in this life, you get what you pay for. Higher quality tires carry excellent treadwear ratings (240+/-) as well as A ratings for traction and temperature.

❑ What's Up With All Those Codes?

Each letter/number stamped in the sidewall is important to someone, but not necessarily to you. From most relevant to least . . .

13

- A pothole ate your left front tire. The other three have plenty of miles left on them. The best alternative for most of you is a duplicate of the other three. You'll need the name of the tire, its manufacturer, and its size, for example: Michelin Defender 225/60R17 99H.

  Fortunately, you don't need to decipher all that numerical gibberish, all you need to know is how to write it down–price shopping without it is impossible. Now if you're going to skip that part, just show up at the tire store with your car on its space-saver, all you'll need is your credit card and sufficient faith to believe that they'll have that tire in stock.

  But for those who who enjoy wandering into the weeds, here goes:

  – The first number (225) is the tire's nominal width measured in millimeters. The term "nominal" implies that the tires could be mounted on different width rims, have different inflations and temperatures, and be mounted on cars of varying weights.

  – The second number (60) indicates the ratio between the height of the sidewall and its width: the height of the sidewall is 60% of its width.

  – R indicates it's a radial tire as opposed to a bias-ply, for example. Remember them? No? Then you're not as old as I am . . .)

  – The number following R (17) is the diameter of the rim (wheel) in inches.

  – The load index follows (99). For this you need a chart. Or not! The number 99 = 1709 pounds, which means that the car, its passengers, and their baggage should weigh less than 1709x4=6836 pounds. The internet has hundreds of these load charts.

  – Finally, we come to the letter H, which refers to the tire's speed rating. The letter H implies that the tire's maximum speed should be no more than 130 mph, V's should carry you safely to 149 mph. For comparison, snow tires typically carry the letter Q (99 mph), space savers (M = 81 mph), . Again, the internet is loaded with these charts.

- It's time to replace some tires. The front tires are worn, the rears are still in decent shape. The tires that came with the car have certainly been satisfactory, but you're looking for something different. Before you start to explore, either online or at a local tire store, you'll need to know the tire size. The only variable you can safely change is the last one–its speed rating. Is it worth the extra $28 per tire to continue to drive at 130 mph all day? Or would 106 mph (R) suit you just fine?

But be advised: Perhaps the higher rated tire is actually a more durable tire; perhaps it has better traction. That'll take you right back to the treadwear, traction, and temperature codes.

- You've been itching to set up your car for some *serious* driving, you know, tripling the legal speed limit! But those V, W, and Y rated tires are expensive! Maybe I could just spring for two now, the other two in six months. Sorry. All of these tires are unidirectional; they corner at vastly different rates than bidirectional offerings. I suspect that danger increases sharply in a curve, and even more so on wet pavement.

- This may never come in handy, but it is an interesting fact: The inner ring of letters/numbers that encircle the rim includes a number of data points:

  – number of plys (layers) and their composition–steel belted, polyester

  – maximum load at the tire's maximum pressure, and

  – the Department of Transportation tire ID: All of which is meaningless to me (useless, too, unless you are a DOT tire inspector or are interested in a warranty claim or lawsuit)–except for the last four numbers. The first two numbers signify the week that your tire was manufactured (01 through 52), the last indicates its year (18=2018).

❑ Price and Availability

The price you pay for a tire has everything to do with where you buy it and how much homework you're willing to do. The absolute worst place to buy a tire is from a new car dealer: Their selection is virtually nonexistent and their prices are, um, pricey. Second on this list are most gas stations. Almost all of them are required to sell their oil company's tire line. These are usually no-name generics, manufactured by various tire companies to be sold in huge lots at a contracted price. Now I ask you: If the manufacturer is not putting its own name on that tire, do you really think that quality is their chief concern? I would suggest that the price per thousand is far more important to both manufacturer and seller than any thought of tread design, traction, or longevity. Furthermore, most gas stations charge full-freight because they know that no one with any intelligence would be buying a tire from them if they had any choice in the matter. By the way, don't think that the sale of generic tires is limited to gas stations; every tire store sells one or more "house brands."

So where should you go for a good tire at a reasonable price? A good place to start is the internet. Do a bit of comparison-shopping. Look for a small to medium-sized, independent tire store. Their sales staff will tend to be well informed. Furthermore, they're free to stock what's good and dump what's bad. They usually offer better pricing than the national chains because they're not saddled with all that overhead. But for those of you who travel incessantly, the regional and national independents are probably a better bet–even if they are more expensive–because they provide the assurance of service in distant cities.

Time was that good tires all came with road hazard warranties, honored across the country by every dealer that carried that brand. If you picked up a nail or tore the sidewall on a curb, the manufacturer would either fix your tire or give you a new one, prorated on the basis of remaining tread depth. This policy is in steep decline. However, road hazard warranties are still available to those who want to pay extra for them. Because most of these warranties are only honored through the selling dealer, the national or regional chains are, again, the place to go for those who drive all over the map.

That being said, I've never been particularly impressed by national chains that only sell their own brand of tire. They're usually expensive, you have fewer options, and the individual stores aren't at liberty to dump a particular line even if that tire is no good. But my biggest concern with these shops is that many of them need to rely upon selling mechanical work–struts, steering components, coil springs and the like–just to make ends meet. I've seen this over and over in the Washington, D.C. area–low-mileage cars that come into our shop with all kinds of aftermarket crap installed, all because their owners were scared into believing that their cars were "dangerous." In all fairness, I'm sure that the nationals have their share of good shops with honest mechanics scattered across this country. Furthermore, the above condemnations can be just as true of those small independents I praised above. As I've said so often in this book, you need to evaluate the shop you're in. If somebody is trying to sell you a laundry list when all you came in for was a few tires, get a second opinion.

Fig.13-7L & R. Tire changers: the new and the old

How about big-box and department store auto shops? A name-brand tire at the right price? Why not! But I would avoid getting any mechanical work done in these shops–including alignments. For the most part, the guys that work in these garages have only received basic training. That means a mount and balance and you're out the door!

13

I know nothing about buying tires online except for the obvious: shipping costs, labor fees for mounting and balancing, what to do if there's a problem with one of them . . . It's just never seemed worth it to me.

❏ **Some Closing Thoughts:**

- Every tire you buy should be spin balanced.

- Never mount a new tire on a bent rim (unless the bend is slight and the wheel is to be used on the rear).

- If you have aluminum alloy wheels, verify that the shop has a tire changer designed specifically for them (Fig.13-7L shows one type). Older tire changers (Fig.13-7R) can bend or distort an alloy wheel, particularly when operated by the wrong hands.

- Whenever a pair of new tires are mounted on the front of the car, the car should have a front-end alignment–unless the tires coming off have been wearing uniformly across their entire tread span. And check out their entire circumference, just to be sure. My point is that there's no sense spending money for something that's demonstrably unnecessary. The same holds true for new tires being mounted on alignable rears.

- If you intend on buying a dedicated set of snow tires, consider buying two (or four) additional wheels. That enables simply swapping them on in the fall, then swapping them off in the spring. It gets expensive if you don't: breaking down two tires, then mounting and balancing the snows. Repeating that process again next spring, this time with your summer tires.

  There are two ways to accomplish this reasonably: Steel wheels are available cheap, especially online. If you need all four tires on matching wheels–your original equipment alloys, for example–search junkyards on the internet. You'll find them, guaranteed; and if they're not local, they'll ship 'em. One caveat regarding alloy wheels: They bend fairly easily; they crack, too. Find out about their return policy before you offer up a credit card number . . .

9. Tire Pressure Monitoring Systems

Low pressure monitoring systems became standard equipment in the US in 2008. There are two versions: those that send information directly from every wheel to their microprocessor, and systems that infer tire pressures from wheel speed rotation.

- Direct systems employ a pressure sensor incorporated into each tire valve, Fig.13-8. These are powered by a battery, which means that it's now standard procedure to replace them whenever you buy new tires. The benefits to this system are that the dashboard can accurately display individual pressures on the dashboard display, although not all of them do. Some cars just offer a warning light if one tire signals that it's low.

- Indirect systems use the vehicle's wheel speed sensors, which have been around since ABS braking systems became standard equipment. A tire that's lost sufficient air will essentially be smaller than it should be, which

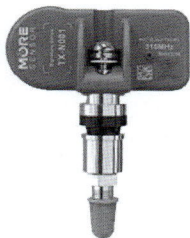

Fig. 13-9. Tire pressure valve (antique)

Fig. 13-8. Tire pressure warning sensors

13

means that it has to rotate faster than the others, just to keep up. ABS systems have a speed sensor (Fig. 13-10) located behind every wheel that sends its data either directly to the ABS modulator (Fig. 12-1, p. 256) or to the car's electronic control unit.

I'm not crazy about these systems. First, they're significantly less accurate than the direct systems. Second, you'll see a warning light, nothing more. (In fairness, the same can be said of the less expensive direct systems.) That's certainly better than nothing, but can lead many of us to ignore yet another check that should be part of our routine. I went into some detail about how easy it is to end up running on chronically deflated tires in §1A, p. 94.

Every time you have a tire replaced, its tire valve should be replaced as well. But what was once a two dollar part Fig. 13-9 could now cost fifteeen bucks. Tire pressure sensors became mandatory equipment in November 2003, but they were standard equipment on a number of product lines prior to that; typically, cars equipped with ABS brakes.

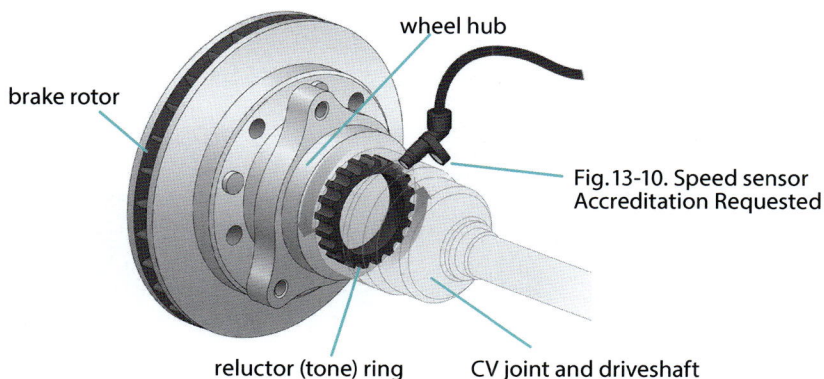

wheel hub

brake rotor

Fig. 13-10. Speed sensor
Accreditation Requested

reluctor (tone) ring

CV joint and driveshaft

10. Alignment

10A. General Considerations

Assuming proper inflation, a matched set of tires and proper alignment, your car should go straight whenever you take your hands just off the steering wheel. This should be true at any constant speed, assuming a straight, level road. Most cars tend to drift a little–this is not a perfect world.

Proper alignment is important for two reasons: It will prolong tire life considerably, and because you won't be continuously tugging at the wheel to keep it going straight, it can greatly reduce driver fatigue. Although fatigue may not seem important to you as you bang around on city streets, it becomes increasingly more so on road trips of several hours or more.

I find it hard to justify a realignment on used tires that are already chopping, particularly when mounted on a commuter car used strictly for low-speed slugfests. On the other hand, when you invest in a new set of tires, it's just common sense to insure them as best you can against premature wear. Alignment work should always follow tire replacement, it should never precede it.

The factories publish a set of alignment specifications for every car they build. These specs are all measured in small angles, many in fractions of one degree. To measure these angles accurately, you need sophisticated equipment–machines that cost $15,000 and up and occupy an entire bay. Most small shops and gas stations aren't willing to commit to this–even if they could afford to–so a lot of them have invested in portable set-ups that can be rolled into place. The main problem with *all* of these units is maintaining their calibration. If you want a good alignment, find a shop that specializes in front-end work.

There are five aspects to wheel alignment:

- inspection of the tires–both for their condition and any tell-tale signs
- inspection of steering components and suspension components
- toe-in adjustment
- camber adjustment
- caster adjustment (when possible)
- at least one thorough road test

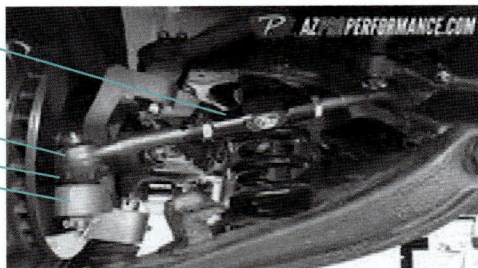

adjustment sleeve
locknut on each end

tie-rod end

ball joint

steering knuckle

Fig. 13-11. Tie-rod
Courtesy of AZ Pro Performance

Fig.13-12. Caster, camber and toe-in
when looking at the leading edge of
the driver's front tire.
Courtesy of Yokohama Tire

toe-in

toe-out

13

camber

positive caster

❑ Toe-In, Toe-Out

Toe-in refers to the direction in which the front tires are pointed. A tire with a toe setting of 0° would be pointed straight ahead. A car with severe toe-in appears to be pigeon-toed, with its tires' forward-facing surfaces positioned noticeably closer to one another than the tires' rearward-facing surfaces. Toe-in is adjusted by rotating that wheel's tie-rod turnbuckle, Fig. 13-11, thereby changing the tie-rod's length. Tie-rods connect the steering rack to the front wheels; they are the intermediary link that actually turns the wheels.

Toe-in problems produce wear patterns on the inner or outer edge of the tire in a band that can chew up as much as an eighth of its tread. Unfortunately, toe settings are the easiest to knock out of whack: Consider the angle at which force is applied to the right front tire when the "braille method" is employed to back into a parking space. Small wonder we see so many right front tires with feathered edges–the rippling that develops on the outside or inside of the tread in response to incorrect toe settings.

❑ Camber

Imagine an rectangular plane that cuts right through the midline of a tire, extending from the road surface to the top of the tire. Camber defines the angle between that plane and the tire. A camber setting of 0° indicates that the tire 's tread will roll flat against the road. Problems with camber produce tires that appears to have lost either the inner or outer eighth of their tread, as if the car were rolling on just that one limited band. That's the same location as the wear you'd see with toe problems, but improper camber settings wear away the tread much faster. None of that feathery stuff, just a uniform "chiselling".

Not all cars have the means to adjust camber, the logic being that you have to smack the front end pretty damned hard to alter it. I personally don't agree with that logic, but . . . In those cases, either a really good alignment shop or a body shop equipped with a frame machine is required. Cars on which camber is adjustable typically employ a cam that's positioned between the top of the steering knuckle and the strut assembly, Fig. 14-15, p. 380, while others locate that cam between their lower control arm and the frame.

A

B

D

C

E

13

F

G

H

I

J

K

L

M

N

O

P

Q

© PT

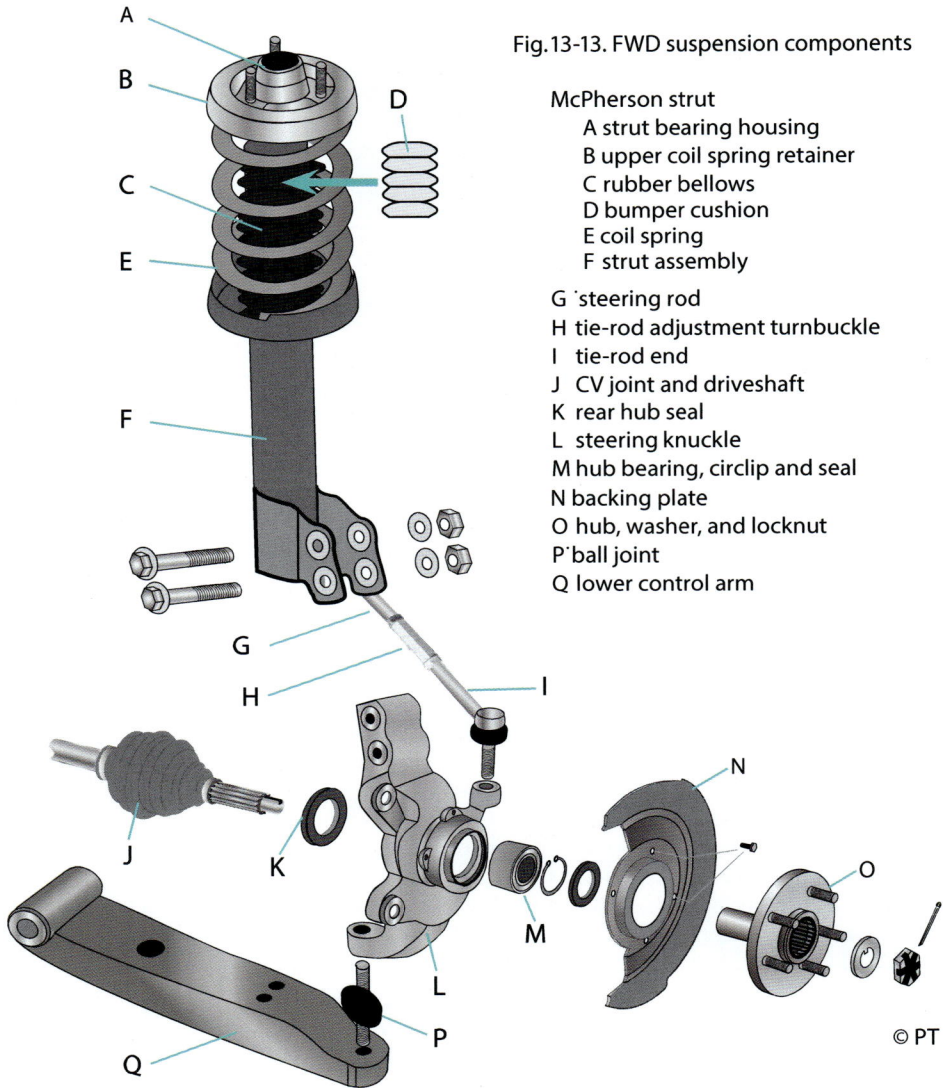

Fig. 13-13. FWD suspension components

McPherson strut
  A strut bearing housing
  B upper coil spring retainer
  C rubber bellows
  D bumper cushion
  E coil spring
  F strut assembly

G steering rod
H tie-rod adjustment turnbuckle
I   tie-rod end
J   CV joint and driveshaft
K rear hub seal
L   steering knuckle
M hub bearing, circlip and seal
N backing plate
O hub, washer, and locknut
P ball joint
Q lower control arm

Fig. 13-13. McPherson strut with associated parts on a FWD vehicle. Typically, the steering assembly is bolted to the firewall – behind the transaxle – which implies that you're looking at the driver's front assembly: Forward movement would roll it left, off the page.

Not shown are the speed sensor and tone ring. The tone ring is mounted on the hub (O), the speed sensor is mounted in the steering knuckle (L). See Fig. 13-10.

Also not shown are the rubber cushions at each end of the coil spring, there to both secure and quiet the coil's motion.

The bumper cushion (D) slides overtop of the strut's interior shaft inside the rubber bellows (C), there to help prevent damage to the car by providing a last line of defense if the shock absorber is totally shot.

❏ Caster

The term caster defines the tire's position relative to another vertical plane–this one perpendicular to the one defined above. Positive caster implies that the wheel's center lies ahead of that plane (just like the front fork on every bicycle is purposefully bent forward to position the wheel ahead of the handlebars). Caster has very little effect on tire wear, but has a great deal to do with where the car is headed. Changing the caster on one wheel can position that wheel ahead of, even with, or behind the tire on the other side. A car with too much caster on the driver's front wheel will pull right, simply because that left wheel is out there leading the charge. Cars with heavy positive caster settings on both front wheels tend toward overly sensitive steering–they steer too easily. The caster for each wheel is adjusted by varying the relative length of its strut rod (where present). These adjustments alter the distance between the lower control arm and a fixed point on the frame, thereby shifting the tire forward or backward. Cars not equipped with strut rods do not have adjustable caster. Should adjustment be required, the front crossmember is loosened, then shifted relative to the frame. But fear not: It's almost impossible to smack the front end that hard without totalling the car. . .

10B. Rear-Wheel Alignment

Alignment used to refer strictly to the car's front wheels. But all that changed with the advent of front-wheel drive and fully independent rear suspensions. There are a number of advantages to this–better handling, improved tire wear–but it leaves owners with a new problem: How to decide whether the car needs to have a complete four-wheel alignment when it goes in for service. Some of the answers follow:

- The majority of front-wheel drive cars have a fully alignable rear.

- All rear-wheel drive cars equipped with a solid rear axle have no means for aligning the rear; it simply follows wherever the front chooses to take it.

- Those rear-wheel drives equipped with a fully independent rear suspension are alignable both front and rear.

The following statements assume an alignable rear.

- You are replacing your two front tires because of wear. The rear tires are a couple of years old, but they've been wearing well. As a general rule, there is no reason to align the rear until the rear tires are replaced.

- However, if the alignment shop points out a wear pattern that you've missed, or if their instruments are measuring a significant misalignment that resulted from some recent encounter with a pothole, then your decision depends on the condition of those rear tires and how long you intend to keep them. How much tread do they have left? How bad is their distortion? If the rear tires are fairly young, protect them with both a tire balance and rear-end alignment. If the rear tires are cupping or noisy (lumpy), skip both the balance and the rear-end alignment until they're replaced.

- If you are replacing your rear tires because of uneven wear, you will need a four-wheel alignment regardless of the condition of the front tires. Because the rear follows the front, the two must roll in harmony.

- If all four tires have just been replaced, you'll need a four-wheel alignment.

13

An alignment should be considered as insurance against premature tire wear. It should always follow tire installation, never precede it.

10C. A Checklist on Alignment Work

Whenever you have an alignment performed, consult the following checklist:

- If you need to replace tires, do your shopping while you still have options, because once you're there, you'll be buying whatever they have to sell at whatever price they want to quote.

- Road-test your car (*§3, p.320*) prior to drop-off so that you can clearly define the problems that you're experiencing.

- Never drop your car off on a rainy day. It's impossible for us to know how the thing will steer on dry pavement if the roads that we're testing on are wet, because a wet road offers little to no traction (friction).

- As with all other drop-offs, set up the appointment on a wait basis so that you can evaluate first-hand any curve balls that might be thrown your way. You might actually have some loose components in the front end, you may have a blown strut. You'll never know if you're sitting at the other end of a telephone.

- When you drop off the car, ask the service writer to include on your copy of the repair order both the alignment settings that the car had upon its arrival, and its settings upon leaving their rack. These numbers might be meaningless to you, but at minimum that request forces the shop to verify that their settings are in accordance with factory spec. Furthermore, any change in a particular number serves to identify the adjustments that they needed to make. Modern alignment machines provide a computerized printout of all that information.

- When you pick up the car, always ask if it was necessary to swap your tires left to right, see *§5, p.328*.

- Following your pick-up, road test the car again, looking for any changes in the way it handles. And finally, once you're around the corner and out of view, check your tire pressures. They should be identical left and right. If you find that they are radically different (50 psi on the left, 25 psi on the right, for example), take the car back riding high on that rush of righteous indignation.

Assuming that you don't smack another curb or crash through another pothole, this process will slow the rate of future tire wear and hopefully, eliminate your car's wanderlust as well.

11. Wheel Bearings: Bearing Noise and Diagnostics

Bearings are an essential component of every wheel, permitting them the freedom to spin with minimal frictional resistance. Wheel bearings begin to make noise as a direct result of wear. A bearing on its way out makes a grumbling or growling sound in motion that typically increases in volume with increasing speed. In their earlier stages of deterioration, these sounds are usually intermittent, rising to the audible level as tiny chunks of steel slough off the ball (or roller) bearings into the greasy track (race) in which they ride. The noise goes away as those loose pieces are either spit out of the track or are beaten into the race–where they hasten the further sloughing of other, larger pieces.

The noise that's produced varies with the bearing's placement on the car. Front bearings tend to growl far more noticeably as the car leans into a corner. In some cases, the sound gets louder when the additional weight of the car is being thrown onto that particular bearing; in others, the momentary absence of weight causes the unloaded bearing to wail–it all depends upon the physical location of the deterioration within the bearing.

13

I've found that the best means to identify a front wheel bearing is on a quiet, smooth road, driving at constant speed (about 30 mph).

Note: In the following discussion, I use the word road for lack of a better single-word term. The parking lot in your nearby mega-mall is a much safer place to perform these tests–early in the morning while the lot is still empty. Choosing a safe location for the following tests–an area with excellent visibility and no traffic–is the essential ingredient to your well-being and the safety of your community.

If there is a noise present in the front end during cornering, try to isolate the sound to one side or the other by traveling in a straight line, then making slight adjustments to the steering wheel–right and left–in order to make the sound come and go. Clearly, accurate observation is essential, but so too is patience–waiting for traffic to clear so that the road grows quiet. (Erratic changes in direction have a way of terrifying other drivers.)

Back in the shop, on jackstands or a lift, grab each front wheel at 12 and 6 o'clock and jiggle in and out (Fig. 13–14) in order to identify the extent of slop (freedom of movement) present on that side–if any. Any front wheel bearing that can rock when jiggled should be repaired at once. Note that a noisy front wheel bearing can show absolutely no sign of any free play; note also that a perfectly silent wheel bearing can be so loose as to scare the wits out of the wrench just back from his road test! A bearing that is intermittently noisy and shows no sign of any slop can probably be put off for the short term–possibly till the next oil change–as long as the car is being driven conservatively at in-town speeds. However, those of you who carry this procrastination beyond the limits of the bearings' endurance will find yourselves buying a hub or spindle as well (the parts that the bearing is pressed onto) for considerably more money. It's hard to push a bearing past the point of severe injury to yourself–they just get too loud to ignore.

Fig. 13-14. Inspecting wheel bearings

13

Rear wheel bearings on a front-wheel drive car growl much the same way as those on the front, although cornering is nowhere near the factor it is with the front bearings.

Rear wheel bearings on rear-wheel drive cars that are equipped with a traditional solid axle typically produce a consistent "woa-woa-woa" sound that increases in volume and frequency with increased speed. Again, cornering is nowhere near the factor it is with the front bearings. To isolate which side is making the sound, you may need an assistant's ears in the back seat. Come over the crest of a hill at 50 mph, then quickly hit the brakes hard, momentarily lifting the rear of the car off the wheels. If the sound disappears for that brief moment, you have a bad rear axle bearing. This is NOT a panic-stop we're talking about here, but more of a split-second rapid deceleration–as if you were pausing for a squirrel that's not yet running.

> Be sure to check your rear-view mirror before locking up your brakes. Keep your hands glued to the steering wheel.

There's no reason to continue braking for more than a second–the noise will only disappear while the weight of the car actually lifts off the rear end. And there's plenty of reason to get off the brakes: Locking up your brakes is dangerous under all conditions . . . If you're good, you can repeat this test two to three times within ten seconds: alternately slamming on, then releasing the brakes.

Three final notes:

- Replacement of a bearing on one side of the car does not necessitate replacement of the bearing on the other side.
- Whenever a wheel bearing is replaced, that job should include a new seal or seal kit. If it does not, you're going to have trouble with that bearing–keeping grease in and dirt and water out. Ask about it at the time of authorization, then check your bill's parts column to confirm that you paid for one.
- In most configurations, front bearing replacement necessitates realignment of the front end.

11A1. Front-Wheel Drive Cars

- Front Wheel Bearings: To identify those bearings that are noisy on the road but tight on the lift, mechanics usually run the car in the air, listening to the bearings spin with a stethoscope or long-bladed screwdriver pressed against the stationary steering knuckle. Should you try this yourself, be *very* careful: An earful of steel is not an inviting prospect.

  Expect to pay 2.5 to 3 hours labor per side. This job typically requires that a front-end alignment be performed following reassembly. Don't try doing this job yourself–it requires a press and several SS tools. None of these bearings can be repacked.

- Rear Wheel Bearings: The inspection of these has much in common with the description above–rocking the wheel at 12 and 6 o'clock, and spinning the tire to listen for the gravelly sound of deterioration. But depending upon their design, repair can be a relatively simple job (Fig. 12–31, p. 303) or it can require a press to remove and install the sealed bearing.

11A2. Rear-Wheel Drive Cars

- Front Wheel Bearings: These are by far the easiest of all the bearings to replace. Labor to clean the hub, replace the bearings, and repack should not run more than 1 to 1.5 hours per side. These bearings can be repacked. An alignment check is not necessary.

- Rear Wheel Bearings: To identify which bearing is making all that racket, mechanics often run the car in the air, listening with a stethoscope or long-bladed screwdriver. Should you try this yourself, be *very* careful: An earful of steel is not an inviting prospect.

  The fixed rear axle (Fig. 12-30, p. 302) carries one axle bearing and seal per side. Labor to replace is usually 1.5 hours. Don't try this yourself; you need an axle puller and bearing press, minimum. These bearings cannot be repacked. These configurations do not need realignment.

- Some of the fully independent rears require a new crush sleeve in addition to the bearing set and seal kit. Costs will run 2.5 to 3 hours. These bearings cannot be repacked. These configurations do need realignment.

11B. Repacking Wheel Bearings

In days of yore, wheel bearings were routinely repacked every 30,000 miles or so. But with the rising cost of labor and the tremendous improvements being made in modern-era greases, wheel bearing repacks are becoming a thing of the past. In fact, every front wheel bearing on every front-wheel drive car is a sealed unit. They can't be lubricated, period. I've indicated above which bearings can be repacked and which cannot. Even with those that can be repacked, I'd advise you to save your money for more essential service work, unless your mechanic can provide some compelling reason for doing so.

12. McPherson Struts and Shock Absorbers

Figure 13–15 illustrates the suspension on a typical late-model car. The front suspension employs a pair of McPherson struts, the rear relies on a pair of shocks in combination with coil springs. There are any number of variations on this theme–wishbone suspensions up front, (Fig. 13–16), McPherson struts on the rear, leaf springs on the rear (trucks), no struts at all (mostly trucks)–but by and large, Fig. 13–15 is a valid generalization because so many front-wheel drive cars look just like what's shown.

The primary purpose of these components is to keep each wheel in contact with the road; smoothing out the ride for passenger comfort is a secondary consideration.

Any car that has reached 80,000 to 100,000 miles without strut or shock replacement would benefit markedly from their replacement.

13

Fig. 13-15. A typical suspension: McPherson struts up front, shock absorbers with coil assist in the rear.

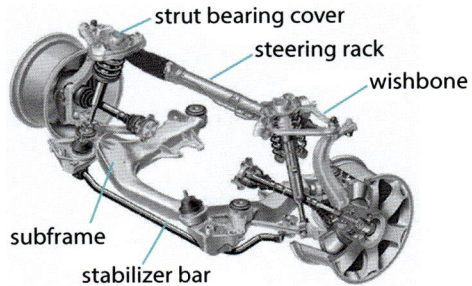

Fig. 13-16. Front-end wishbone suspension, power steering rack, and engine/trans subframe. Attribution requested

Both shocks and struts tend to start leaking as the direct result of collisions, typically within a year of the impact. The word collision can refer to an accident with another car, but it can also refer to a significant encounter with a pothole. A leaking strut or shock is one small step away from becoming completely useless. Not only will your car's ability to handle be compromised, but also that tire will suffer from ply separations and severe wear patterns.

A completely blown strut will produce a "tock" or "tock-tock" sound when run through a pothole or over a bump. Blown front struts often produce a strong steering wheel wobble when run over a bump.

A blown strut or shock is dangerous and should be replaced.

12A. Inspection

Start one corner of the car moving up and down in a deep rocking motion, using your knee to push downward on that corner of the bumper as it reaches the upper extent of its travel. Once you have it rocking as much as it's going to, stop to observe its motion. If the strut or shock on that particular corner is sound, then at the end of the last downward stroke you've applied with your knee, that corner will rise to the top of its travel–though not quite so high as when you were working it–then start back down, stopping with a slight upturn. If you have more than one full oscillation in addition to that, that shock is either weak or shot. The less dampening (shock absorption) it provides, the more oscillations you'll observe.

Look for signs of fluid leakage down the outside wall of the strut, Fig. 13-13F, p. 340. Ditto for shock absorbers. Leakage appears as a trail of darkened, wet-looking dirt sticking to what was originally just leaked oil. Any leakage whatsoever indicates the need for replacement.

A less common problem is presented by the coil springs. These can weaken with time; the car very gradually sinks lower . . . In severe cases, the car can start to bottom-out over speed bumps, scraping its undercarriage as it drags itself over them. (A low-hanging exhaust can create a similar problem.)

To determine whether your car needs coil springs, park your car on a flat piece of ground and bounce the bumpers at each corner as described above. Now compare the relative height of R/F versus L/F, and R/R versus L/R. (Measure the distance from the ground to the top of each fender, directly through the center of each wheel.) Does the height of the R/F fender vary from the L/F by more than an inch? How about the rears; are they significantly different left to right?

If your answer to either question is yes, you could benefit from coil spring replacement, particularly if you have a blown strut to deal with anyway. If you don't need strut work and the car doesn't bottom out going over speed bumps, save your money.

A blown strut, by itself, will not cause its corner of the car to sit any lower than its more healthy partner. However, over time a blown strut can take the sproing out of any spring.

12B. Strut Replacement

In cases where one strut has blown at 40,000 miles due to a collision, replacement of that one strut should suffice. However, by 80,000 to 90,000 miles, struts should be replaced in pairs. Because shocks cost so much less to replace than struts, shocks are typically replaced in pairs.

Depending on definition, strut repair can mean replacement of the complete assembly (strut and spring), but in common parlance the term refers to replacement of just the strut. In some cases, that can mean substitution for the "guts" of the original unit with an insert that fits into the strut's exterior tube.

Whenever you authorize strut replacement, find out whether the design of your particular model necessitates an alignment. Certain struts can be removed, then reinstalled without altering the alignment one iota. However, on many cars, wheel alignment is intertwined with the strut, and an alignment check becomes an absolute necessity following strut repair.

Coil springs are cushioned top and bottom in preformed rubber retainers. The lower cushion, especially, is likely to become soaked by the oil that bleeds out of a blown strut. This oil causes the cushion to swell and become distended as soon as the coil is removed. When reassembed, it no longer fits the coil. This can lead to all kinds of threatening noises–squeaks, grinds, clacks. The rubber bumper (C in Fig. 13-13, p. 340) can literally be beaten to death because a blown strut can no longer cushion the blows it's taking. Not replacing the bumper will probably go unnoticed by the owner in the short term, but its absence can become a significant issue down the road after the strut has lost some of its "bounce." Unfortunately, neither of these parts is commonly stocked by the dealers–much less the corner garage. And to further complicate matters, once the strut has been removed, the lift on which that car is sitting becomes tied up until the strut assembly has been reinstalled. (Cars don't move particularly well on three wheels, even with a jack). The economics of this dilemma usually dictates the outcome: namely, the customer's car goes out the same day with what's left of his old parts. So before anybody takes your blown strut apart, assure yourself that all the parts required for repair will be on hand. Typical labor costs for strut replacement range from 1.5 to 2.5 hours.

12C. Shock Replacement

Fortunately, shocks aren't that complicated or expensive–unless they're rusted in place. Typical labor fees run 1/2 to 1 hour each, depending upon accessibility and rust. (An occasional shock needs to be torched off the car–very smoky, very smelly.) In most cases, the new shocks just bolt in. New rubber cushions come with the shock. Shocks should be replaced in pairs.

13. Steering Components

13A. Inspection

Aside from leakage, the most common problem associated with steering is wear to any one of the ball joints that unite its various components. Because looseness is due strictly to wear you usually won't see these problems cropping up before 60,000 miles. An evaluation of these components is part of every good alignment check. Because it's possible to spend lots of money replacing parts that are still intrinsically sound, it's important to find an alignment shop worthy of your trust.

The simplest test for (very) loose steering components requires nothing more than a stationary car, front wheels pointing straight ahead, engine running (on cars equipped with power steering). Rock the steering wheel back and forth approximately ten degrees in either direction. If any of the components are loose enough to shift, they give themselves away as slop in the wheel, a distinct "clunk," or both. (By most definitions, these parts can be loose enough to change long before you begin feeling slop in the steering.)

The next step requires the front end to be raised and supported on jackstands (§7D, p.140). With the key in the ignition (engine off) unlock the steering column. Grab either front tire at 3 and 9 o'clock and try to jiggle it in and out, Fig.13-17, p.351. If any play exists, you'll feel it as slop in the wheel (hub). Next, try your hand on the opposite wheel.

Note that looseness on either side could easily be caused by a loose wheel bearing. So should you detect freeplay at 3 and 9, compare the amount of freeplay with your hands placed at 12 and 6 (Fig.13–14, p.343). A comparable amount of play here indicates a loose front wheel bearing. No slop at 12 and 6 indicates that the looseness is limited to the steering. Have an assistant observe each joint as each front wheel is rocked (hands at 3 and 9 o'clock). This inspection is done both by eye, and by the laying on of hands. Loose joints give themselves away by their abnormal (klunky) movements.

To inspect lower ball joints, the front end must be elevated, supported on its frame, §7D, p.140. A hydraulic jack is placed beneath the lower control arm; the lower control arm is raised just far enough (3 inches or so) to "unload" that side, but not so far that the car starts lifting off the jackstand. "Unloading" the suspension positions the ball joint much like it would be with the car on the road. Grasp the wheel at 12 and 6 o'clock, then jiggle in and out. If you feel slop to the wheel that was not present prior to the unloading, then you have a worn ball joint. Repeat the same test on the opposite side. It is unnecessary to replace ball joints in pairs.

13

Once upon a time, every joint in the front end had a grease fitting, and every year or so the lucky ones saw a shot of grease. With the exception of certain heavy-duty trucks, this practice is long gone. Virtually all of these joints are now sealed and should remain that way. The only union that could conceivably takes a shot of grease is the lower ball joint. Some of these come from the factory with a bolt in place rather than a grease fitting. At some point in the car's life–30K is an appropriate interval–these bolts should be replaced with grease fittings. More frequent lubrication is unnecessary.

Just under the hood, at the top of the strut, is a cap that pops off to reveal a bearing. On rare occasions, these bearings need replacement to correct for binding in the steering. However, in most cases this problem can be resolved simply by greasing the bearing from above without disassembly. You can do this with a standard issue caulking gun. Grease cartridges are available from any parts jobber.

13B. Power Steering

The two most common problems associated with steering racks involve fluid leakage and worn tie rod ends. Collisions are the most likely cause of leakage; leaks typically develop within a year of the impact. The steering rack has a seal at each end. If one of the wheels takes a shot during the collision, the impact can initiate a leak in that seal. A distended or soggy bellows is a dead giveaway, Fig. 13-18, p. 351.

The power steering pump supply and return lines are connected to the steering rack with both flex lines and solid piping. Fig. 13-19, p. 351 shows one potential source of leakage: a flex line's end cap. Often, these are concealed by rubber tubing meant to reduce vibration, but vibration can wear that tubing away, especially when a line is left unsecured during collision repair reassembly (some of those clamps are near impossible to see).

Every power steering system uses pressurized hydraulic fluid delivered to the steering rack to assist in turning the wheels. The source of that pressure is a power steering pump located at the front of the engine, off to the side, Fig. 9-6, p. 159. The loss of fluid from any part of the system results in a moaning sound when the steering wheel is turned, usually first thing in the morning. The sound emanates from the pump, and can often be temporarily eliminated by adding more fluid, §3F, p. 106. Be sure to use the correct type and Do Not Overfill! Add fluid promptly, because running these systems low on fluid can cause significant damage both to the pump and the steering rack.

In the short term, a small leak can be dealt with by routine addition of fluid. But in the long run, most of these leaks get to be unmanageable because of the pressures being applied by the pump.

Front seal replacement on the power steering pump is a fairly reliable fix. Resealing the end of a power steering rack has little chance of success. Leakage in a line requires replacement of that line.

I'm not big on rebuilt power steering racks; I've seen too many of them bind right out of the box or leak within the first year. But new factory racks are often outrageously expensive. That leaves you with OEM aftermarket racks and junkyard pickings. You pay your money and you take your choice . . . Buy an OEM rack if you can.

13C. Manual Steering

Beyond the aforementioned problems with looseness, there's not much to these systems. The gear oil level within the steering box should be checked annually. On occasion, a slight adjustment (tightening) to the steering box is required to eliminate slop in the wheel. First, back off the steering box locknut a revolution or two while holding the set-screw (backing off the locknut without doing so will just create more slop in the box. Now have an assistant rock the steering wheel back and forth through the "slop" while you tighten down the set-screw. Once it bottoms out, back it off just a bit to provide a bit of freeplay. Maintain that while you tighten down the locknut.

13D. Intermittent Binding

The ball bearing at the top of each McPherson strut can bind for lack of adequate lubrication. See Fig. 17-26, p 461. Parts jobbers sell pocket grease pens. Steering columns employ a universal joint to transmit steering wheel rotation to the rack. Occasionally, these joints can suffer from binding, requiring their replacement. Then there's the occasional steering rack that binds in a particular spot.

- Start by lubing the two McPherson strut bearings. Road test, cornering a lot to work the grease into each bearing. If the problem remains, raise the front of the car, put it up on jackstands, then disconnect one of the McPherson struts from its outer tie rod end, Fig 13-13(I), p.340. Rotate the strut through its full range of motion. Any binding? If not, move on to the other strut. These bearings can be replaced, but first the strut must be removed, §12B, p.347 and §14 below. You need only replace the one binding bearing.

- To isolate the u-joint, it needs to be removed from the car for examination on the bench.

- The rack comes up guilty in those cases where the above two turn out to be fine.

> Binding in the steering can be extremely dangerous and should be repaired without delay.

14. Inspections and Servicing – Some DIY Details

Much of this work requires a considerable amount of muscle and a 24" breaker bar–possibly even a pipe to slide over top for added leverage, or a compressed air gun (1/2" drive) with plenty of air to back it up. You'll also need some SSTs (special service tools), all of which can be rented. The question is, will you save any money doing this yourself, or would it be better to have the pros do it?

First, the lightweight stuff:

Speed Sensors: To diagnose speed sensor problems, you'll need a moderately sophisticated OBD 2 code scanner (one that will identify which sensor is faulty), your shop manual, a spray can of electrical cleaner, a pair of goggles, and an ohmmeter. Or the problem may

Fig. 13-17. Checking for tie-rod freeplay

Fig. 13-18. Steering rack seal leakage

Fig. 13-19. Flex line end-cap

13

be totally obvious: Before wading into the weeds, check for loose electrical terminals, corroded terminals, or a broken wire. A badly worn or loose wheel bearing can produce both the damage (chipping a tooth on the tone ring) and the diagnosis for you.

As for the rest of it, you'll need a jack and 24" stands, a pry bar, impact sockets, wrenches, your factory repair manual, rust buster spray, wheel bearing grease, plenty of rags, a few cans of brake cleaner and some goggles.

Wheel Bearings (FWD): You'll need a hub puller (rental) and a friendly machine shop to pull off the old bearing and press on the new, and a 1/2" drive torque wrench. You'll also need an inner hub seal and a new locknut. Now is a great time to replace the transmission's driveshaft seal as well. Bearings do not need to be replaced in pairs.

Pressed Rear Bearings (FWD): You'll essentially need the same equipment as you would for the front – minus all that muscle.

> To avoid chipping a tooth off a tone wheel, do not use a hammer or pry bar to free the axle from the hub . . . Use a hub puller!

Rear Spindle and Drum configuration (FWD), Fig. 12-31, p. 303 These bearings should be packed every 60,000 miles or so. They'll rarely cause any problems unless they've been overtightened. To avoid that, spin the drum while tightening the castle nut against its thrust washer. The drum will slow as the nut nears too tight. Back it off a skosh, spin the drum once more. Repeat until you've found a happy medium–free spinning but snug. Reinstall the cotter pin. Should the bearings need replacement, have a machinist install them for you to avoid screwing up the races.

RWD Axle Shaft Bearings: A slide hammer to pull the axle, a machine shop to remove and replace the bearing and a 4 lb. hammer to drive it all home, plus a bit of gear oil to top off the differential. You'll also need an axle seal, possibly a paper gasket.

McPherson Struts: You'll need most if not all of the above tools (OBD scanner and ohmmeter need not apply), plus a line wrench to disconnect the caliper. Other, more necessary tools–with certain qualifications–include a ball joint splitter and a coil spring compressor. As to the qualifications: Not every one of you will need to pop the ball joint free. I think it's easier because it offers considerably more freedom of movement, but doing so can destroy the ball joint's rubber boot, necessitating ball joint replacement.

And then there's the whole question of loaded strut assemblies vs a basic strut, Fig. 13-20. Buying a pair of loaded struts is certainly faster and safer: No need to swap over all of those parts, and no need to mess with that coiled spring which, if it comes loose, could put a hole in your garage door–or you, for that matter. But the reason I say a pair is that even if you only have one leaking strut, the variance in coil spring height and compression can vary between original equipment and OEM replacements, which can lead to handling problems.

That being said, struts are typically replaced in pairs. And no matter how careful you are in marking your camber settings, an alignment check is always a good idea whether that be now or immediately following tire replacement.

Fig. 13-20. Loaded vs unloaded strut assemblies

> Coil springs must *always* be locked down in a coil spring compressor prior to disassembly, and should remain so until the strut's top locknut has been re-secured. *Always* use a new locknut, and *always* make damned sure it's tight.

Chapter 14. Transmissions, Differentials, and Driveshafts

14

Chapter 14. Transmissions, Differentials, and Driveshafts

14

Notes

Notes

Transmissions, Differentials, and Driveshafts

1A. Fluid Level and Condition

Over the past twenty years, an increasing number of manufacturers have been eliminating the automatic transmission's dipstick in favor of an inspection port (plug) on the side of the transmission case. There is one primary reason for that: It prevents unskilled owners from screwing up their *very* expensive gearboxes. Overfilling or adding the wrong type of fluid are primary concerns. It happens all the time, particularly in cases where the owner doesn't take the time to read his owner's manual . . . But for those of you who do (read and have an automatic transmission with a dipstick), use your owner's manual to locate it!

❑ Dipstick Access

Unless your manual says otherwise, the fluid level is inspected with the transmission fully warmed (driven for a few miles, run through all the gears), the engine idling, the gearshift in Park. You need to be on level ground. Pull the dipstick, wipe it off, reinsert it, then pull it again. The level should lie within the hot range, at or near its full line at the top of the cold-hot grid, Fig. 14-1.

Wipe off the dipstick with white paper toweling. Clean fluid is clear and ruby red in color. The older it is, the browner it is. (Transmission fluid contains cleaning agents that pull dirt and worn clutch pack material right into solution.) Fluid that has turned opaque brown should be replaced sooner than later, *§*1B, below. Fluid that's turned black should be changed now!

Fig. 14-1. transmission dipstick
Courtesy of Mercedes Benz

hot range   cold range

If you need to add fluid, check the specifications page in your owner's manual for the type of fluid required. Most will find that Dexron or Mercon is called for, but there are others, and all have evolved through a series of formulations (Dexron I, II, III, VI and ULV for example); each improvement providing better cleaning agents, better lubrication, and longer service intervals.

For those of you who own transmissions that require a different fluid, be advised: These fluids are not compatible and are not interchangeable. The composition of the rubber used in transmission seals is one critical issue here: Some seals swell in the presence of the wrong fluid, others dry out and crack. Just one more reason to be alert in the presence of lube jockeys! Better safe than sorry: Consult your owner's manual!

To avoid making a mess, insert a small- to medium-sized funnel into the opening, hold it in with light pressure to create a seal between funnel and tube. Those of you playing with half a deck (that includes me) should wrap or tie a rag around the tube directly below the funnel as a further precaution against yourself. Fluid should be added a bit at a time to avoid overfilling, because once it's in, there's no way to get fluid back out short of opening the transmission drain plug or using your wife's turkey baster (bad idea!).

Run the engine while you're topping off the system; doing so speeds the filling process by eliminating the need to repeatedly restart the engine to measure level. It takes thirty seconds or more for the fluid to work its way down the dipstick tube and into the transmission. Measurements made immediately after adding fluid are meaningless because the stick will be wet all over.

The low-full markings etched into the transmission's dipstick do not necessarily represent a full-quart gradation. You could be making a big mistake if you add a quart just because you find that the fluid level is about one full notch down. Drive the car first to assure that the transmission is fully warmed. Transmission fluid expands much more readily with heat than oil does; in fact, you'll often find two sets of markings on the stick, the lower pair is for a cold transmission, the higher pair is for hot ones, Fig.14-1. If you find that fluid is needed, add it sparingly so as not to overshoot the full line. Note that overfilling by a little is no big deal, but drowning the thing can cause excessive internal pressures which can lead to leaking seals and/or hard shifts. A transmission that has been significantly overfilled should have some sucked out with a hand pump or drained.

❏ Inspection Port / Plug (No Dipstick Access)

You'll need a lift or four jackstands. The car must be level. Filling or topping off a transmission with no dipstick requires a hand pump whose outlet can be inserted into that port – and will stay there once you look away . . . You will need the manufacturer's service manual or its online equivalent (alldatadiy.com) to continue safely.

The only reason you might want to go to all this trouble is because you've been seeing fluid on the street. If so, read §5A, p.133 first . . .

Most manufacturers provide an inspection/fill port located roughly halfway up one side of the main case. This port is in addition to the drainage port located at the transmission's lowest point. Both of these typically carry threaded plugs that require a hex- or star- wrench to open them.

Others employ just one port, always at the bottom of the transmission, that doubles as both, Fig.14-13, p.378. As long as the inspection tube is threaded into the case, the trans can't drain anything other than the dribble you'd expect to find exiting any properly filled unit. Replacing the transmission fluid requires that the inspection tube be removed.

The manuals I consulted all referenced their manufacturer-specific OBD scanners. Each of these cars have transmission fluid temperature sensors and all specified that the fluid temperature be in the 110º-115º F range – plenty warm, not blisteringly hot.

Clean fluid is clear and ruby red in color. The older it is, the browner it is. (Transmission fluid contains cleaning agents that pull dirt and worn clutch pack material right into solution.) Fluid that has turned opaque brown should be replaced sooner than later, §1B, below. Fluid that's turned black should be changed now!

If no fluid drains out, you will probably not need to add more than a few ounces before you'll see some overflow exiting the inspection port. More than that and you already know that you have a significant leak. If you need to add fluid, check the specifications page in your owner's manual for the type of fluid required. Most factories specify their own proprietary brand. I'm no chemist and am not about to suggest that you use anything different. The composition of the rubber used in their transmission seals is one critical issue here: Some seals swell in the presence of the wrong fluid, others dry out and crack.

If the fluid level is more than six ounces low, read §D4, pp.362 and inspect Fig.14-3, p.363 before proceeding further.

1B. Automatic Transmission Servicing

Back in the day, most manufacturers recommended a transmission fluid change every two years or 30,000 miles, whichever occurred first. But over time, vast improvements in the makeup of transmission fluids has allowed that interval to double, even triple. So consult your owner's manual and follow their recommendations. Don't ignore the frequency differences between normal and severe driving (mountains, towing, etc.) Remember that your car's warranty becomes a useless piece of paper if you fail to keep up with the factory's maintenance requirements.

Transmission servicing comes in three flavors:

❑ A simple drain and refill: Most Speedy-Quick lube joints would suggest that this constitutes adequate servicing. I wouldn't necessarily disagree if it's done with sufficient frequency and the proper fluid, at least on low-mileage cars. But before you go dashing off to your local Speedy-Quick, please read §1A, p.357 and §1C, p.360.

❑ A more tradional approach to servicing involves removing the transmission pan following drainage. That's every one of those 15 bolts ringing the pan (Fig.14-2). Doing so provides access to the transmission filter, which can be holding back an additional quart or two of dirty fluid.

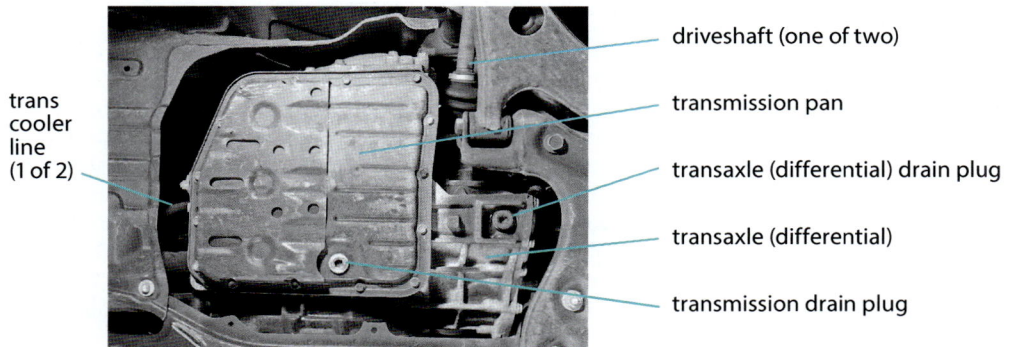

trans cooler line (1 of 2)

driveshaft (one of two)
transmission pan
transaxle (differential) drain plug
transaxle (differential)
transmission drain plug

Fig. 14 –2. FWD transmission viewed from below

The pan's drain plug contains a magnet to collect any loose metal shards that might be floating around. One or two small pieces might not be a problem; numerous larger pieces could be. Hopefully, you'll just find a thick paste of metal dust adhering to the magnet. Clean it and forget about it. The interior of the pan will typically be covered in dust from a different source: worn clutch pack material. If you're doing this work, use varsol, brake cleaner, or something similar.

■

14

The pan gasket should always be replaced. In cases where the filter gasket gets torn during filter removal, it too must be replaced. Unfortunately, very few shops – including many dealerships – carry these gaskets in stock, which means that in most cases the car leaves with a piece of it missing. Since this gasket only costs a few bucks, it makes sense for you to stock it – assuming of course, that you remember to leave it on your passenger's seat with a note, "Feel free to use this should you have the need."

Even when the more thorough approach is used, the fluid that remains in the torque converter[1] stays there when the engine is turned off, ready to contaminate your new fluid as soon as the engine is fired off. Because of this, those transmissions that have been neglected over the years will not derive tremendous benefit from just one transmission service. On these, getting your fluid to remain clean over time requires two or three additional drain and refills at intervals of several thousand miles, or

❑ A transmission flush: Every automatic transmission cycles its fluid through a dedicated portion of the radiator. This box resides within the base of the radiator but its contents are separated from the engine's coolant for as long as the car is not involved in a severe front-end collision. Two flex lines connect the transmission's cooler to piping that leads back to the trans. One line is visible in Fig. 14-2.

Flushing the transmission involves a dedicated machine that pumps "fresh" fluid through the transmission's cooling lines. Fluid passing through the machine is filtered, presumably "refreshed". For some shops, that's the extent of their service. Better shops finish the job by then draining the transmission and adding new fluid. The trouble is, it's impossible to know whether that last step is actually included. It's also conceivable that the fluid being pumped through your trans is not the correct type; in all probability, it's a mish-mash of fluids from various makes, models and years.

1C. The Transaxle – Special Considerations

Virtually all front-wheel-drive cars house their differential in the same case with the transmission. In many (most?) of these designs both the transmission and differential share the same fluid, passing through shared openings within the case. If you drain one you are in essence draining a portion of the fluid for the other. But just as tidal pools are left behind on the beach when the tide goes out, so too, a puddle of dirty transmission fluid is left behind in the differential's portion of the case unless it is also drained. And not all front-wheel drive transmissions share their fluid with the differential – some run Dexron in the transmission, gear oil in the differential. To add another wrinkle, some front-wheel drive transaxles (both transmission and differential) can be refilled through one opening, others require that the transmission and differential be filled independently, even though they both use the same type of fluid.

Lube jockeys can make close to minimum wage working at Perfect Service Quickie-Jiff. Even if their employees have been trained to consult a chart of fluids, you assume a lot on faith about Louie's abilities and his interest in your car. You need to know what type of fluid your transmission takes, what type of fluid your differential takes, and whether the two are connected or independent. You then need to communicate this information

---

1 A large donut-shaped structure, bolted to the flywheel at the rear of the engine. Its primary function is the smooth transfer of power from engine to transmission. It holds a whole lot of transmission fluid.

to the service writer when authorizing the work. To get this type of information reliably, you need to consult the factory's shop manual or on-line equivalent (alldatadiy.com), an experienced mechanic, or the shop foreman at a dealership.

On differentials that run gear oil, a drain and refill every 60,000 miles should be sufficient.

No one can tell you that your differential fluid is dirty without first draining it or siphoning some off – it's simply not visible through that wall of steel.

1D. Problems with Automatic Transmissions

1D1. Slipping

Slipping refers to a car's refusal to move when it's given gas in gear. Automatic transmissions can slip in Drive, Reverse, or both. By far the most common cause of this is a low fluid level, usually due to leakage, §D4 p.362. Any slippage causes wear to the transmission's clutch packs, so check that level NOW before doing any further damage, §1A, p.357. Should you find that the transmission level is down, top it off, verify that the slippage has ceased, then get thee to a mechanic to determine the cause of the fluid loss.

14

If you find that the fluid level is fine, then clearly, the transmission is not . . . Hopefully, the slippage is a cold-only problem. If that's the case, try letting the car warm up in Park for three to five minutes before starting off. If you can sidestep the issue of slippage by allowing the transmission those few minutes to wake up, you could be saving yourself a bundle: In a fair number of cases, the problem will not get appreciably worse for periods of up to a year. Naturally, the colder the weather, the longer it takes to complete the warm-up.

In certain cases, you'll find that putting the transmission into Low and shifting up manually through D1 to D2 eliminates the slippage altogether. The joys of a manual transmission without that pesky clutch!

If the above suggestions are fruitless, you are probably looking at a major bill, §1E, p.364. But before casting hope aside, pay to have a full transmission service, to include pan and filter screen removal. The shop may find that the filter screen is torn, that its gasket is torn or missing, that a pipe transporting fluid from A to B has worked its way loose, or that you have some problem with the kickdown cable. In and of itself, dirty fluid will not cause slippage.

1D2. Sluggish Shifting Between Gears

Sluggish shifting is almost always due to dirty fluid. The dirtier it gets, the thicker it becomes. Its rate of flow drops, not only through the filter, but everywhere. In consequence, less pressure is delivered to the valve body – an automatic's hydraulic brain. Try a complete transmission service, then reinspect the fluid 2,000 to 3,000 miles later. If it's muddy again, change it again, this time by means of a simple drain and refill.

1D3. Problems with Engaging the Correct Gear at the Correct Time

This used to be governed exclusively by fluid pressure, using a hydraulic "brain" known as the valve body. Even now, the valve body is an essential component of shifting, but increasingly, the control of the transmission's behavior has been usurped by the transmission control unit (TCU), an electronic black box that governs shift points. The TCU acts through solenoids (electronic switches) housed in or near the valve body. As with the engine's emission-control system, information detailing problems with a particular component can be retrieved from the TCU (with some scanners, not all; your odds will improve dramatically at the dealership). You'll need a specialist for this type of work; the problem could have a simple solution. Then again, it could be internal to the transmission.

1D4. Leakage

Fluid loss can be major or minor; the urgency of repair depends upon its rate of loss. The procedure for discovering its source is identical to that for engine oil leaks: a careful cleanup, topping off the level to its full line, then reinspection 250 to 500 miles later (§ 5A, p.133). Possible sources follow. Refer to Fig.14-2, Fig.14-3 & Fig.14-6.

❑   Both Automatic and Manual Transmissions

- Transmission or differential drain plug – It's either loose or its gasket has been crushed beyond oblivion. If tightening won't suffice, buy both a new drain plug and a new gasket – both from the dealer – so you won't have to do this twice. Fig.14-2, p.359.

- Driveshaft seals (front-wheel drive (FWD) only) – There are two, one for each shaft, coming out of opposite sides of the differential. They can be replaced independently of one another; each requires removal of its driveshaft. (Labor: 1-1.5 hours). Fig.14–3.

- Transmission rear seal (rear-wheel drive (RWD) only) – Replacement requires driveshaft removal (1-1.5 hours). If the leak stops briefly, then resumes, it's because the tailshaft itself is worn, requiring bushing or tailshaft replacement (2 to 3 hours). Fig.14-6, p.365.

- Transmission front seal – Replacement requires transmission removal (5 to 6 hours). Note that an oil leak from the engine's rear main seal exits from the same seam as the transmission's front seal. The only means to distinguish between the two is to examine the color and smell of the fluid. Should you ever need to have either of these seals replaced, have the other addressed at the same time (1/2 to 1 hour additional). For that matter, whenever you're paying to have the transmission put on the floor, having all of its seals replaced is money well spent. (3/4 to 1 hour additional). Fig.14-3

- Speedo drive seal, speedo drive o-ring (antiques) – Easy and cheap to fix.

❑   Automatic Transmissions

- The transmission pan gasket – These usually leak only if the transmission pan has recently been removed for servicing, then reassembled using either the original gasket or glue. (That's right, glue!) Fig.14-2

- Transmission cooler lines (not shown) – These two lines run between the trans and the bottom tank on the radiator. You'll find flex tubing on both ends with metal tubing between them. If the line has split at a hose clamp, you'll have a BIG leak. Inspect the hose; it might be long enough to permit

14

shortening, then reclamping – at least as a temporary fix. Another possibility? Old clamps in need of tightening or replacement. If new cooler lines are needed, wait for factory replacements.

- Neutral safety switch seal (not shown) – A small seal in the transmission case directly behind the switch.

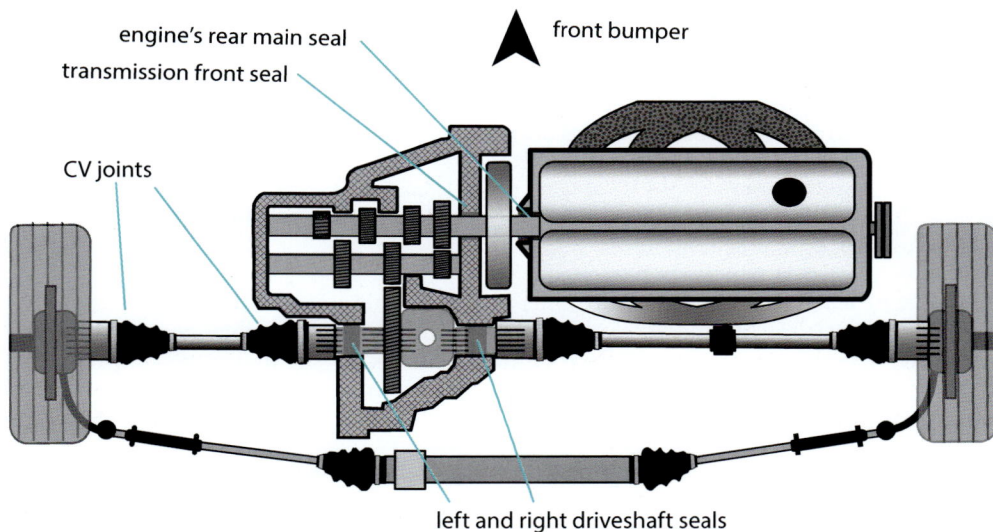

Fig. 14–3. Transmission seals, driveshafts, and steering gear

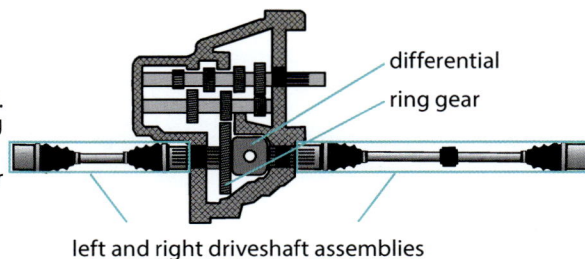

Fig. 14–4. Transmission & differential. Note that the two are co-joined, making this a transaxle.

All other FWD vehicles have similar structures.

left and right driveshaft assemblies

1D5. Metallic Noises

The whirring sounds of a noisy bearing or the grumbling sounds of a bearing that's self-destructing are unsettling enough. But when they can be clearly linked to your engaging the transmission in Drive, or to acceleration within a particular gear range (L, D1, or D2), they take on a whole new dimension. If you are not particularly enamored of your car, now is a good time to trade it in – *before* the transmission comes unglued. While it makes sense to have an experienced wrench listen to the noise before jumping to conclusions, it might not be cost-effective to remove the transmission pan for internal inspection – you both already know what you're going to find.

1E. Internal Transmission Damage – Where Do We Go From Here?

In virtually all cases where internal problems are suspected, removing the transmission pan for inspection of the interior can be a worthwhile endeavor, if only to confirm your worst fears. On those occasions in which the transmission has been coming apart, it will become immediately apparent to the mechanic – clutch pack material and/or loose pieces of metal will be lying loose in the pan, or will be stuck to the magnets that reside there.

All transmissions slough off clutch pack material over the years; the magnets lying in the pan are there for the express purpose of attracting this stuff as it floats past. Don't let anyone sell you a transmission just because the magnets have brown, powdery stuff adhering to them. That's normal. However, distinct chips of shiny metal lying loose in the pan are unmistakable proof of internal damage.

If the car drove in, you should be able to drive it back out. Now comes the moment to decide: Do you switch horses, using what's left of your car as trade bait? Or do you invest $2,000 to $3,000 in a resurrection?

The least expensive approach is to install a junkyard unit, but be advised: Transmissions that have been sitting around for a period of six months or more might give you more grief than satisfaction: Transmission seals dry out over time when not in use, causing problems with slippage, chatter, you name it. This concern is less of an issue on popular, late-model cars where the junkyards' turnover of stock is much faster.

The best approach is to replace the transmission with a brand-new, factory unit or a factory remanufactured one – straight out of the box. Several companies within the aftermarket specialize in rebuilding; these transmissions usually come with a one-year warranty with the option to buy additional coverage. Transmission specialty shops offer rebuilding in-house. My own experiences with these shops have often been underwhelming, I suppose because they're tied to repairing what you hand them. Scrap metal being pumped around inside the box – even for a short period of time – can cause tremendous damage, often unseen. The result can be a transmission whose behavior is a qualified success, one that might only last a year or two.

Front-wheel drive cars only – Transmission removal typically requires that both driveshafts be removed. Either a castle-nut and cotter pin, stake nut, or a self-locking nut secure each driveshaft at the outside of its front hub, Fig. 14–5. Check that these pieces have been reinstalled. Their absence allows the front wheel bearing to run loose, leading at best to

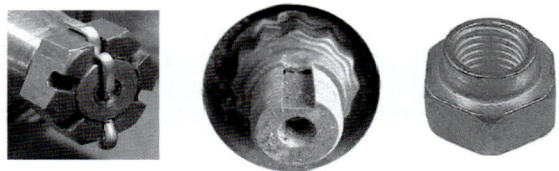

Fig. 14-5. Driveshaft lock nuts (L>R): castle nut, stake nut, self-locking nut

its premature demise, at worst – to yours. Inspection requires wheel (or hubcap) removal. Driveshafts secured with a self-locking nut should always be reassembled with a shiny new one. All must be torqued to factory spec during installation.

2. Manual Transmissions

2A. Servicing

Aside from timely clutch replacement, the only other requirement of a happy transmission is a gear oil change every 50,000 to 60,000 miles. The same holds true for the differential. If either your transmission or differential runs automatic transmission fluid, you'll need to swap the fluid more frequently, possibly once every 30,000 miles. See §1C, p.360 for further information.

2B. Leakage: See §1D4, p.362.

2C. The Clutch Assembly

The clutch assembly is located between the engine and transmission; it enables the transmission to disengage from the continuously spinning engine to permit gear changes. The most common clutch assembly is comprised of four pieces, Fig. 14-6: the clutch disc, pressure plate, throw-out bearing, and pilot shaft bearing. The pressure plate – otherwise known as the clutch cover – is bolted directly to the flywheel (the engine's rearmost component). The two spin as one. The clutch disc lies between these two; and is connected to the input shaft of the transmission by deeply grooved splines. The pilot shaft bearing is located in the center of the flywheel and supports the end of the transmission's input shaft. It allows the input shaft to decelerate, spinning independently of the flywheel when the transmission is disengaged. Under normal driving conditions – with the clutch pedal resting undisturbed at the top of its path – a heavy, circular spring within the pressure plate acts to lock the clutch disc hard up against the flywheel. In consequence, the transmission's input shaft turns at the same speed as the engine's crankshaft. Depressing the clutch pedal to the floor moves the throw-out bearing forward where it contacts the release forks of the pressure plate, which releases the pressure plate's grip on the clutch disc. This action frees the disc from its bondage to the flywheel, disengaging the transmission from the engine.

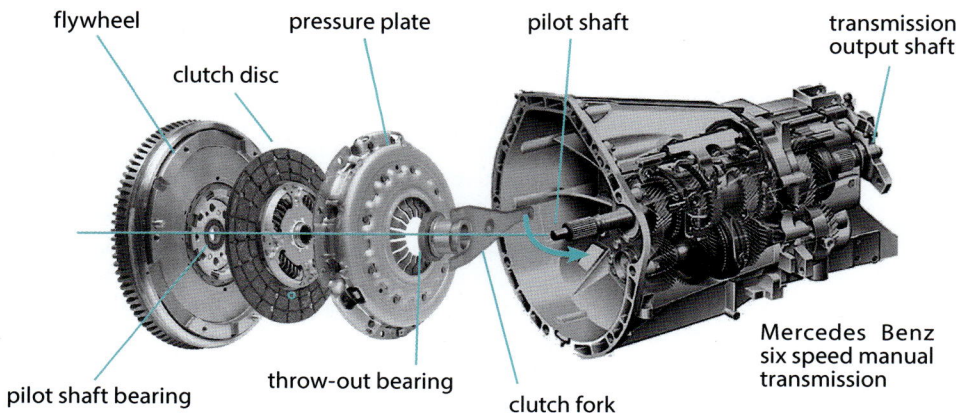

Fig. 14-6. Manual transmission with clutch assembly

As the clutch disc wears thin, the pressure plate begins to lose its ability to hold the disc tightly against the flywheel. As a result, the disc is free to slip, allowing the engine to spin faster than the transmission.

The results of a slipping clutch are lousy mileage, poor acceleration, and difficulty engaging and disengaging gears.

2D. How to Test for a Slipping Clutch

Approach a long, fairly steep hill in second gear at 25 mph. As you get into the steeper portion of the incline, rapidly shift to third gear, flooring the accelerator as you do so.

- Clutches in good condition might permit the engine to race momentarily as the disc begins to "bite," but almost immediately the engine's RPMs will drop into sync with the rear wheels, producing the low RPMs of a car that's straining uphill in the wrong gear.

- If the clutch is slipping, the engine's RPMs will just race away while the car loses power.

For those of you living on flat terrain, there's a variation of this test that requires nothing more than a slight incline. Traveling in third gear at 50 mph up that excuse for a hill, quickly slam the transmission into fifth gear while flooring the accelerator. A slipping clutch will produce the same result – an engine that's screaming, yet producing no power.

2E. Some Driving Tips: How Not to Use a Clutch!

You and your clutch will get along just fine if you remember the following four points:

- The only time that the clutch disc is subject to wear is when your foot is on the pedal. The greatest frictional losses occur while the clutch pedal is moving toward the floor or returning from the floor.

- Sitting with the clutch pedal to the floor, transmission in first, waiting for the light to change is terrible for the disc. That flywheel is spinning past while the disc is essentially motionless. That's a lot of frictional heat to deal with! It's far, far better to put the transmission in Neutral, take your foot off the clutch pedal, and relax . . .

- The very worst thing you can do to a clutch is to use it as a brake while you're sitting on a hill, waiting for the light to change. I know it provides the timid with the sure and certain knowledge that they won't roll backwards when the light turns green, but by holding the clutch pedal partway out with the transmission in first, you are actively engaged in slipping the clutch – not for a matter of moments, but possibly for minutes at a time.

- Try this: Use the handbrake to hold the car. That way, you can work the clutch and accelerator pedals with no fear of rolling backwards. When you feel the car starting to surge forward, simply ease off the handbrake.

- Another egregious habit is to use the clutch pedal as a footrest. By so doing, you have eliminated the freeplay so crucial to keeping the clutch fully engaged.

2F. Clutch Replacement

The vast majority of clutches are buried between engine and transmission, requiring the removal of one, the other, or both. In most cases, the transmission is the easiest piece to yank. Labor costs generally run five to six hours. Those designs that require the removal of both engine and transmission can run twelve hours or more.

As always, when the parts are buried deep inside the engine or trans, I strongly urge factory replacement parts. If presented with the option of buying brand-new versus remanufactured clutch parts, don't even blink! You'll be into these guys for plenty of money anyway, so buy yourself every assurance that you won't have to repeat this misery anytime soon.

Your bill should reflect a four-piece clutch assembly (*S*2C), fresh gear oil (certain cars use automatic transmission fluid instead), possibly an exhaust pipe gasket or cooling system hose. If the transmission has any leaks, now is the time to get them repaired – seal replacement is a simple matter with the transmission on the floor.

If your car has high mileage, you might want to spring for resealing the transmission regardless of whether you currently have any leaks. The same argument can be made for the engine's rear main seal, because the flywheel is the only component that stands in the way of its replacement. Believe me, it's a lot cheaper to get it done now than it would be once that transmission is back in place! If the logic of all this appeals to you, take it up with the service manager or shop foreman when you're scheduling the work so that they can be prepared with the appropriate parts. It's hard to find fault with wisdom – you might just find that your willingness to pay a bit more translates into increased respect for both you and your car.

Front-wheel drive cars only – Transmission removal usually requires that the driveshafts be removed. Either a castle-nut and cotter pin, or a self-locking nut secure each driveshaft at the outside of its front hub, Fig. 14-5, p.364. Check that these pieces have been reinstalled. Their absence allows the front wheel bearing to run loose, leading at best to its premature demise, at worst – to yours. Inspection requires hubcap removal.

Driveshafts secured with a self-locking nut should always be reassembled with a shiny new one.

2G. The Link Between Pedal and Clutch: Hydraulics or Cable?

2G1. Anatomy

Your clutch is activated by a cable mechanism or by hydraulic cylinders, but not by both. To determine what you have, look under the hood just to the outside of the brake master cylinder, directly in front of the clutch pedal. If you find a fluid reservoir there that is not connected to the brake master, then your clutch is activated by hydraulic pressure. The component you've discovered is the clutch master cylinder, joined by metal tubing to the clutch release cylinder (Fig. 14-7 and Fig. 14-8), located on the side of the transmission's bell housing.

If you find no secondary reservoir, look for a cable-within-sheath assembly exiting the firewall directly in front of the clutch pedal.

14

All later-model hydraulic and more sophisticated cable systems are self-adjusting, but both the cable and older hydraulic systems have versions that require routine adjustment to maintain some freeplay between the pressure plate and throw-out bearing. The clearance between these two changes over time as the clutch disc grows thinner. Failure to maintain this clearance results in a slipping clutch. So before anybody sells you a new clutch, you need to determine whether your clutch is adjustable, and if it is, whether an adjustment is all that's needed.

The point at which the clutch engages the transmission with the engine – that is, the height of the clutch pedal when the car first begins to move away from a standstill – changes as the clutch disc wears, rising ever higher off the floor until you'd almost think that your foot is off the pedal before the car starts to move. If that seems familiar, find a hill and test the clutch, § 2D, p.366.

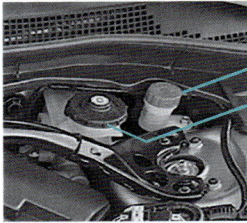

Fig. 14–7. Clutch master cylinder location

clutch master cylinder
brake master cylinder

Fig. 14–8. Clutch release cylinder

bleeder valve
hydraulic line inlet
clutch release fork piston

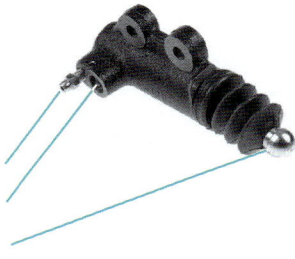

2G2. Problems with the Cable Systems

Cables can stretch, they can fray, they can break. If your clutch pedal engagement height starts to change rapidly, or if it suddenly becomes difficult to engage or disengage your gears, have your mechanic inspect your cable *before* it snaps.

Self-adjusting cable mechanisms can stick in cold or hot weather, signaling the need for lubrication (CRC 5-56 or WD-40) applied to its pivots. These cogs and cams are commonly mounted to the clutch pedal, high up under the dash. Alternately depress, then release the clutch pedal while standing on your head – you'll see 'em move.

As mentioned above, systems that require routine adjustment can produce a clutch that slips badly when that freeplay is gone. If not promptly corrected, the slippage will ruin the clutch disc in short order: Slipping produces lots of friction, lots of heat, and rapid wear. A standard measure of freeplay is about 1/2 inch, measured as a drop in pedal height as the pedal is depressed with just one fingertip. Consult your service manual for instructions on how to make adjustments

2G3. Problems with the Hydraulic Systems

If either the clutch master or release cylinder starts to leak, air gets sucked into the hydraulic line. Since air is vastly more compressible than brake fluid, the result is a reduction in the distance that the release cylinder travels in response to depressing the clutch pedal. That's like trying to shift the transmission with the clutch pedal depressed only partway. What you get is difficulty engaging and disengaging the gears. Jamming the transmission into gear can do real damage, expensive damage.

To verify leaking hydraulics, look at the fluid level within the clutch master cylinder's reservoir. Is it down an inch? If so, open the driver's door, kneel down on the pavement, and stick your head under the dash to enable viewing the firewall directly in front of the clutch pedal. Is that area wet? Is there a stain trailing down toward the carpeting? In either case, the master is on its way out. If the firewall is bone dry, then the clutch release cylinder is leaking. Its inspection requires pulling back its rubber boot (usually from beneath the car); a dribble of muddy fluid is all the evidence you need.

Many shops insist on selling both the master and release as a pair, their reasoning being that if one is leaking, the other will soon follow. There's a certain ring of logic to that, but it ain't necessarily so. I'd go with the realities of your pocketbook.

As mentioned above, systems that require routine adjustment can produce a clutch that slips badly when that freeplay is gone. If not promptly corrected, the slippage will ruin the clutch disc in short order: slipping produces lots of friction, lots of heat, and rapid wear. Adjustments to correct for freeplay are performed at the clutch release cylinder. A standard measure of freeplay is about 1/2 inch with its return spring removed (where present).

The clutch master has two adjustments, but unless you've had some clown under there, neither one should ever need to be altered.

- Both the brake pedal and clutch pedal should sit at the same height relative to the floor. The at-rest height of the clutch pedal can be adjusted by a positioning bolt and locknut that are located in sheet metal, up high in the dash on the driver's-seat-side of the pedal. Don't go fooling with brake pedal height!

- The clutch master has its own internal freeplay adjustment that should never need attention unless the master has just been replaced. Adjustment of the master cylinder's freeplay should not be performed until you've established that the at-rest height of brake and clutch pedal are the same. A standard measure of clutch master freeplay is 1/8 to 1/4 inch, measured as a drop in pedal height as the pedal is depressed with one fingertip.

2H. Problems with Gear Engagement

The three most likely causes of this have already been discussed:

- Worn clutch: See *§*2C, p.365 and *§*2D, p.366 above
- Leaking clutch hydraulic cylinder or defective cable: *§*G2 and *§*G3, above
- Clutch in need of adjustment: *§*G2 and *§*G3, above

14

Other possibilities include:

- Front-wheel drive linkage mechanisms: Since the gearshift lever sits right there beside you and the transmission is out there under the hood, some means is necessary to join the two. Most cars employ a pair of cables that are housed within fixed sheaths, anchored both at the gearshift and the transmission. Look for a sheath that's lost its moorings, or a frayed cable. On those cars that employ linkage rods, look for a worn bushing[2] or loose support bracket.

- Rear-wheel drive linkages: Unfortunately, most of these are located internal to the transmission and are therefore inaccessible without transmission removal and disassembly. For the lucky few who sport exposed linkages, look for worn bushings or a loose support bracket.

- Something internal to the transmission: First ask yourself whether the problem is isolated to one specific gear, or to a pair of gears that share a rail (1st and 2nd, 3rd and 4th, 5th and R).[3] If so, the problem is almost certainly due to something gone foul inside the transmission case. If, however, the problem is manifest in all gears, resume your search among the first three possibilities and consider yourself lucky!

2I. Kicking Out of Gear

This problem is almost always due to wear inside the transmission. Replacement of the three detent springs that tension the shifter rails offers hope to the hopeful. If they're easily accessed, why not try? Consult your service manual.

2J. Transmission Overhaul or Replacement?

Because of the cost of skilled labor, you will probably end up paying more to rebuild what you have than you would if you just replaced it. All factories sell brand-new transmissions, some offer remanufactured units as well. Very expensive, very nice . . . In the aftermarket, several companies specialize in rebuilding; these units usually come with a one-year warranty and the option to buy additional coverage. Get it! The least expensive option is the junkyard. A used manual transmission is a lot more user-friendly than its automatic counterpart, because sitting around for months at a time is of much less concern – assuming there's been no water collecting inside the case . . . Ask your mechanic whether he's seen any trends with your particular model: Have a fair number of these transmissions been self-destructing? If so, a junkyard replacement becomes a much less attractive option. The next question is whether the factory has improved its original design.

If you need to replace your transmission, consider clutch replacement. The labor should be minimal if the transmission is sitting there on the floor. A clutch with more than 50,000 miles of use should definitely be replaced.

---

2 A washer or collar used to reduce friction and provide a better fit.

3 This assumes a standard five-speed gearbox with the accompanying layout: 1   3   5
                                                                             2   4   R

❏ **Front-wheel-drive cars only:**

Transmission removal usually requires that the driveshafts be removed. Either a castle-nut and cotter pin, or a self-locking nut secure each driveshaft at the outside of its front hub, Fig. 14–3, p. 363. Check that these pieces have been reinstalled. Their absence allows the front wheel bearing to run loose, leading at best to its premature demise, at worst – to yours. Inspection requires hubcap removal. Driveshafts secured with a self-locking nut should always be reassembled with a shiny new one.

3. Differentials

When you turn your car to the right, the tires on the left side have to travel significantly farther than those on the right – in the exact same amount of time. That's no big deal for a tire that's free-wheeling, it just rolls that much faster to make up the distance. But if the wheel is connected to a driveshaft, that's another story. Power is being applied to both, and yet one is turning faster than the other. How do they do that? By means of an intricate cluster of gears known as the differential. (In rear-wheel drive parlance, you might hear it called the "rear end" – they're one in the same.)

14

3A. Servicing

A happy differential gets a gear oil change once every 60,000 miles. If your differential uses automatic transmission fluid, you'll need to swap it every 30,000 miles. Check the specifications page in your owner's manual and see §1C, p.360 for further information.

3B. Problems with the Differential

3B1. Leakage

❏ Front-Wheel Drive: There are only three possibilities: the drain plug, a driveshaft seal, or leakage past a seam (gasket) in the case. See §D4, p.362.

❏ Rear-Wheel Drive: Again, three possibilities: the drain plug, the differential pinion seal, or leakage past a seam (gasket).

- See §D4, p.362 regarding drain plugs.
- Leakage past a seam requires removal of the differential from the axle housing to replace the paper gasket, a 2- to 3- hour job.
- Personally, I don't like changing pinion seals. The very act of disassembly can alter the physical relationship between the pinion drive and driven gears, with the result that after a month or two of driving, the differential gets noisy – a whine that usually occurs only under acceleration or deceleration – not both. This noise grows louder with time but seldom creates any real damage. Still, that's hard to explain to a customer after the fact. It's better for the owner to know of this possibility before the work is undertaken.

3B2. Noise

Differential noise is due to wear, either in the gearing or in the bearings that support those gears. It can also be due to chipped or broken teeth. Draining the differential permits inspection of its fluid. It is not at all unusual to find flecks of metal pouring forth – differentials take a lot of abuse! Nor is it unusual for the drain plug to have tiny chunks of steel adhering to it (these drain plugs are magnetic). However, large chunks are a costly sight.

A differential whine can go on for years with no great consequence as long as the fluid is kept fairly clean. However, those diffs that change their tune from a whine to a noisy clunk – particularly in corners – have reached Tombstone. Get it replaced before it takes you there too!

3C. Differential Rebuilding and Replacement

Forty-five years ago, I worked with a mechanic named Frenchie who could rebuild a differential in a matter of hours. He used to say that the alignments inside – of which there are plenty – were easy once you had the "feel" of the thing. That may be true, but I have not met a mechanic since who even pretends to rebuild them.

14

- Front-Wheel Drive: Because the differential is so intimately tied to the transmission, the only sensible approach is to replace the entire transaxle (transmission plus differential). See *§* 1E, p.364 (automatic transmissions) or *§* 2J, p.370 (manual transmissions).

- Rear-Wheel Drive: A brand new differential is an expensive proposition. If the factory offers remanufactured units, that would certainly be more palatable. I don't know of companies in the aftermarket that rebuild differentials in any kind of volume. The junkyard is your cheapest alternative. Find a low-mileage replacement, preferably one with 60,000 miles or less.

4. CV Boots, CV Joints, and Front Driveshafts

❑ Applicable to front-wheel drive and four-wheel drive vehicles only:

Each of the front wheels has a driveshaft passing through its center from the inside, Fig. 14-3, p.363 and Fig. 14-9. To provide the flexibility required for steering, each of these driveshafts has two constant velocity joints (CV joints) – a pair of universal swivels that can transmit power from the inboard to the outboard shaft while allowing a wide range of motion in any direction. This is accomplished using a lovely design of extremely tough metal running in a gooey mass of grease. Each CV is encased in a rubber boot that keeps the grease in and dirt and water out. Once this boot tears, the life of that CV becomes limited as its grease is rapidly slung out by centrifugal force. Routine inspection of these boots is essential to proper maintenance, and should also be performed immediately following any towing. The outer CV joints are located just inside each wheel and are quite visible with the wheels turned hard over. You don't have to squeeze under the car in search of the tear to know that a boot is shot; grease that has been slung in a circle surrounding the boot is all the evidence you need to work up a good case of the blues.

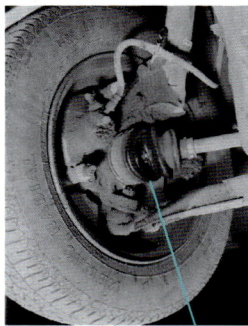

Fig. 14-9. Torn CV boot. Dark areas surrounding the boot are grease mixed with dirt.

The two inboard joints lie on either side of the transaxle and can be quite hard to see – if not impossible – without having the car on a lift. Fortunately, their inaccessibility usually protects them.

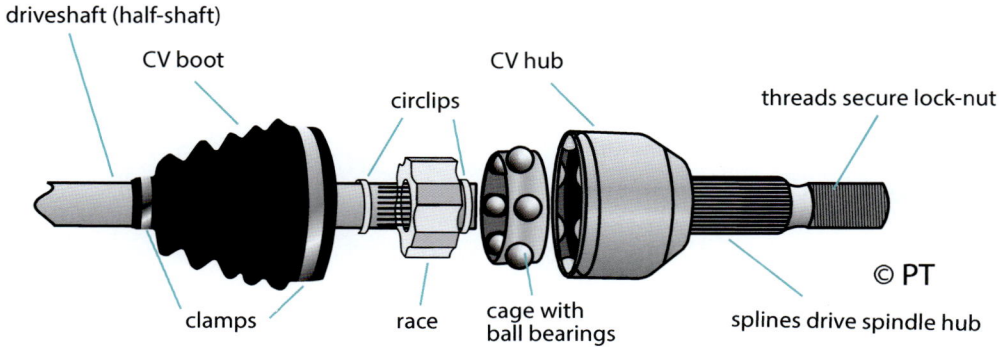

driveshaft (half-shaft)
CV boot
circlips
CV hub
threads secure lock-nut

© PT

clamps
race
cage with ball bearings
splines drive spindle hub

Fig. 14-10. Outboard CV joint
The hub & boot are loaded with grease

As CV joints age (90K and up), wear causes some slop (freeplay) to develop, usually within the outer joint. This slop can translate to noise, a repetitive b-d-d-d-d-d sound when cornering under hard acceleration. In and of itself, this noise is no big deal unless (until) it reaches a decibel level that can no longer be ignored. However, since the sound of clicking can also be produced by a joint that is rapidly running out of grease, you or your mechanic need to check that the boots are not torn. Severely worn CV joints require that the entire driveshaft be replaced. You have three options: a junkyard replacement, a remanufactured unit, or brand-new.

5. U-Joints

❑ Applicable to rear-wheel drive and four-wheel drive vehicles only:

On rear-wheel drive cars, the driveshaft that connects the rear of the transmission to the front of the differential must rise and fall with the movement of the rear wheels. This is accomplished by inserting two U-joints – a type of swivel that can transmit power down the shaft – into the equation. With increasing age, U-joints can wear to the point that they clunk when starting from a standstill. They can also bind, reducing their range of motion, producing a teeth-chattering vibration throughout the car when driving at higher speeds.

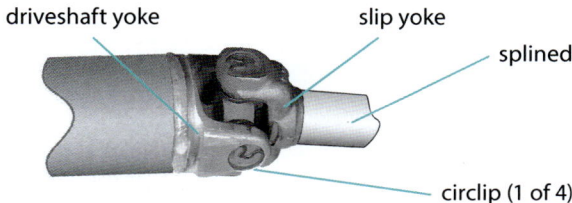

driveshaft yoke
slip yoke
splined output shaft
needle bearings
cap
circlip (1 of 4)
grease seal

Fig. 14-11. U-joint mounted in driveshaft

Fig. 14-12. U-joint assembly

The clunking kind are a lot simpler to diagnose: First of all, it's obvious what's wrong; second, when grasped between your two hands and twisted, the joint has slop (too much freedom of movement) to it.

In order to prove that binding within a U-joint is responsible for that vibration, the driveshaft has to be removed for inspection.

If your car needs to have one U-joint replaced, get 'em both done. Why go through this twice?

6. Wheel Speed: Speedometer Cables

Speedometer cables are relics from a bygone era, having no use in cars with electronic fuel injection (EFI), anti-lock braking systems (ABS), all-wheel drive (AWD), and/or traction control. But for those still driving cars built before the early 1990s, this section is for you.

These cables are driven off a nylon gear in the transmission's tailshaft and wend their way to the backside of the dashboard's speedometer gauge.

A speedometer that suddenly stops working can usually be resurrected by replacing the speedo cable. The cable rides within a protective sheath; they are sold as a unit and should be replaced that way. (The "quick and dirty" approach is to leave the original sheath in place, slap a bunch of grease onto the new cable and insert it into the old sheath. It's easy to tell whether the sheath has been replaced – it will be clean.) All speedo cables enter the passenger's compartment directly in front of the driver.

A flickering speedometer needle is due to a kink in the cable or to inadequate lubrication within the sheath. Replace it.

A speedometer that races to 100 mph while the car is travelling 25 mph is caused by transmission gear oil that has worked its way past the speedometer drive seal, ultimately ending up in the speedometer head. Replacement of the seal plus removal of the speedo head to clean its gearing (*Lectra-Motive* by CRC) can often fix this problem without having to resort to speedometer head replacement. But the trip odometer often fails as a result of this procedure. The main odometer should continue to operate.

7. Transmission Inspections and Servicing – DIY Detail

A. Automatic Transmission Servicing (with dipstick only)

Reread §1A, 1B and 1C, pp. 357-360. Refer to Fig. 14-2, p. 359. before you begin. Also, consult your owners' manual for the type of transmission fluid required and the transmission's fluid capacity. Most of you will find two capacities listed: The smaller volume corresponds to a "wet refill" – the maximum amount your trans will take if you are performing a complete trans service as described below. The larger number represents a "dry fill" – the amount you'd be adding if the transmission is being completely replaced – torque converter and all. At least that's how I always understood it . . .But many a lube jockey will look at his chart, then add that smaller volume while performing a basic drain and fill. I think that's why we saw so many transmissions drive through our door with fluid leaks, shifting problems, or worse. To be certain, follow the directions below regarding initial filling and topping off to the proper level.

FWD Only: Consult your manufacturer's service manual or its online equivalent regarding the fluid in the differential–its type, its scheduled service interval, and its connection (if any) with the transmission.

❑ Tools and Supplies

- Jack and four jackstands (24"), *§4C, p.118*
- 3/8" ratchet and the appropriate socket to address the drain plug (hex, star, or standard)
- To avoid overtightening, use a 1/4" ratchet to reinstall both the transmission pan and its filter.
- Varsol or brake cleaner
- Drain pan
- Rags
- Single-edged razor blades
- Transmission pan gasket
- Transmission filter gasket
- Drain plug o-ring
- Transmission fluid
- Long-necked funnel

I strongly recommend that you purchase both gaskets and the o-ring from a dealership or on-line factory supplier. Consult your owners' manual for the correct type of transmission fluid.

❑ Procedure

Clean the exterior of the transmission pan and the area directly above its gasket seal to eliminate any chance of getting dirt into the transmission's interior. Drain the trans for as long as practical before you remove the pan. The fluid should be warm, not hot. Remove all but two adjacent pan bolts, then loosen (but don't remove) those two as well. Pry the pan loose in such a way that the spillover will end up in the drain pan, not on your face.

Access to the transmission filter may be blocked by piping that connects passageways on one side of the transmission's valve body to other points within the unit. If so, take a picture of their configuration before you proceed. Some of you will find that one or more of these pipes have a flange (flat tab) and bolt holding them in place; if so, you'll need to remove them – and only them – before proceeding. For the majority of you, just a simple twist with a wide, flat-bladed screwdriver, applied near each end, will pop the pipes loose. Wiggle them out from their center.

All clear? Then remove the filter just like you did the pan: Pull all but two bolts, loosen them, etc. Be advised: Dirty filters can hold back a lot of fluid.

Use varsol or brake cleaner to clean the trans pan's interior. Use a rag to wipe off as much metal dust from the magnet(s) as seems reasonable; reinstall them exactly where you found them (their position will be clearly outlined – just like a bathtub ring).

After you remove the transmission's filter, use varsol or brake cleaner to clean it. Brake cleaner is advantageous because its spray will dislodge most crud clinging there, and it evaporates quicky. Do not spray any solvents into the transmission! The cleaning agents in transmission fluid were engineered to clean without harming the transmission's seals and solenoids.

The majority of both gaskets will just peel off; brushing varsol onto the residue helps to loosen the rest. Surrounding each bolt hole you may find "tougher sledding". If so, hold a single-edged razor blade perpendicular to the surface and scrape rapidly back and forth. (A steel razor blade, held at an oblique angle to an aluminum surface, can gouge that surface quite easily, so keep it vertical to the plane you're scraping.) An occasional shot of brake cleaner will soften the material, speeding the process.

14

Once you're satisfied, install the filter, the pipes, then the pan. Do not use glue! Finger tighten a few bolts on opposite sides, then install the rest, tightening them *evenly*. The filter should be torqued tight but not overwhelmingly so; the pan takes nothing more than light to light-moderate torque. I always use a 1/4" drive ratchet to avoid snapping off any of these bolts. If all that seems vague, consult a reliable source regarding tightening torque. Install the drain plug with a new o-ring.

Tie a rag around the transmission's filler neck before inserting a long-necked funnel.

Recheck your owners' manual for the transmission's fluid capacity. If you find that it provides two measures, one for a standard refill and another for a dry fill (that includes the torque converter), then:

- Almost all of you will be using the smaller capacities. Add all but one quart, *slowly*. If you have fluid standing stationary in the funnel, read the text following the second dot. The rest of you took my advice and poured your quarts deliberately. Congratulations!

  Remove the funnel, insert the transmission's dipstick, then start the engine in Park. Let it idle for a brief period before moving through every gear *with your foot firmly planted on the brake pedal*. (Shifting gears with the wheels spinning freely can destroy a transmission.) Leave it idling in each gear for 3-10 seconds or so to provide time for fluid to fill every passageway. Return to Park, leave it idling, and check your fluid level. Don't freak out if the level seems grossly overfilled; just wipe off the stick and look again. Since the trans is essentially cold, use the dipstick's cold markings as your reference. If there's nothing on the stick, add 1/4 quart and check again.

- If fluid is standing stationary in the funnel, have an assistant crank the engine in Park or Neutral until the fluid level drops, while you (1) hold the funnel securely in place and (2) call out "Stop Cranking" as soon as the funnel is empty. Add more fluid, *slowly* this time, then return to Dot 1.

- If your torque converter has a drain plug, your manual will tell you so, but it probably won't tell you how to deal with it. For that, you'll need the vehicle's factory shop manual or its equivalent.

B. Automatic Transmission Servicing (with no dipstick)

Every manual I consulted referenced their manufacturer-specific OBD scanners, specifically to provide a read-out of transmission fluid temperature. Each of those cars come equipped with temp sensors; all specified that trans temps should be in the 110°-115° F· range – plenty warm, not blisteringly hot. Due to the expense and complexity of these transmissions, I would strongly urge that you begin with the manufacturer's service manual or its online equivalent (alldatadiy.com).

FWD Only: Speaking of service manuals, consult it regarding the fluid in the differential – its type, its scheduled service interval, and its connection (if any) with the transmission.

Reread §1A, 1B & 1C, pp.357-360.

❑ Tools and Supplies

- Jack and four jackstands (24″), §4C, p.118
- Hand pump suitable to your particular vehicle
- 3/8″ ratchet and the appropriate sockets (hex? star?) to address the drain plug and inspection port plug
- Varsol or brake cleaner
- Drain pan
- Rags
- Drain plug o-ring
- Trans fluid filter when specified
- Transmission fluid

Numerous models now have a spin-on fluid filter just like an engine oil filter. Consult your shop manual for its replacement interval. I strongly recommend that you purchase the o-ring(s) and filter from a dealership or on-line factory supplier. Check your manual for the correct type of transmission fluid.

❑ Procedure

Rule #1 – Follow the procedures outlined in the manufacturer's service manual!

A few thoughts . . .

- If the only reason given for the manufacturer-specific OBD scanner is to read out temperature, feel free to proceed with a thermometer. Two points:
  - Don't wish away any other reasons given!
  - It's time to head to the dealer, not your local Perfectly Quick . . .

- Some of you will be inspecting the fluid level with the engine running, others will be told to turn it off. But in either case, before opening up the inspection port, run the engine through every gear *with your foot on the brake*. (Shifting gears with the wheels spinning freely can destroy a transmission.) Leave it idling in each gear for 3-10 seconds or so to provide time for fluid to fill every passageway. Unless otherwise specified, put the gearshift into Park before proceeding.

- The car must be raised high enough to enable working comfortably underneath. It has to be level. The transmission fluid must be warm, but not so hot that it could burn you.

Most manufacturers provide an inspection/fill port located roughly halfway up one side of the main case. This port is in addition to the drainage port located at the transmission's lowest point. Both of these typically carry threaded plugs that require a hex- or star- socket to open them.

Others employ just one port, always at the bottom of the transmission, that doubles as both, Fig. 14-13. As long as the inspection tube is threaded into the case, the trans can't drain anything other than the dribble you'd expect to find exiting any properly filled unit. Replacing the transmission fluid requires that the inspection tube be removed. Not to state the obvious, but it also requires that you turn off the engine!

14

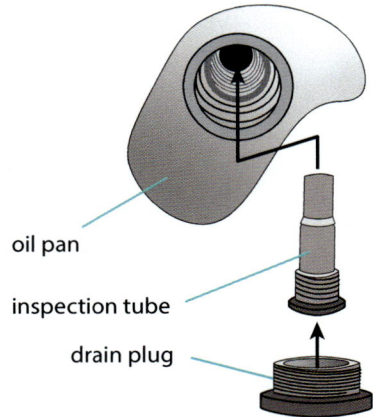

oil pan

inspection tube

drain plug

Fig. 14-13. Combination inspection port and drain © PT

- Follow the shop manual's direction regarding the transmission filter with regard to both mileage and time. If you are removing a transmission pan similar to that seen in Fig. 14-2, p. 359, read *7A, p.374 before proceeding.

- When refilling, add the amount specified in your shop manual. It will be sufficient to fill the transmission to slightly overfull; a light stream should be trickling out. Install the inspection port plug or combination drain plug. Unless the shop manual states otherwise, start the engine and let it idle while you run through the gears *with your foot on the brake*. Don't linger in any particular gear, 3 - 10 seconds is plenty to fill each passageway.

- Unless the shop manual states otherwise, check your fluid level with the engine running. Top it off if necessary.

7C. Differentials

Nothing much to add here other than what's already been said in *1C, p.360 and *3, p.371. Consult your shop manual for service interval, type of fluid, and service procedures.

7D. Clutch Replacement

This can be a tough job on the street. You'll need help: Someone to hand you tools once you've crawled under the car, someone strong enough to help you pull the transmission away from the engine, then get it to the ground without dropping it, then repeating the process: lifting, holding the thing above their head – and yours – while lining up the transmission's input shaft with the clutch assembly and pilot shaft bushing, pushing it "home", then installing two bolts (loosely) to hold it in place.

In a shop you can manage the whole process at a convenient height, with plenty of elbow room, a skilled assistant, and compressed air guns (ratchets), in a third the time.

I've replaced RWD clutch assemblies five or six times on the street, both as a much younger man during my "shade-tree" days and once since I've retired. It was never fun, but it is doable. However, with regard to FWD transmissions, I'd rather pay my mechanic to do the job: too many special service tools, too many chances to screw it up, and a warranty!

7D1. Front-Wheel Drive

Front-wheel-drive transmissions add much more complexity to the job: much tighter working quarters, steering components, two driveshafts . . . stress. You'll need your manufacturer's shop manual or its online equivalent (alldatadiy.com). You'll be referencing it frequently.

❑ **Tools and Supplies**

14

- 1/2″ drive air ratchet and air compressor. Rent a big one; smaller tanks run out of air very fast.
- 1/2″ drive impact socket for the driveshafts; 1/2″ drive impact sockets for the bolts that secure the trans to the engine block; 1/2″ drive impact swivel
- 3/8″ air ratchet, hand ratchet and socket set
- 1/2″ drive torque wrench; 1/2″ drive breaker bar, 2′ long
- Pilot shaft bearing puller; ball joint (tie-rod) separator
- Seal installer kit (round discs of various sizes)
- Jack and four jackstands (24″), *§4C, p.118*
- Hand pump suitable to your particular vehicle
- Varsol or brake cleaner
- Drain pan for oil, perhaps a drain pan for coolant
- Rags
- Emery cloth, medium
- Mechanic's wire
- Wheel bearing grease
- Clutch assembly (4 pieces)
- Clutch centering tool, Fig. 14-14
- Driveshaft seals
- Drain plug o-rings
- Gear oil or automatic transmission fluid – check your shop manual
- Creeper or two

Fig. 14-14. Clutch centering tool
Courtesy of Dorman

❑ **Procedure** – Some random thoughts:

- This job requires a lot of patience, perhaps a few extra parts, maybe even another trip to the tool-rental store. If you get rattled, walk away before you break something.

- Use factory parts throughout: a four-piece clutch assembly, driveshaft seals, drain plug gaskets, whatever else your shop manual mentions during disassembly.

- Depending upon the car's mileage and how much longer you intend to keep it, you might want to consider replacing the engine's rear main seal and the transmission's front seal.

- The wheels have to be on the ground to enable you crack loose the axle nuts. You'll need a 1/2" drive air ratchet just to bust them loose, but *don't back the nuts out while the car is still on the ground* – you run the risk of damaging the wheel bearings. Keep both hand on the ratchet at all times – a 1/2" drive ratchet carries a lot of torque.

- Disconnect the battery's positive terminal. To maintain your car's memory settings, use an OBD-II scanner with that capability.

- Drain the transmission and differential, possibly the coolant.

- Mechanic's wire is used to suspend the brake caliper assemblies to eliminate any strain on their hydraulic lines. Also to suspend the free end of each driveshaft whenever necessary. Take care to position the wires' ends so they can't possibly blind you.

- Certain suspensions require that two bolts on the front strut be removed to free the driveshaft's outboard end, Fig. 14-15. The upper bolt has a cam that defines the wheel's camber. *Mark its position before disassembly.*

- Once the outboard side of the driveshaft is free, temporarily suspend/support it with mechanic's wire while you work to free its inboard side. Most are secured with a spring-loaded circlip. There are pullers for that, but if you have a pry bar or two, have at it.

Fig. 14-15. Corolla front hub

Fig. 14-16. Pry bar, 18" long

- Emery paper should be used to sand any glaze off the surface of the flywheel. Replace the flywheel if it has hot spots (Fig. 12-10, p.265). If you don't, the chatter will continue every time you engage lower gears.

- Assuming the flywheel remains in place, you'll need a pilot shaft bearing puller to remove it, and an appropriately-sized seal installer disc to put in the new one, Fig. 14-17.

- Most seals come pre-lubricated and won't require any attention, but to ease their entry, apply a thin film of grease around their circumference. Same goes for the pilot shaft bearing. If the axle shaft seals were packaged dry, fill the interior groove (where the "teeth" are in Fig. 14-18) with wheel-bearing grease. The figures below show appropriately sized seal installers. Note that in both cases, the installing disc is just slightly smaller than the outside circumference of the bearing or seal, large enough to drive the piece home without distorting their interior space.

- Follow your manual's disassembly and reassembly instructions to the letter, particularly with regard to sequence and torque specifications. Always reassemble pieces evenly: Remember that you're dealing with aluminum in most cases, and aluminum *loves* to crack – as does mis-aligned steel, for that matter.

Fig. 14-17. Pilot shaft bearing and installing disc

Fig. 14-18. Driveshaft bearing and installing disc
Bearing courtesy of AC Delco

7D2. Rear-Wheel Drive

Compared with FWD transmissions, a RWD is like a walk in the park! You'll still need a manual, but some of the special service tools (SST) required on Page 379 are unnecessary. That being said, having a 1/2″ air gun on site is a blessing.

Depending upon the car's mileage and how much longer you intend to keep it, you might want to consider replacing the engine's rear main seal and the transmission's front seal.

❑ **Tools and Supplies**

- 1/2″ drive air ratchet and air compressor. Rent a big one; smaller tanks run out of air very fast.
- 1/2″ drive impact sockets for the bolts that secure the trans to the engine block; 1/2″ drive impact swivel; a long 1/2″ extension (3′)
- 1/2″ drive breaker bar, 2′ long
- 3/8″ air ratchet, hand ratchet and socket set
- Pilot shaft bearing puller
- Seal installer kit (round discs of various sizes)
- Jack and four jackstands (24″), *S4C, p.118*
- Hand pump suitable to your particular vehicle
- Varsol or brake cleaner
- Drain pan for oil, drain pan for coolant
- Rags
- Emery cloth, medium
- Wheel bearing grease
- Clutch assembly (4 pieces)
- Clutch centering tool, Fig. 14-14, p. 379
- Transmisssion rear seal
- Drain plug o-rings
- Gear oil or automatic transmission fluid – check your shop manual
- Creeper or two

❑ **Procedure** – Some random thoughts:

- Breaking loose the transmission bolts is doable with a breaker bar. I would much prefer to do it with an air ratchet. Keep both hand on the ratchet at all times – a 1/2″ drive ratchet carries a lot of torque.

14

- Speaking of safety . . . Using a breaker bar? When those bolts snap loose, the end of the bar will move very fast. Envision where it will end up: Your gut can absorb a blow far better than your head!

- Disconnect the battery's positive terminal. To maintain your car's memory settings, use an OBD-II scanner with that capability.

- Drain the transmission and the coolant. Disconnect one end of the upper radiator hose to relieve the stress on the radiator's upper neck when the rear of the transmission is freed from its rear crossmember.

- The clutch release cylinder has to come off. Start by cracking loose its hydraulic line while the cylinder is still bolted to the tansmission case.

- Emery paper should be used to sand any glaze off the surface of the flywheel. Replace the flywheel if it has hot spots (Fig. 12-10, p.265). If you don't, the chatter will continue every time you engage lower gears.

- Assuming the flywheel remains in place, you'll need a pilot shaft bearing puller to remove it, and an appropriately-sized seal installer disc to put in the new one, Fig. 14-17. App;ly a thin film of grease around its circumference to ease entry.

- As for the tailshaft bearing: Most seals come pre-lubricated and won't require any attention other than a thin film of grease to ease its entry, but if the seal were packaged dry, fill the interior groove (where the "teeth" are in Fig. 14-18) with wheel-bearing grease. You'll need an appropriately-sized seal installer disc to put in the new one. Note that the installing disc is just slightly smaller than the outside circumference of the seal, large enough to drive the piece home without distorting its interior space.

- Always reassemble pieces evenly: Remember that you're dealing with aluminum in most cases, and aluminum *loves* to crack – as do misaligned steel flanges, for that matter.

Chapter 15. Heating and Air Conditioning

15

Notes

Heating and Air Conditioning

1. Anatomy

Automotive heating and air conditioning units share a common ventilation system buried behind the dash that includes a fan motor, its speed control system, control doors/drums and their actuators. These doors act as gates to provide separation between fresh and recirculated air, to blend heated and cooled air, and to control its destination (vent, defrost, and floor). Most of these "boxes" extend from the driver's right knee to the far wall, just to the right of the glovebox.

Heat supply is simple: Engine coolant is piped from the rear of the engine (its hottest zone) into a small "radiator"–the heater core–that's housed within the ventilation box, Fig.15-1, p.386. The fan motor blows air through its coils on its way to the passenger compartment. The coolant is pulled back to the engine through the suction side of the water pump, Fig.9-1, p.148. The amount of heat delivered is regulated by the heater control valve, essentially a faucet controlling the amount of coolant allowed entry to the loop, §3, p.397.

The real beauty in air conditioning design is that it works at all . . . Imagine taking a simple physical truth–that evaporation produces cooling–and transforming that knowledge into a continuously cycling machine. The trick was to find a substance that could flip-flop between its liquid and gaseous states within a reasonable range of temperature and pressure. Easier said than done!

By definition, evaporation is the conversion of a liquid to a gas. It creates a cooling effect. You can demonstrate this by first standing in front of a fan on a hot day, then wetting your skin and trying it again. You'll find that the evaporation of that water makes your second pass far more chilling.

By turning the process of evaporation into a continuous cycle, air conditioning systems take this principle and transform it into a technological work of art. A gas is compressed into its liquid state, then driven through a teeny(-ish) hole into a much larger chamber that provides it with the space to expand rapidly into gaseous form. The pressure drop of this transition produces a very strong cooling effect.

The composition of that gas has changed markedly over the years, from ammonia in early trials (the 1890's) to Freon (R-12) throughout most of the 20th century to R-134a since 1987. Freon was developed because ammonia is, among other things, explosive . . . R-134 superceded R-12 because reon was burning ozone holes into our atmosphere. These days, the word freon is used in much the same way as the word kleenex is: It depicts a product, not necessarily a brand.

While general principles and physical layouts are for the most part identical, the means of controlling internal pressures has been approached in two different ways. One system employs an expansion valve and receiver-dryer (Fig. 15-1), the other uses an orifice tube and accumulator (Fig. 15-2). More on that in a moment.

The expansion chamber is known as the evaporator. It's finned much like a radiator so that its coldness can be quickly transferred to the air blowing through it. The now-gaseous freon moves through the evaporator and connecting piping into the compressor–the workhorse of the system. The system's pressure between the evaporator coil and the compressor is typically in the vicinity of 30 psi (pounds per square inch) at idle, lower when revved. System pressure from the compressor to the expansion valve is typically in the vicinity of 250 psi at idle, higher when revved.

The compressor derives its power from a drive belt, typically the serpentine belt. The amount of energy consumed by the compressor is significant: Turn on your air conditioner as you accelerate up a hill; you'll notice a distinct drop in power as the compressor kicks in. This power drain accounts for a drop in fuel economy of roughly two miles per gallon.

The compressor takes in low-pressure gas and sends it out as high-pressure gas. This creates a tremendous amount of heat in addition to the pressure, so much so that the freon exiting the compressor must be cooled to keep it from overheating and bursting the system. This is accomplished by the condenser, a radiator-like structure located directly in front of the engine's radiator. Within the condenser, freon cools to its liquid state and the technological marvel of turning a liquid into a gas and back again is complete!

5

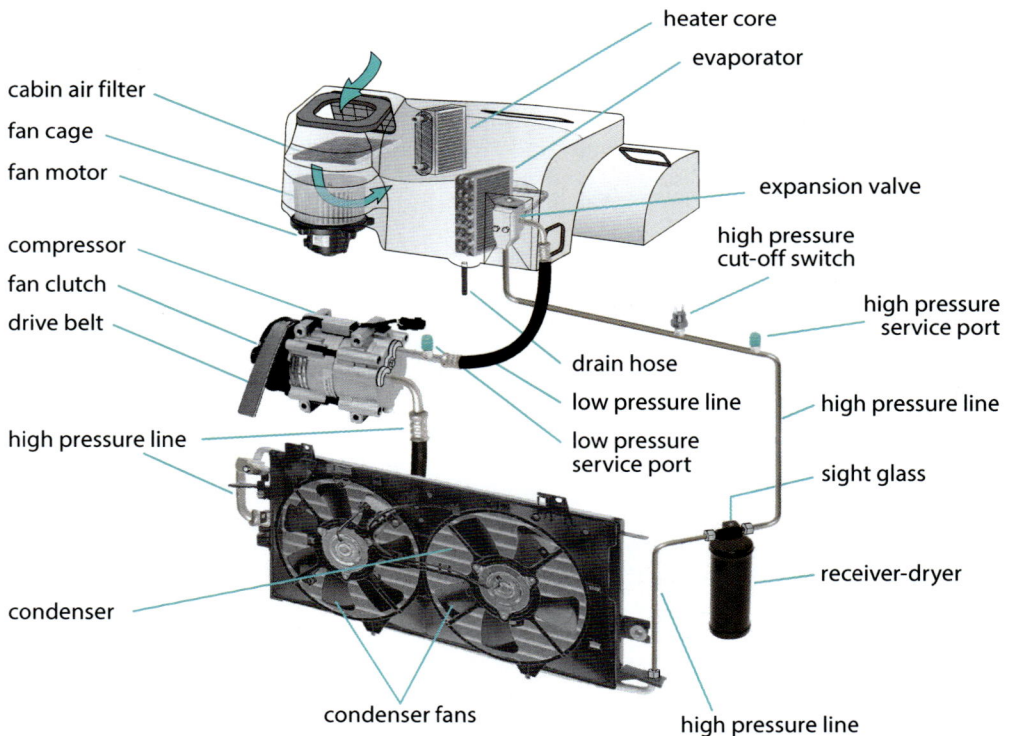

Fig. 15 - 1. AC system with expansion valve and receiver-dryer. (Type 1) © PT

Receiver-dryers and accumulators share a few similarities:

- They look similar–they're both cans with an inlet and an outlet–except that accumulators are roughly twice as large.

- Both act as storage compartments for excess freon, held in reserve to accomodate changes in engine speed, ambient temperature, humidity, and system efficiency.

- Both contain a filter for capturing crud–oil residues, corrosion particulates, and in worst case scenarios, metal flakes from a failing compressor.

- Both contain a supply of dessicant to absorb moisture trapped in the system.

Three differences:

- They're at opposite ends of the system. The receiver-dryer lies between the condenser and the expansion valve on the high-pressure side of the system. The accumulator resides on the low-pressure side between the evaporator and compressor.

- The primary purpose of an accumulator is to prevent liquid freon from finding its way into the compressor. Liquid freon won't compress; instead, it will weaken and ultimately destroy the compressor. The receiver-dryer does exactly what its name suggests. Many also provide a sight-glass for inspecting the condition and amount of freon.

- New accumulators come with a supply of compressor oil that gets sucked up through a pinhole at the bottom of its canister.

15

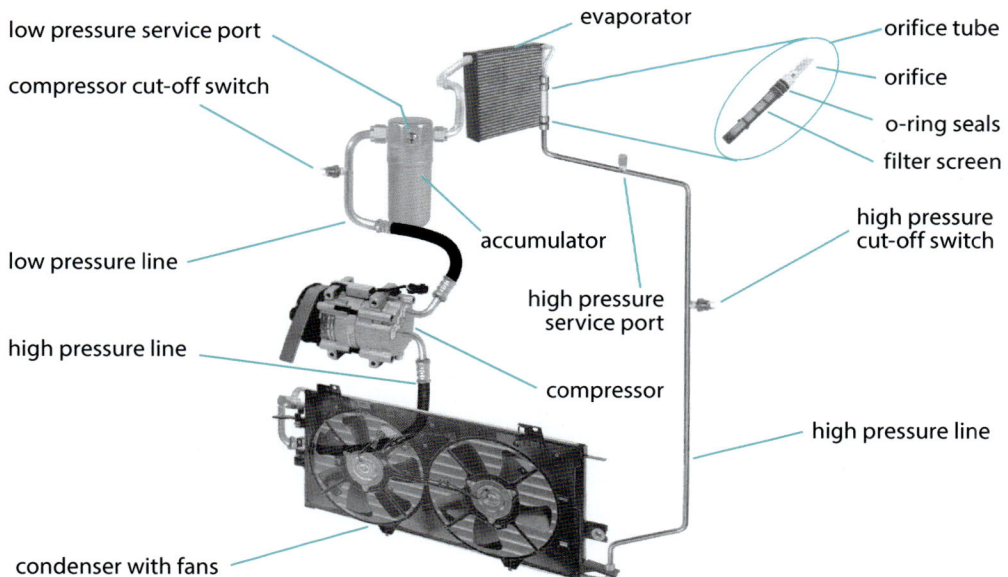

Fig. 15 - 2. AC system with orifice tube and accumulator. (Type 2) © PT

Expansion valves regulate the amount of freon passing through them by varying the size of their opening–again, just like a faucet. The size of that opening is governed by a thermostatic bulb which, in the old days, used to mount right up against the evaporator coils. Some still do, but in most modern applications the metering is done within the expansion valve itself. A properly funtioning expansion valve is very efficient, converting liquid to gas almost completely under all conditions.

An orifice tube has no moving parts; the volume of freon passing through it remains constant regardless of downstream temperature. Because of that, a compressor cut-out switch is employed to regulate the pressure of the system at the accumulator. In a perfect world, all of the freon entering the accumulator would be in its gaseous state, the evaporator having captured all of the coolness available to it. But in reality, because the orifice tube can't respond to changes in engine speed, ambient temperature, humidity, and system efficiency, there will always be some liquid entering it's chamber. Liquid freon drops to the bottom where it will boil off into its gaseous state; freon gas just moves on through after it's passed through a filter and drying agent.

The compressor is activated by an electro-magnetic clutch assembly (Fig. 15-3) that sits just to the outside of the compressor on its "belted" side. The AC drive belt spins the clutch pulley continuously while the engine is running, but it's "idling" as long as the air conditioning is off. Turning on the AC energizes the coil; its electromagnet pulls the clutch plate hard up against the pulley's front face which locks it to the compressor. The compressor's pump-like innards begin spinning in sync with the drive belt.

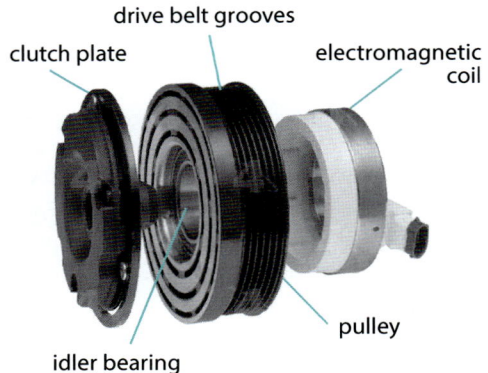

Fig. 15 - 3. AC compressor clutch
Courtesy of Four Seasons

Modern condensers have one or more electric cooling fans that kick in when the AC system is turned on. Without these fans, the internal temperature of the system climbs sky-high, causing internal pressures to build beyond the system's capacity to contain them. A high-pressure cut-off switch is used to disengage the compressor should pressures climb too high. Condenser fans typically run continuously whenever the air conditioning is on.

It takes a great deal more energy to cool down a droplet of water than it does to cool down a corresponding amount of dry air. For this reason, the AC system works far more efficiently when it is operated in recirc mode (recirculated air rather than fresh). The AC system acts to bolster its own efforts by dehumidifying the car's interior—this by the very process of cooling. Moisture condenses on the evaporator coils' fins, then drips its way down to a drain at the bottom of the evaporator case. This drain leads through the firewall to the ground. In humid weather, the amount of water that puddles beneath the car can be astonishing. But unlike a real leak, this water is clear and not slippery to the touch.

2. Problems with the Air Conditioning System

2A. Freon Loss (Leakage)

Most AC system failures are caused by leakage. Likely sources include:

- the compressor's front seal, Fig. 15-4
- at line fittings where two components are joined together, Fig. 15-5
- past the valve in either one of the service ports, Fig. 15-6
- through a pinhole in the condenser, Fig. 15-1

Freon loss is more likely to occur on the high pressure side of the system.

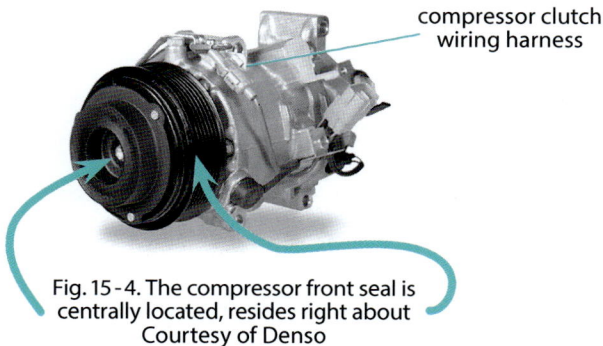

compressor clutch
wiring harness

Fig. 15-4. The compressor front seal is centrally located, resides right about
Courtesy of Denso

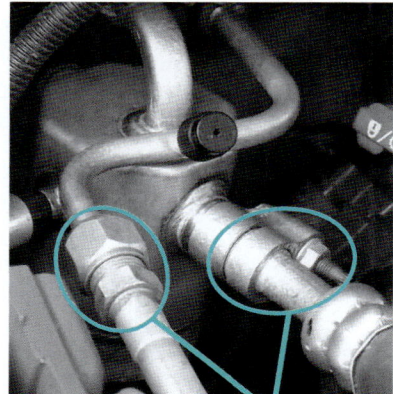

Fig. 15-5. Two types of line fittings and one service port cap
Courtesy of rechargeac.com

Schrader valve

Fig. 15-6. Service port

Front seals can simply wear out over time, but a lack of sufficient compressor oil will certainly hasten the process. A defective cut-out switch on Type 2 systems can overwhelm the compressor with liquid freon, washing away its lubricating oil. I've never had any luck resealing compressors, and the labor to try is high. Spring for a new compressor or a factory remanufactured unit.

Most line leaks develop simply as a matter of aging and vibration, but they can also result from a stress fracture. (All of these components are made from aluminum to enable rapid heat dissipation. Unfortunately, aluminum tubing can be very soft and subject to metal fatigue.) Line fitting leakage can also result from the human touch—whenever any component requires replacement, the fittings can be overtightened, undertightened, or stripped. Any time a line gets stripped, both the male and female sides of that union must be replaced to assure a leak-free joint.

Servicing AC systems requires attaching a set of gauges to a pair of service ports, one on the low pressure side, the other on the high side. Each fitting contains a Schrader valve–identical to those in your tires but capable of withstanding much higher pressures, (Fig. 15-6, p. 389). On older cars, these valves should be replaced whenever the system requires repair.

Any freon leak permits the entry of moisture from the surrounding air. In the car's early years, this is usually not much of an issue, but the process is cumulative. If you live in a humid climate and your system has been apart more than once for repairs, be sure that the receiver-dryer/accumulator is replaced as insurance against corrosion. Once the receiver-dryer has absorbed all the moisture it can, whatever moisture remains starts to corrode the system from the inside out. When moisture mixes with freon under high temperatures and pressures, it becomes highly corrosive. If your AC system has been opened and is to remain that way for a period of days while special order parts come trickling in, insist that a replacement bottle be installed. It should be the very last component installed before the system is evacuated. Failure to do so may just cost you a condenser sometime down the road. Developing a pinhole within the condenser is far more likely than finding one in the evaporator (high pressure vs low pressure), but leakage can occur in either place.

Another all-too-common problem is the overfilled system, usually caused by adding freon to a partially filled system. Overfilling can create tremendous internal pressures and produce a multitude of leaks. Each one requires time to track it down, some form of repair, and a recharge. All too often, the freon level is low again two weeks later, an ugly process that repeats itself until all leaks have been uncovered and repaired.

Most aftermarket kits provide the means to add freon without first evacuating the system. Unfortunately, just adding freon *does not* displace either the air or the moisture that have seeped into the system. That requires a vacuum pump. Skipping that part can end up costing thousands of dollars. I'm not saying that's true in every case, because it's not. What I am saying is that vacuum pumps are inexpensive to rent, and a cheap set of gauges can be purchased for thirty bucks.

How can you tell whether your system is low on freon?

- Your first clue will always be that the cabin simply doesn't cool off as well as it once did. It'll take longer to cool down and won't get as cold, regardless of how patient you might be.

  To verify, pick up an AC service thermometer (Fig. 15-7) and insert it into any one of your vents. Start the engine, turn on the AC. It should start to read 55-60 degrees within a matter of minutes. Anything higher than that indicates that trouble is brewing. It doesn't necessarily mean freon loss; but it does suggest that the system is not properly functioning.

- If your receiver-dryer or accumulator has a sight glass (Fig. 15-8), see §4, p. 398.

- If not, you'll need a set of AC gauges or a mechanic. See §5A, p. 400 for a primer on gauge readings and what they signify.

Fig. 15-7. Thermometer with probe
Courtesy of FJC

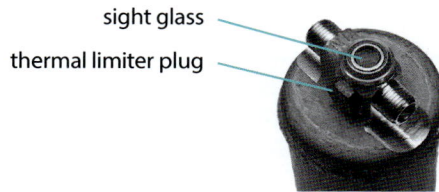

sight glass
thermal limiter plug

Fig. 15-8. Receiver-dryer, upper face

2B. Leak Detection

Since freon is an odorless, colorless gas, it leaves no trace other than a hint of oil–subtle evidence at best. But by adding an ultraviolet dye to the system, large leaks become immediately apparent–they offer a yellowish-green or purple glow whenever an ultraviolet (UV) light source is pointed in their direction. Smaller leaks can take some time to appear, requiring a return visit for reinspection.

> Ultraviolet light will damage unprotected eyes.
> *Always* wear UV goggles in the presence of a UV light source.

These dyes can be added in one of two ways:

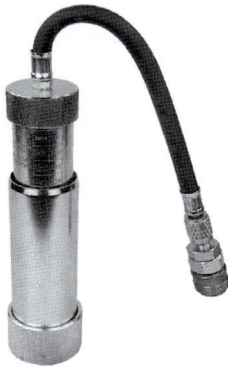

Fig. 15-9. Dye injector
Courtesy of Mastercool

- If there's still enough freon in the system for the compressor to kick on normally (without your help), a dye concentrate can be added using a dye injector tool, Fig. 15-9.

- If not, cans of R-134a are available both with and without dye. One can with dye is installed first. The balance of a full charge–typically one more can of regular R-134a–is then added.

There are some good electronic freon detectors out there, but they're expensive, and the mechanic has to be lucky enough to place the probe directly into the freon's vapors to make it sound its alarm. Compounding this problem is the fact that many leaks only occur under pressure, requiring that the air conditioning system be up and running. That means that the engine's cooling fans are running as well, blowing away many a hope for a clear signal. As an added bonus, accessibility to many line fittings is atrocious.

So, no, I seldom use mine. However, there's one spot that an electronic detector comes in handy: Because the evaporator is buried deep inside the dash, uncovering leakage there can be a nightmare without one. Simply inserting the detector's probe into the drain tube that exits the climate control box–typically, it's directly under the evaporator–will indicate a problem there. The engine should be off. You'll probably have to access the tube from beneath the car. See ❑ Gurgling Sounds, p.393.

2C. Freon Recycling

> It is illegal to allow freon to escape to the atmosphere. That's not to say that those of you with leaking AC systems will be hauled off to jail, but if you knowingly open up your system without having some means to capture whatever freon is left in there, that's a crime.

Until the late 1980's, the freon remaining in any system that required repair was simply allowed to escape into the air, where it would waft away on the breeze, floating off to the sky. Over time, cancers, sunburns and cataracts began to spike. Scientists traced these phenomena to a large hole in the atmosphere that was allowing the sun's rays to penetrate unfiltered all the way to the ground. Careful analysis revealed that this "ozone hole" had evolved from the interplay of oxygen, sunlight, and man-made chlorofluorocarbons, R-12 freon being one of the principal offenders.

Believe it or not, governing bodies worldwide actually got together in 1987 and agreed to the Montreal Protocol, which eliminated the use of R-12 and required that all freon from all repairs be captured and recycled. Because it is significantly less chemically active, R-134a was phased in to replace R-12 by the early 90's. The ozone hole has shrunk dramatically since then, but because its presence in the atmosphere contributes to global warming, R-134a must still be recycled.

Recovery systems take two basic forms: those that simply trap the freon for someone else to recycle, and those that recycle on site. Theoretically, the latter should be cheaper for the customer, because whatever comes out of your system can be cleansed, then reinstalled following repair of the leak. Naturally, the quality of the equipment and the integrity of its filters determines whether your car is recharged with clean, dry freon, or not . . .

> Remember: It's extremely important to have the receiver-dryer or accumulator replaced whenever the system is opened up for repair. This is especially true on older cars, on systems that have been opened up more than once, and on systems left open while parts are being ordered.

2D. Noises

❏ The AC system puts a considerable load on the engine. When you turn on the AC, a worn or loose drive belt can squeal, chirp, or otherwise protest. So start there, *§*2, p.444.

❏ The compressor is the only system component that's prone to mechanical failure–it's the system's only moving part. Signs of failure almost always start with bearing noise before degrading to the crunch and grind of disintegrating metal. There are any number of bearings associated with the drive belt(s); see *§*18, p.175 to tease out which component (AC compressor, water pump, power steering pump, etc.) is producing the sound. A noisy bearing can last for quite awhile before it starts to come apart; however, because every one of those components can produce serious problems, locate the source before it gets worse.

Compressor noise can originate from its innards (bearings or pump components), the fan clutch, or its moorings. Turn off the AC.

- If the noise goes away entirely, there's little doubt that the compressor is the problem. If the noise was "just" the whirr of a worn bearing, then the compressor probably has some life left in it–enough to figure out what to do next. But if it's been grinding or crunching, **don't turn it back on!** Odds are that metal fragments could get stuck inside the condenser and/or further down the line. Once lodged there, the AC might never work properly again. Regardless of the severity of the sound, you'll be looking at replacement, because compressor overhaul requires special tools, special skills, and a considerable amount of practice.

- But first . . . Verify that the compressor is securely bolted to its brackets, and that those brackets are bolted hard up against the engine. A loose compressor will produce more of a chatter than a grind, and will almost always create belt noise as well. Simply tightening its retaining bolts can cure the problem.

- If the grinding continues once the AC is turned off, it is probably not the compressor–at least not its innards. But a compressor clutch can make noise whether the AC is on or off. Its idler bearing could be coming apart, or its shims[1] could have worn down to such an extent that the inner face of the clutch plate is in constant contact with the pulley's outer face, grinding both down to dust. The compressor clutch can be replaced without removing the compressor or evacuating the freon. If caught early enough–while the sound is still more of a scaping sound than a grind–you could conceivably get away with just replacing the shims and sanding any roughness off those two faces.

15

❑ I lied when I said that the compressor is the *only* system component that's prone to mechanical failure. Pressure switches and relays certainly qualify, but noises aren't part of their symptomology. Condenser fan(s) on the other hand . . . A bad fan bearing will produce a whirring, then a grinding, just like any other. Later stage issues involve plastic fan blades warping from excess heat, or whapping against their cage as the bearing comes apart. In every case, the entire fan motor assembly will have to be replaced.

Ditto for the cabin fan and fan cage, §2E, p.394.

❑ **Gurgling Sounds, Particularly While Cornering** – Heater-AC fan motors often fail as the direct result of being forced to run in water. The air conditioning system acts as a dehumidifier: Water in the air (humidity) condenses on its frosty coils. These droplets are supposed to flow down and out through a rubber drain tube that runs from the bottom of the evaporator box forward through the firewall, then downward a few inches toward the street. (The water that you see when you leave your car idling on a hot summer's day–AC on–passed through that tube.)

With increasing age, crud can accumulate to block this tubing, creating a backup within the system. This problem gives itself away by the sound of water being thrown around as you make a turn (only when the fan is running). That water is on its way to the bearing at the bottom of the heater fan motor. Unblock the drain as soon as possible.

---

1 Figure 15-3 (p.388) doesn't show them, but one or more shims lie between the central shaft of the compressor and the central shaft of the clutch plate.

❑ If you ever have the misfortune of experiencing the sound of an explosion emanating from beneath your hood, coupled with a cloud of oily gas (you're seeing the oil, not the freon), you'll know that your AC system's high side pressures went through the roof, resulting in either a melted thermal limiter plug (Fig. 15-8, p. 391) or a split high pressure line. Causes include:

- the expansion valve/orifice tube became plugged with contaminants
- the condenser cooling fan has failed
- the high-pressure cut-out switch has failed.

Both system corrosion (internal) or metal fragments from a disintegrating compressor can plug the expansion valve/orifice tube.

2E. Electrical Problems

Time was, AC systems had a fan motor switch, a temperature (cold-hot) lever, a high-pressure cut-off switch, and a thermal limiter plug to back-up that switch. We still have the fan speed switch and the high pressure cut-off switch, but most of us now enjoy climate control modules on which we set our temp of choice: Set it and forget it! . . . It's true that we have other buttons and dials we can play with, but there's a whole lot more going on behind the scenes that can render the entire system useless: air temperature sensors–one for evaporator output, one for the cabin, occasionally one on the dash, one for ambient (outside) air; then component sensors–some for temperature, some for pressure; oftentimes a freon flow sensor, and a computer or two to run the whole thing. Did I mention relays?

> Chapter 16 covers rudimentary electrical diagnostics. It won't make you a genius, but it might help you stay out of trouble.

Fortunately, most (all?) of these systems have self-diagnostic programs built into them. If you have an OBD-II scanner and a factory service manual (oe equivalent), you might be lucky enough to find how to extract those codes.

Still, I would always start with the basics: drive belt condition (§2, p. 444 and Fig 17-25, p. 461), compressor clutch(below), condenser fans (below), low-high gauge readings (§5A, p. 400), and fuses (§11, p. 422),before wandering into the weeds.

❑ **Compressor Clutch:** A defective compressor clutch can render a totally functional AC system completely useless. You'll find its electrical plug attached to the compressor, (Fig. 15-4, p. 389); it will have one wire or two. Disconnect the plug.

- One wire – When the AC is switched on, the incoming wire should light up a testlamp (§10, p. 421). If it does, then either the clutch coil is bad or its ground wire is faulty. Most find their ground on the compressor case itself, check that it's secure and unbroken.

- Two wires – One supplies power, the other goes to ground. The incoming power lead will light up a testlamp (§10, p. 421). Section 15, p. 429 describes the procedure for testing the ground circuit.

In both cases, there's a relay involved in supplying power to the clutch, (§8 & §9, pp. 418-421).

- One wire connectors – You can hot-wire the clutch to determine if it's functioning–engine *off*, ignition key *off*, clutch coil connector disconnected. Connect a jumper wire (*S*26A, p.440) to the coil's wire terminal. Tap the jumper's other end against the battery's positive terminal. The clutch plate should jump toward the pulley–not much more than 1/16″ or so–you'll hear a clack as the two slap together. Release the jumper and it will shift back to it's original position. If the clutch plate doesn't move, then either the coil is fried, it's not grounded, or the entire assembly is bound up, probably toast.

  > Note that I said "tap the jumper" – **not** "connect". It's conceivable (but not likely) that the coil has a short-circuit to ground. If so, that jumper will get hot instantaneously.

- Two wire connectors – You'll need two jumpers, one for the positive battery terminal, the other to any good ground. Set up the ground first; it doesn't matter which of the coil's terminals you choose as long as the coil's connector has been disconnected from the car's wiring harness. Once that jumper is secured at both ends (establishing a solid ground for the coil), connect the "hot" jumper to the other terminal inside that plug.

  > *Verify that the two jumpers cannot touch each other,* because when you tap the "hot" jumper to the battery's positive terminal, any accidental contact would produce a direct path to ground. At best, you'll burn your hand; at worst, you'll torch your car. Seriously.

  Tap the jumper's other end against the battery's positive terminal. The result should be identical to what's described in One wire connectors, directly above.

15

❑ **Condenser Cooling Fan(s):** In most cases, these should start spinning as soon as the AC is turned on. They are controlled by a relay that activates as soon as you press the AC switch. This relay can have any number of names–AC, fan, or condenser fan relay. Section 9, p.419 covers relay inspection.

Testing the fan follows the same procedures cited above under **Compressor Clutch**. The same Cautions apply as well.

A few systems switch the condenser fan(s) on intermittently–only when the condenser temperature reaches "too hot". Good luck with that . . . You'll need a wiring diagram and the skills to read it.

❑ **Cabin Fan Issues:**

- **Losing a Speed** – The fan circuit is composed of a fan speed switch, the heater-AC fan motor, and a fan resistor wired between the two. The setting of the speed switch (low, medium, high) determines the pathway for current flow through the resistor. The amount of resistance in that portion of the circuit determines how much current the fan motor sees, thereby defining how fast it will spin–the Low circuit passes through the highest resistance coil, the High circuit bypasses the resistor coil altogether. Systems that have lost a particular speed usually suffered a meltdown within the heater fan resistor (which always sits directly downwind from the fan) or at the electrical contacts within the switch. Since the low-speed setting runs current through the coil of highest resistance, the low-speed coil is the one most likely to burn out (higher resistance = more heat).

- **Not Blowing at All** – Power is applied either through a fan-motor fuse, relay, or fusible link, §19, p.435. Your service manual should contain the specifics. Check those first; you might get lucky.

  If you find a blown fuse, then discover that its replacement blows shortly or immediately after switching the system back on, you have a short circuit. The short could be located anywhere within the circuit. Possibilities include the fan speed switch, the heater-AC fan motor, the fan resistor, or the wiring between. See §19, p.435.

  To test the fan motor, first disconnect it from the rest of its circuit. Test it just like you would a condenser cooling fan (previous page).

  If the fan quit running following a protracted period of slower than normal spinning, it's likely that the motor was failing. This process is usually accompanied by noise, produced either by a bearing that's seizing up, or because of wear that has allowed the fan motor cage (the fan "blade" itself) to start beating itself against its adjoining housing. In either case, the smart money is on replacement of the motor and cage at the same time. Unfortunately, just replacing these two components will not necessarily cure all. A failing motor usually burns out one or more coils (circuits) through the fan resistor, eliminating one speed for each melted coil.

2F. Other Problems

❑ **Frost Exiting the Vents:** Either too much humidity is in the air or the evaporator is drowning. First, check your Recirc control. If it's off (set to Fresh), you've found your problem–assuming it's a humid day. Recirc doors do get jammed on occasion, usually by a faulty controller. Both reside behind the glovebox.

As for a drowning evaporator: Let the car idle on some dry pavement; you should start seeing a puddle of water forming beneath the car. The muggier the day, the bigger the puddle. No? Read the next paragraph and ❑ Gurgling Sounds, p.393.

❑ **Mold and Mildew:** Mold and mildew can flourish within the ventilation system, normally developing due to blockage within the drain tube that leads off the evaporator box, Fig.15-1 and ❑ Gurgling Sounds, p.393. To cure the problem, start by clearing the drain. Most mechanics just blow compressed air into the drain tube's outlet—a backflush of sorts. While this might solve the immediate problem of blockage, it drives the crud that blocked the drain right back into the evaporator box where it can trickle back down to repeat the process all over again. A better solution is to feed a flexible bottle brush (available in most aquarium stores) up through the drain hose.

To kill mold and mildew, first locate the ventilation system's fresh-recirc door, usually located directly behind the glove compartment. (Flip the fresh-recirc control lever back and forth: The slap-slapping of the recirc door should give away its placement.)

If opening the glove box door provides no access, look upward from beneath the dash.

Do you see a grille up there? That's the inlet for recirculated air. Slide the door to its recirc position. With the air conditioning up and running and the fan speed set to low, spray an aerosol can of unscented *Lysol* directly into the grille area for a period of 30 seconds or more. Now shut off both the car and the fan and let the spray do its work for at least a half-hour before firing up the system again. Repeat this procedure routinely to keep the problem from recurring.

If the problem started because of an open window on a rainy day, you've got another problem entirely. First remove any carpet mats, let them dry in the sun. Use a wet-dry vacuum cleaner to suck as much water out of the carpets as possible. Then towels. Ventilate as much as possible–open windows, AC on (using the floor vents). Pick up a can/bottle of mold killer and a tub of dessicant granules as soon as possible; use the first liberally on any surface that was wet as often as is necessary, leave the second in the car till the problem goes away. There's a great deal of padding (soundproofing) beneath the carpets and it soaks up water like a sponge, so go after the moisture hard and often.

3. Insufficient Heat to the Interior

Is your cabin fan is working? (p.395) If so, any of the following could be to blame.

- frozen heater control valve
- defective thermostat, allowing the engine to run too cool, or
- air pockets within the engine's cooling system, blocking the passage of coolant to the heater core

Where you start depends on whether the engine is hot or cold. If it's cold, start by inspecting the coolant level, §1B, p.95 and §3A, p.100. If it's hot, observe your temperature gauge with the engine running. Is it reading lower than it used to? If so, wait till the engine is stone cold, then inspect its coolant level. All good? Then replace the thermostat, §10, p.161.

Virtually every heater control valve used to be mounted on the engine side of the firewall. Easy to find–two coolant hoses and a cable gave that away–and easy to service. In a perfect world, that's what you'll find. For the rest of you, note where the heater hoses pass through the firewall. That's where you're headed . . .

Wherever it's located, observe it while someone else changes the temperature from cold to hot, then back. The heater control valve's lever should move freely through its full range, doing whatever its control cable or servo directs it to. No? Refer to Fig. 15-10, p.398.

- Cable – Disconnect the cable. Have your friend repeat his/her hot-cold switching. The cable will either move or it won't. To check the valve, simply move its lever. It'll move or it'll bind.

- Vacuum Servo – Disconnect the vacuum hose, put your thumb over the end of it. Have your friend repeat his/her hot-cold switching. The hose will either produce strong suction or it won't. Check the valve/servo with a brake bleeder/vacuum tool or a piece of hose and your cheeks. To distinguish between a binding valve or a leaky servo, disconnect the two and test each individually.

- Electric Servo – You'll need access to the factory service manual. Notice that the valve shown has two servos and three hose connections. That's to accommodate driver/passenger temperature control.

A binding valve might just need some lubrication. Ditto for a frozen cable. Vacuum tubing can be pinched, cracked, or disconnected.

Fig. 15-10. Heater control valves
Courtesy of Four Seasons

4. Freon Level, Visual Inspection

Back in the day, most air conditioning units had a sight glass located on top of the receiver-dryer bottle, Fig. 15-8, p. 391. Some accumulators did as well. Unfortunately, that's no longer true, but on cars that do, peering through its window allows you to easily distinguish between a clean system and one that's been contaminated: The latter clouds over the sight glass from the inside out, making it virtually impossible to see anything within. To assure yourself that what you are seeing is on the inside of the system, use some glass cleaner or, if necessary, a rag soaked in solvent to clean the exterior of the glass.

With the system running, you can evaluate the level of freon as well: A system that's low on freon displays a stream of air bubbles moving rapidly through the sight glass.

- Turn on the air conditioning and let it run for five minutes
- Turn off the AC (engine running) and let the freon "rest" for a minute. During this period, you should see liquid freon pouring past the glass, down into the chamber of the receiver-dryer. Once that process has ceased,
- Have a friend switch on the AC while you observe the sight glass.

In a system that's completely full, the freon that had collected within the dryer will rise up all at once and zip right past on its way to the outlet tube. Following that, you'll see nothing but clear glass. Systems possessed of a slight amount of air exhibit a stream of bubbles that climb more gradually out of the bottle, then disappear out the exit over a period of 10 to 30 seconds. You'll then see the bubbles cycling back through from time to time. Systems with more air than that display a continuous stream of bubbles. The more air in the system, the more dense that bubbling appears.

Consider an occasional cluster of bubbles or a very thin stream to be no big deal as long as they haven't affected the system's ability to cool the car. However, a more dense stream of air bubbles will definitely detract from that capability: While the car might eventually cool

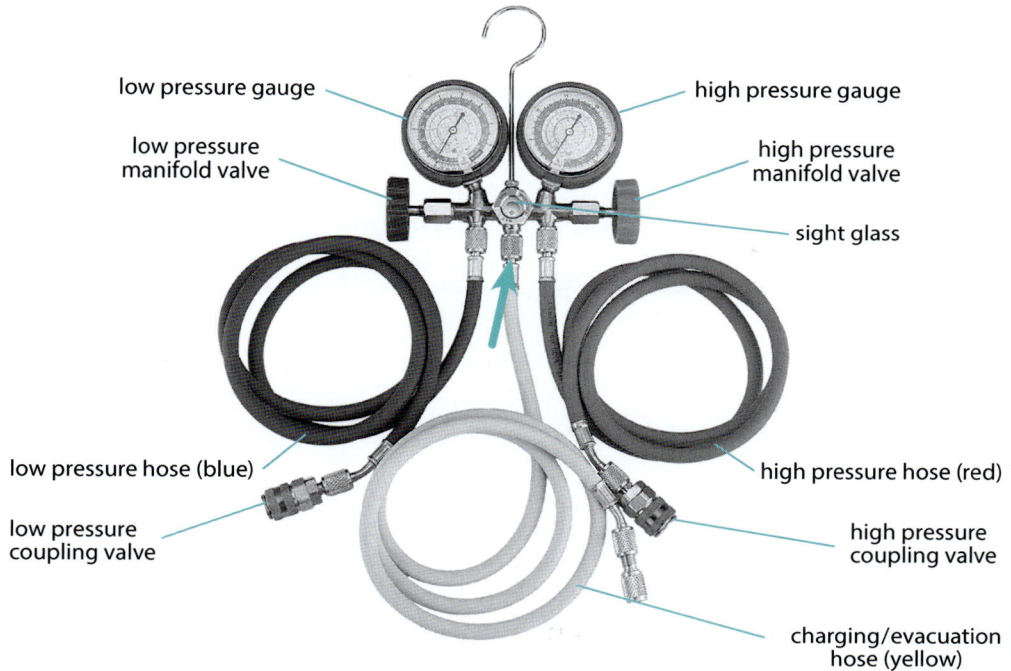

low pressure gauge

low pressure
manifold valve

high pressure gauge

high pressure
manifold valve

sight glass

low pressure hose (blue)

low pressure
coupling valve

high pressure hose (red)

high pressure
coupling valve

charging/evacuation
hose (yellow)

Fig. 15-11. R-134a gauge set
Courtesy of OrionMotorTech

down, the process takes a lot longer and requires a great deal more energy to accomplish. You need to weigh the cost to evacuate the system, repair the leak, and recharge it against the system's present ability to cool the car.

Note: A system that is just being fired up will commonly display some bubbles over the first minute or two. That's why I suggest letting the system warm to operating temperature.

5. Diagnosing and Repairing AC Systems– DIY

Most of you won't be blessed with a sight glass to peer into your system, *§4 above*. That's OK, because just looking at the freon won't fix anything. For that, you'll need a set of AC gauges, Fig. 15-11. Gauges are easy to connect: The valves used to couple them to the system can only go on one way–the high side valve is significantly larger than the the low side valve. Each valve has a slide collar; just retract it, place the valve, then let the collar snap back into place.

> But first! Verify that the manifold valves are closed!

5A. Evaluation

I said in *§1* that normal low side pressure should be in the vicinity of 30 psi at idle; high side pressure should be in the vicinity of 250 psi at idle. That's true in the Mid-Atlantic region on a humid, summer day. Those numbers vary somewhat with temperature.

Table 15-1. Pressure Variations with Temperature		
Engine at Idle	**Low Side**	**High Side**
Summer Day – 80°-90°	25-30 psi +/-	210-250 psi +/-
Hotter	30-40 psi +/-	235-285 psi +/-
Cooler	22-27 psi +/-	150-210 psi +/-

Don't get hung up on slight differences between your car's readings and values in the table above. If your AC is not performing properly, one or both of those gauges will indicate a wide divergence from the numbers cited above. For example, a low pressure reading that's hovering around 5-10 psi or a high side reading of 320 psi or higher signals trouble, particularly when coupled with warm air blowing through your vents.

When you rev the engine, the low side pressure will drop toward zero, the high side pressure will rise dramatically. The reason? The compressor spins in sync with the engine; the faster it turns, the higher its output. Pressures rise downstream because the freon still has to pass through that itty-bitty hole (the orifice tube/expansion valve). Stronger suction on the low-pressure side produces the opposite effect

Table 15–2. Abnormal Gauge Readings & What They Signify – *Engine at Idle*		
Low Side	**High Side**	**Likely Causes**
Low	Low	• low on freon, probable leakage
Low	High	• plugged / defective expansion valve, or • plugged orifice tube • some other blockage on the high side
High	High	system is overcharged, may have air in the system
High	Low	compressor failure
Normal	Normal	But Not Cooling – moisture in the system

❑ Low/Low: You want to find the leak before you evacuate the system, *§*2B, p.391.

❑ Low/High: An expansion valve is more likely to suffer blockage than an orifice tube. Two reasons:

- Expansion valves are mechanical devices with moving parts; orifice tubes are little more than a filter and a tube.

- Expansion valves are capable of creating a significantly smaller hole, making them more efficient but also easier to plug.

Blockage will always occur on the high-pressure side of the system. Two reasons:

- The interior diameter of the piping that connects the condenser to the expansion valve/orifice tube is dramatically smaller than everything on the low-pressure side, so any corrosion in the system will just tool around till there's enough of it to plug the smallest opening.

- Compressor failure begins with steel dust and graduates to progressively larger fragments, all travelling toward the condenser and high-pressure components beyond.

Because any blockage will simulate an orifice tube, there's a distinct pressure and temperature drop across the obstruction. So a temperature probe (Fig. 15-7, p.391) applied anywhere near the condenser outlet and anywhere near the high side of the orifice tube/expansion valve inlet will tell you whether the problem lies in the line (dramatic temperature drop) or beyond (roughly the same temperature in both places).

15

❑ High/High: You will never see these readings unless someone worked on the AC recently. The most common cause is adding freon to a system without first evacuating, drawing out the moisture, etc. See *§*2A, p.389. Every overcharged system will ultimately start leaking.

❑ High/Low: That's right; the low side is reading too high, the high side is reading too low. The reason is that the compressor is not creating sufficient pressure on the downstream side, which also means that it's not pulling freon out of the evaporator/accumulator particularly well.

Every system cuts in and out routinely, primarily to manage system pressure and cabin temperature. When that happens, the low and high gauges equalize somewhere in the middle. To distinguish between the compressor and fan clutch, verify that the fan clutch is engaging the compressor. If it is, the compressor is shot. If it isn't, you need to differentiate between a defective clutch assembly (*§*2E, p.394) or a non-existent system signal.

❑ Normal/Normal, but Blowing Hot Air: You'll need to evacuate the system, replace the receiver-dryer or accumulator, then pull a vacuum for an hour or so to remove any vestige of moisture from the system. Charge it following leak testing.

5B. Freon Removal

It's illegal to allow freon to escape to the atmosphere, *§*2C, p.392. Naturally, if a line blew, there'll be nothing in the system anyway; both gauges will be reading zero and you won't have to worry about our planet's future.

For the rest of you, if at all possible, find a shop that will capture it for you. (Odds are, that'll be an expensive visit for ten minutes work.) But with a bit of ingenuity and some brass fittings, an empty propane tank can be coupled to your charging/evacuation hose. Problem solved.

5C. Repairs

Every fitting has an o-ring seal. These have a distinctly different composition than both engine and fuel line seals. All AC-compatible seals are green. Every fitting that comes apart requires a brand-new seal, identical in size to what came out. Soak the seal in refrigerant oil–yup, you'll need a small bottle of that too–and coat the fittings' threads for good measure.

Some of these line fittings can be beastly hard to loosen. As you place your wrenches, orient them so that you can squeeze their free ends together with both hands (as opposed to pushing one wrench away from the other). Figure 15-12 illustrates the idea without the furrowed brow.Use line wrenches on smaller fittings to avoid rounding off their heads. For larger fittings, a decent open-end wrench is perfectly suitable.

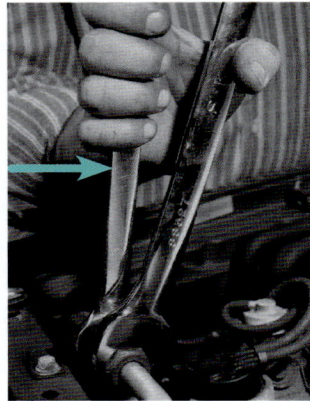

Fig. 15-12. Wrench placement for stubborn fittings. Stand facing the wrenches edge-on, as shown by the arrow.

Just because they were hard to get apart doesn't mean that you've got to crush them on the way back together. Lubricating those threads greatly reduces the amount of torque required to get a good seal. Snug them up, then apply moderate torque. Aligning your wrenches as in Fig. 15-12 works for tightening too.

Don't tip over a new compressor; it's full of oil. The same goes for accumulators.

Never recharge an older system or even a newer one that's been out of commission for awhile without replacing the receiver-dryer/accumulator. Because they both contain dessicant material, they should always be installed after everything else has been addressed.

Replace both Schrader valves on older cars, as well as on newer models that have been worked on frequently.

Fig. 15-13. AC vacuum pump
Courtesy of Zeny

5D. Evacuation

Proper procedure requires a vacuum pump capable of pulling a vacuum of minus 28-32 in-Hg
(the green arrow, Fig. 15-14). Both manifold valves should be wide open for this. Let the pump pull for five minutes, then close both valves before immediately shutting off the pump. Assuming a leak-free system and a decent set of gauges, the low-side needle shouldn't budge. I wouldn't be overly concerned with a slight drop (1 in-Hg, for example) within the first minute or so, as long as it stops there and remains steady for the next ten minutes. If it does, move on to the next paragraph. But if the vacuum gradually drops off toward zero, then you've got another leak in the system. Go to §5F, p. 405.

Fig. 15-14. AC gauges
Courtesy of Mastercool

15

Assuming all's right in your world, turn the pump back on and let it pull for another half hour or more to entirely dehumidify the system's innards. Double-check your work by repeating the waiting game–valves closed, pump off. If all is well, move on to §5E.

5E. Charging the System

Check the Specs pages in your service manual to determine the system's capacity. The same information can be found on a tag (usually yellow) in the engine compartment. Most of you will be using 12 oz cans, most manufacturers measure system capacity in pounds. There are 16 ounces to the pound.

I think it's always a good idea to use one can of dye-infused freon as insurance against a future leak. Use standard (clear) R-134a for the rest. To attach a can of freon to the charging hose, you'll need a coupler, Fig. 15-15. Since freon will exit the can as a freezing gas, eye protection is critical.

Verify that both manifold valves are closed. Remove the vacuum pump from the charging/evacuation hose.

The can with dye goes in first. To prevent freon loss as you thread coupler to can, open the coupler valve before you begin. Thread it onto the can, then onto the charging hose. Now close the coupler valve; doing so will puncture the can. Open the valve again; the charging line will fill with freon. But there's air in that line that must be removed: Crack open the charging line's connector directly below the manifold (green

Fig. 15-15. R-134a can tap/coupler
Courtesy of Mastercoo

> Freon is *always* added through the low-pressure side of the system. Opening the high-pressure side can cause the can to explode.

arrow, Fig. 15-11, p. 399). You'll be shutting it almost immediately as freon begins rushing out. (Some gauge sets offer a Schrader valve located centrally on the manifold for this purpose. Use it as you would a tire valve, but be advised: that gas is freezing cold!)

Open the low-pressure manifold valve.[2] You'll see both gauges climb as freon enters the system. Lay the can down someplace safe while you start the engine and turn on the AC.

The can will get quite cold as it empties, and lighter, of course. As it gets near empty, close the coupler valve, close the manifold valve, then swap in a new can. Because you've closed both sides of the charging hose, there's no need to purge air from it this time. Just open the coupler valve, open the *low-side* manifold valve, and continue with your charging.

I'll leave it to you to figure out exactly how much freon to add. The obsessive among you can bring a scale out to the car that can measure in ounces. Connect a full can to your gauges, weigh it, then subtract the number of ounces in the can. That's what an empty can would weigh. From there, you're on your own.

Or you could assume that most manufacturers have a pretty good idea how much freon actually comes out of one of those cans under normal circumstances, and engineered their system correspondingly, so that the number of pounds/ounces they specify as full capacity is pretty close to what the average bloke would get out of x number of full cans. But please don't trust me on that, because I've made a few assumptions in this life have gotten me into a *world* of trouble.

All of which gets us back to sight glasses–those within the AC system, not those provided in certain manifolds. For those of you that have them, once the bubbles stop coursing past the glass, the system is full.

For the impatient among you, revving the engine will hasten the process of emptying the cans. Body heat will also help–and it might just help cool you down . . .

Once the system is full, close the low-pressure manifold valve, close the coupler valve. Compare your gauge readings with Table 15-1, p. 400. Reread the first three paragraphs of §5A, p. 400 and see if you agree. Stick a thermometer into a central dashboard vent; expect a temperature in the 50°-60° range. All good?

Turn off the engine and disconnect the gauge set. Be advised, some freon will escape as those couplings are removed. No big deal if you haven't used dye, a potentially enormous deal if you have and intend to start tracking down a leak. If you have, wrap the connectors in a rag as you pull them off, get the gauges clear of the car, then mop up the mess. Clean those areas with brake cleaner, use clean rags for drying.

---

2   This assumes that the gauges have remained in place with both valves closed since the system was evacuated. If the gauges were removed for any reason, then both low and high side lines have air in them and must first be purged using the vacuum pump.

5F. Tracking a Leak

Ok, so maybe you're frustrated by now. There's a leak somewhere, but you have no clue where it might be. Start by purchasing a UV flashlight–the smaller the better–a pair of UV glasses, and a 12 ounce can of R-134a with dye. Oh, and a small mirror with telescoping handle, p.117. Charge the system, *5E,p.403.

> Ultraviolet light will damage unprotected eyes.
> *Always* wear **UV goggles in the presence of a UV light source.**

With the engine off *and your goggles on*, trace the system's high-pressure side from the compressor around to the firewall. The ultraviolet light will illuminate any leak as a green or purple glow. Big leaks are obvious; small leaks are subtle. Focus on the unions between components and piping to start with. As with any other leak, the trail will gravitate downward–that's where the mirror comes in. Since you can't see the space between the condenser and the radiator, any leakage on that side of the condenser will require removal of the splash pan beneath and patience for the dye to gravitate south.

Sometimes it takes a week or two before anything turns up.

Leaks occasionally spring up at the seam between a flex hose and its steel connector. In Fig. 15-2, the green line associated with the high pressure line points right at that spot. Since many of those unions are covered in a foam sheath, the dye might never see the light of day unless you peel it back.

A compressor front seal leak can be quite a challenge. You'll definitely need a mirror.

I've saved the worst for last . . . the evaporator, buried deep in the dash. Many (most?) manufacturers hide their expansion valves in there as well. The only way to infer that a leak is lurking there is to look for dye in the drain tube, accessed from beneath the car. Leakage over a considerable period of time will be easy to spot; newly sprouted leaks take time to form. The good news is that because the evaporator is a low pressure unit, it is far less likely to spring a leak.

15

Notes

Chapter 16. The Electrical System

16

Notes

The Electrical System

Starting Troubles–Including Battery Problems– are contained in Chapter 11. Engine Troubles: ∫4, pp.219-230.

Sluggish Cranking (p.219)

Jump Starting a Dead Battery (p.221)

Battery Testing (p.223)

Battery Servicing (p.224)

1. Fundamentals

The electricity required to power your car is supplied by two sources: the battery and the alternator. If the engine is not running, the battery is the sole provider, regardless of the position of your ignition key. But as the engine roars to life, so too does the alternator, supplementing the battery, raising the system's available voltage to approximately 14.6 volts rather than the battery's resting state of roughly 12 volts, and in fact restoring the battery to its fully-charged state over the next 5 to 45 minutes, depending on the battery's initial state of charge.

The cable that connects to the battery's positive terminal–the B+ cable–is connected to the:

- starter motor
- alternator
- junction block (fuse panel), engine compartment
- junction block, passenger's compartment Fig. 16-1 and Fig. 16-2, p.411

16

Every other electrical component in the car finds its current source in one or the other of those two junction blocks.

> If your code reader or scanner is connected to the car, disconnect it before doing any invertigations into B+ wiring, including any searches for open or short circuits, etc. this to ensure that you don't fry the scanner.

Of those four, the fattest wire by far leads directly to the starter. The cabling that leads to the alternator is significantly thinner, but still quite thick, a bit larger than what you might see on a heavy-duty kitchen appliance. The B+ wiring that feeds the junction blocks is of comparable size. These four cables are the functional equivalent of the battery's positive terminal and should be approached with considerable respect.

I read somewhere that Benjamin Franklin was the gentleman responsible for confusing mechanics across the globe for the past 130 years when he decided that electrons must originate from a battery's positive terminal. We have been dealing with that misconception ever since. And all instruments reflect that "truth" – at least in this part of the world – by labelling all batteries' positive terminals as if they were that source: the B+ terminal.

In fact, the electrons that populated your battery originally have long since found their way to the chassis, supplanted by electrons that originated in the car's chassis and travelled into the battery by way of the chassis ground. So the B+ terminal is actually supplying negative electrons to your car, but we view them as positive, and so do our meters.

2. The Starting System

Never attempt to replace the starter, alternator, or ignition switch without first disconnecting the battery's positive (B+) cable.[1] Never remove the back cover of either junction block without doing the same.

Never permit a wrench, screwdriver, or other metallic object to bridge the gap between one of these (exposed) terminals and an adjoining metal surface. Pay attention to the free end of any tool you might be using in their immediate vicinity. (Both the alternator and starter B+ terminals are encased in a rubber cap which protects against disaster. The same is true for the B+ terminal on the battery.)

Any metallic bridge between one of these and an adjoining metal surface will short out the battery to ground, creating an intensely hot spark with sufficient energy to melt (weld) that tool right into place. The most likely result is a severe burn to your hand. Other possibilities include a meltdown of the wiring harness, a fire, or explosion of the battery.[2]

- Following the reconnection of the battery's positive terminal, be certain to reinstall its rubber cover.

- For obvious reasons, never use the battery as a tool tray. To avoid the possibility of explosion, never smoke in the vicinity of a battery.

- If you intend to remove the battery, always disconnect the battery's negative cable first, then the positive, this to avoid the consequences of arcing. On reassembly, reconnect the positive terminal first, then the negative.

1 Doing so will delete any memory stored within the ECU, the radio, seats, etc. Higher-end scanners will save this information for you. Consider this if you're in the market for one.
2 I've only seen this once, but I have seen 6-10 melted wiring harnesses. Electrical fires? Several.

For a car to start, the ignition key must be turned to the Start position and the transmission must be in Park or Neutral. The brakes should be engaged. These three, in tandem, will open the circuit to the starter solenoid, which, in turn, should engage the starter, Fig. 16-3, p. 412. For problems with starting, see §4, pp. 219-230.

Cars with manual transmission have no neutral safety switch; in its place you will find the clutch neutral safety switch, identical in function, but requiring that the clutch pedal be moved all the way to the floor, thereby insuring that the transmission is disengaged.

to the starter

to the alternator circuit by way of a
fusible link, circuit breaker, or fuse

to the engine comparrtment fuse panel
(includes fusible links, circuit breakers,
relays and fuses)

to the interior fuse panel (near driver's knee) –
includes circuit breakers, relays and fuses

B+

battery

ground
to chassis

ground to
engine block

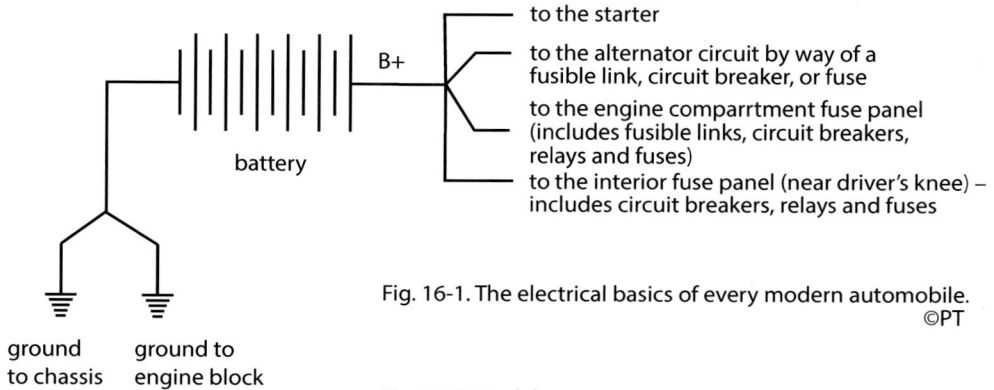

Fig. 16-1. The electrical basics of every modern automobile.
©PT

Fig. 16-2. To elaborate . . .

With regard to the relays, don't hold too much stock in what
I've drawn here, specifically:

Each relay has four or five distinct wires connecting it to the
circuit it's in, all of which are buried in the bottom of the fuse
panel. All of those relays share the same power source (B+)
and ground circuit.   ©PT

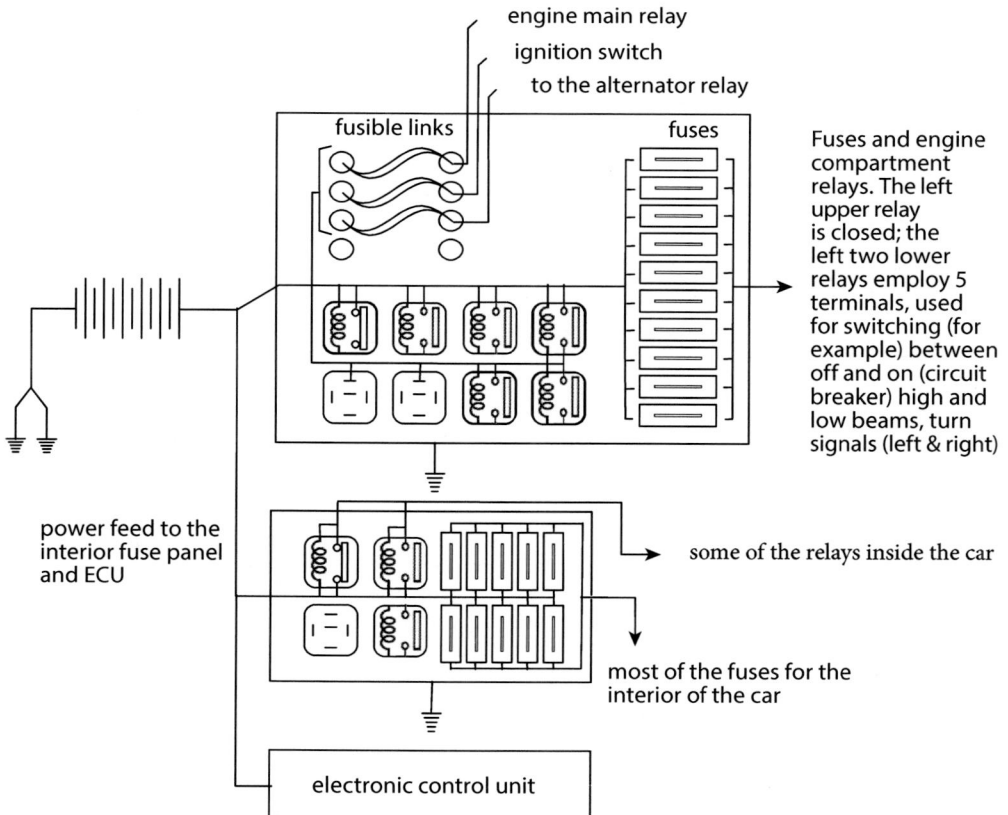

engine main relay

ignition switch

to the alternator relay

fusible links

fuses

16

Fuses and engine
compartment
relays. The left
upper relay
is closed; the
left two lower
relays employ 5
terminals, used
for switching (for
example) between
off and on (circuit
breaker) high and
low beams, turn
signals (left & right)

power feed to the
interior fuse panel
and ECU

some of the relays inside the car

most of the fuses for the
interior of the car

electronic control unit

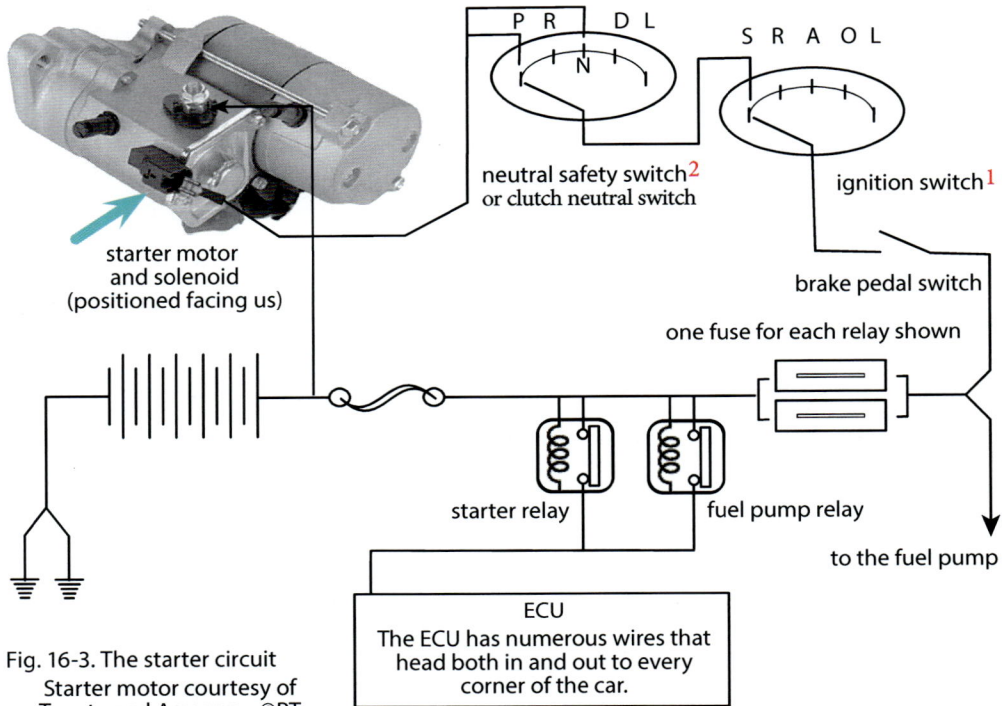

P R D L   S R A O L

N

neutral safety switch[2]
or clutch neutral switch   ignition switch[1]

starter motor
and solenoid
(positioned facing us)

brake pedal switch

one fuse for each relay shown

starter relay   fuel pump relay

to the fuel pump

ECU
The ECU has numerous wires that
head both in and out to every
corner of the car.

Fig. 16-3. The starter circuit
Starter motor courtesy of
Toyota and Amazon   ©PT

16

1 The ignition switch is shown in tbe Start postion. From the left, those poitions are commonly
listed as Start, Run, Accessories, Off, and Lock (Key Removed).

2 The neutral safety switch is shown in tbe Park position. From the left, those positions are commonly
listed as Park, Reverse, Neutral, Drive, and Low. Note that the starter can engage from both the Park
and Neutral Positions. Your foot must be on the brake to close the brake pedal switch.

To differentiate between the starter itself and the starter circuit:

- Nine times out of ten, the starter motor or the battery is the cause of your
  problem. Start with the battery, $\mathcal{S}$A2, p.223.

- **If the battery tests fine,** use a jumper wire to isolate just the starter and
  battery: The starter has two wires leading to it–the heavy B+ cabling from
  the battery's positive terminal (Don't touch!), and a thinner wire from the
  ignition switch by way of the neutral safety switch. Disconnect the thinner
  wire at the starter (as illustrated by the green arrow in Fig. 16-3) and connect
  your jumper to that (starter solenoid) terminal. You'll use the jumper to tap
  the battery's B+ terminal for a second or two. If the starter engages, then the
  problem must reside in the starter's cicuitry. To build a jumper, see $\mathcal{S}$25, p.016.

If the starter does not fire, then you'll be in the market for a new starter. If you wish to
test the wiring, read on . . .

- To test the circuitry: Connect the test lamp's ground clamp to any convenient ground on the chassis, near the starter. Put the probe from your 12V test lamp into the starter solenoid's now disconnected terminal end (the wire, not the solenoid itself). The lamp should not be lit. Now have an assistant turn the ignition key all the way over to the Start position. (Gearshift in Neutral or Park, or clutch pedal pressed to the floor).

- If the circuitry is sound, the test lamp will light, just as if power were being applied to the starter. The starter is defective.

- If the test lamp does not light, there's an open circuit between the battery and the end of that solenoid's wire. You'll need the service manual to diagnose this problem; there's a lot of wire with numerous connector plugs involved (Fig. 16-3) plus each of the components to investigate: ignition relay, ignition switch, neutral safety or clutch neutral switch, and the starter relay to name a few . . .

Before you begin that search, check to see if other systems are involved. With the ignition key turned to the On position, check the heater fan, headlights, wipers, etc. to determine whether the failure is limited to the starting system. If the problem encompasses more than just the starter, then the search proceeds from the battery's positive terminal through the fusible links and on to any individual relays or fuses that have been shown (by component failure) to be involved. Use a testlamp to search for power (§10, p.421) at each point along the way from the battery to the dead spot. As a general rule, the more things that don't work, the easier the problem will be to locate, simply because the open circuit (meltdown) will be closer to the battery.

With multiple systems involved, I'd be looking at terminal ends within the junction blocks that contain some or all of the failed systems. The problem smells suspiciously like a melted harness plug. You'll need to remove the back cover of the engine's junction block to gain access. See §23, pp.438.

3. The Alternator

Think of an alternator as an electric motor operating in reverse: Rather than plugging the motor into an electrical outlet in order to produce mechanical work, the automobile engine puts mechanical energy into the alternator to produce electrical power. The alternator is coupled to the engine by means of the serpentine drive belt (Fig.9-6, p.159); it begins to generate electricity as soon as it starts to rotate. Due to the much smaller diameter of the alternator's driven pulley relative to the crankshaft pulley (the drive pulley on the engine), it spins roughly five times faster than the speed of the engine.

Assuming a healthy alternator and a properly tensioned drive belt, the battery's reserves become virtually irrelevant to the electrical functioning of the car once the engine has been started–aside from backing up the alternator to meet demand when, for example, the air conditioning has just been turned on. An extreme example: If the battery has just been jump started, the air conditioning and any other accessories that might be on could overtax the alternator and cause the engine to stall.

16

If the alternator fails while you're driving, the charging system simply quits, the charge warning lamp should light up. The engine plus any accessories that happen to be on begin to pull their electrical needs from the battery's reserves. The charge warning lamp or gauge is designed to alert the driver to malfunctions within the system, and in most cases it will. Should that light come on, most of you have 30 miles or so before the engine begins to lose power (assuming an initially well-charged battery). A short while later the engine stalls, the battery stone dead. Naturally, those accessories that draw large amounts of current (air conditioning, heater fan, rear window defogger) greatly reduce a battery's range.

If your charge warning light comes on while driving, be sure to pull over within the first quarter mile to verify that your alternator has not thrown its drive belt. Since this belt also drives the water pump, you want to make absolutely sure that continued driving will not overheat the engine. This inspection should be performed with the engine off, so that the condition and tension of the belt can be appraised. But since turning off the engine might leave one in every fifty cars stranded (due to an already weakened battery), you might view this as a risk that's not worth taking. Even so, pull over: An inspection with the engine on is certainly better than no inspection at all. Consider the possibility of having to shut down anyway, three miles down the road with an overheated engine.

The charge warning lamp (battery discharge light) indicates a problem with the charging system. Possibilities include the:

- alternator
- its voltage regulator
- alternator drive belt (broken or very loose)
- charging system fuse, or
- charging system relay, or
- a melted fusible link
- a broken wire or melted terminal plug at the rear of the alternator (where the harness plugs into the alternator)

Because of the number of possibilities, the cost of each component and the reluctance of parts suppliers to accept returns on electrical parts, it should be clear that some testing is in your future. But you might want to cover your bases by inspecting the following first:

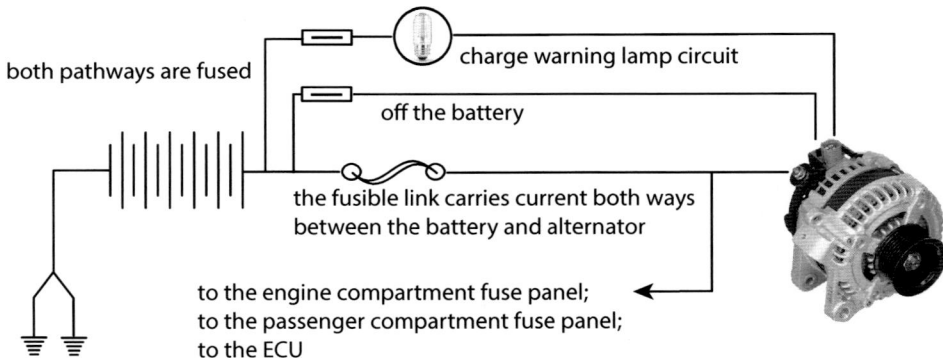

both pathways are fused

charge warning lamp circuit

off the battery

the fusible link carries current both ways between the battery and alternator

to the engine compartment fuse panel;
to the passenger compartment fuse panel;
to the ECU

Fig. 16-4. The alternator circuit
Alternator courtesy of
Toyota andAmazon   ©PT

- Look at your charge fuse. Better yet, replace it. In a pinch you could use a less essential fuse of equivalent amperage.
- Check that your alternator belt is in place and properly tensioned, engine off. See *§2, p.444*.

Finally, save the cost of calling in the hook by addressing the problem before it leaves you stranded. Remember that at best your range will be 30 miles with no accessories turned on.

There is a set of electrical contacts (carbon brushes) inside the alternator that are designed to slide in toward the rotor as they wear. Toward the end of their lives (80K and up) when they're getting quite thin, the brushes occasionally get stuck cockeyed in their holder and Bingo!–the warning light goes on. If you're feeling lucky, a light smack delivered centrally to the back side of the alternator's case will occasionally bring it roaring back to life. Unfortunately, the cure is usually just temporary.

> That thick wire coming off the alternator's case leads straight to the battery's B+ (positive) terminal. It must never come in contact with any adjacent metal parts. Accidentally doing so can cause a meltdown in the wiring harness. See Caution, p.410.

Certain kinds of charging system malfunctions don't trigger the warning lamp. Furthermore, since many owners seldom check their gauges, their first awareness of the problem often comes when the car won't start. A jump start (Chapter 11, p.221) may enable you to drive the car to the shop, simply because running the engine takes very little current–it's the starting that makes the demands.

16

How will your mechanic tell whether the battery or charging system is at fault? First, applying a quick charge to the battery will enable him to start and run the engine. A brief glance at a voltmeter connected across the battery's two posts then tells the story–readings of 13.8 –14.3 volts indicate a sound charging system pumping surplus current toward the battery; a problem with the charging system produces readings of 12 volts or less (the voltage drop exhibited by the battery itself). While there are a number of more sophisticated tests that can help to isolate the offending component, this "quick and dirty" approach will always distinguish the battery from the charging system. Any voltmeter will suffice; a battery load tester or a multimeter is usually the closest one at hand.

Note that a weak alternator or failing voltage regulator[1] can produce just a trickle of current heading toward the battery. In certain cases, this output is sufficient to prevent the charge warning light from switching on, but not sufficient to carry the car's demands over the long haul. This scenario produces voltmeter readings closer to 12.5 volts rather than the 13.8 –14.3 volts one might ordinarily expect.

Note also that the exact opposite can be true, particularly on those cars with a hybrid set-up–aftermarket alternator and factory voltage regulator, or vice versa–the charge warning lamp can be on dimly and yet the battery never goes dead.

---

1 The voltage regulator is mounted to the rear face of the alternator. It usually includes the plug into which all of the alternator wiring is connected. A new one comes installed on virtually all new/replacement alternators.

Be certain that a mechanic well-versed in testing has been consulted before you authorize any significant amount of parts swapping.

Should you need to replace an alternator or other charging system component, be absolutely certain that the shop will be using an OEM part. There is so much junk on the market, a large percentage of which is marketed to fit any number of cars by employing an absurd little adapter harness (a plug, some wiring, another plug) so that your car's wiring will connect electrically to their unit. The likelihood of loose or broken wiring is more than doubled, harness plugs can end up laying against exhaust manifolds, dipsticks, etc.–not to mention the fires and meltdowns along the wiring harness that can result from the wrong connector being employed–all these possibilities increase exponentially when the aftermarket boys start modifying your car. The OEM replacement should look identical to the original. Factory remanufactured components are a suitable alternative, saving money while providing essentially the same level of quality.

> Never remove an alternator without first disconnecting the battery, See Caution, p.410.

Some shops prefer rebuilding alternators rather than replacing them, their profit margin is much higher if your bill includes their labor to disassemble and rebuild rather than simply swapping one part for another. In all fairness, because some alternators are so outrageously priced, rebuilding may be an attractive alternative. If your shop recommends rebuilding, first ask them why, then request a quote both ways.

4. When You Park and Walk Away

Whenever the ignition switch is turned off, the battery takes over and tends to all of the following, where relevant:

- Memory circuits within your radio that recall your favorite stations
- Clock
- Memory circuits within certain on-board computers/controllers; for example, the EFI (electronic fuel injection) or emission-control computers.
- Any relays that remain on when the ignition key is removed, such as the one that controls those radiator cooling fans that continue to run on hot days, as well as the cooling fans themselves (while they're running). Another example: Anti-theft devices–door and window relays, as well as their controller.
- Any light circuits that remain active, such as the brake lights, parking lights, and hazard flashers.
- Any aftermarket installations that were wired directly to the battery (hopefully through some kind of fuse holder), such as a radio, fog lamps, etc.

16

5. V=IR: Ohm's Law

Voltage = Current x Resistance: An Analogy

Consider a mountain stream. The speed of the water's flow–its energy–is directly related to the steepness of the grade over which it flows. Its velocity (energy), therefore, depends upon the height through which it falls. In an electric circuit, voltage is the measure of energy, the battery and alternator are its providers.

The amount of water flowing in the stream can be likened to the flow of current through the circuit–the I of V= IR. All measures of current are linked quite literally to the number of electrons flowing through the wire.

The course that the water takes on its trip down the mountain is the path of least resistance. When water confronts a fork, the greatest volume shoots down the widest path to the valley, lesser amounts trace the more torturous routes. This simple observation explains how resistance to flow defines the inverse relationship between the I and R in the equation above.

Finally, when the stream joins the river meandering through the valley below (chassis ground), its pace slows to the point of becoming almost imperceptible.

My apologies for dragging you away from the pastoral scene above to the mundane world of household plumbing, but if you were to substitute piping of different diameters for your mountain stream, you'll get a much better feel for resistance, particularly as it relates to lighting: The filament within any bulb is very, very thin, and so possesses high resistance. If that filament had low resistance, you wouldn't have any light!

One final analogy: Within a house you can have precisely regulated water pressure; likewise, the automotive electrical system has precisely regulated voltage (and resistance, and current).

16

6. Some Basics of Automotive Circuits

The battery's B+ (positive) terminal, the B+ terminal on the starter, and the B+ terminal on the alternator are all joined continuously, and together they feed the ignition switch, the fuse blocks, and other major players in the car through some very heavy wire, otherwise known as the B+ wiring (battery positive). You will never find this wiring in contact with the body or frame of the car. For that matter you will never find it contacting anything other than an insulated switch, fuse, circuit breaker, or relay; because if it did, the amount of unresisted current flowing through that wire would be sufficient to melt the wiring harness and possibly cause a major fire.

So current flows from the alternator to a particular fuse or relay, from there to a switch, then on to a bulb, wiper motor, or what have you; and from there to the body of the car. Why don't you have a meltdown through that pathway? A physicist would point to the equation V= IR to begin her explanation. My own view is that vehicle electronics are very carefully calculated; that the amount of current required by a particular pathway is all that will find its way down that wire.

Another of those immutable laws of physics states that current will never flow from one place to another without having a pathway back; in short, you must have a closed loop–a completed circuit–before current will flow. How does current get from, say, the brake light bulbs in the rear of the car back to the alternator/battery? The engineers say that the chassis of the car provides the return half of the circuit. We mechanics tend to view the chassis as the end of the line, and refer to it as body ground. Technically, the engineers are correct: You'll notice that your battery's negative terminal is connected directly to the body of the car, thereby completing a closed loop with anything that's grounded on the car. But practically speaking, we mechanics have good reason to view the chassis as the end of the line, because *a test lamp does not indicate the flow of any current on the ground (chassis) side of an electrical device*, whether that be a bulb, motor, or the battery for that matter. More on this later.

7. Open Circuits, Short Circuits

These are the two gremlins that haunt all electrical problems. An open circuit means that there is an interruption to the flow of current: a blown fuse, a blown relay, a broken wire, a loose or dirty terminal, a plug left disconnected, a meltdown within a bulb–these are all examples of open circuits that produce electrical failures. A switch turned to the Off position defines a normal circumstance in which you'll find an open circuit.

> See Caution, p.410 regarding safety issues relating to battery work.

There's no such thing as a normal short circuit–all shorts result from a circuit finding some alternate path to ground, either through a wire being pinched in a door jamb–typically the driver's door (window controls, door locks)–or cut in a collision, or having its insulation gradually rubbed away with the vibrations of the car. When this results in an exposed wire actually contacting a metallic portion of the chassis, it produces a pathway having no electrical resistance (*as long as the break occurs on the "hot" or B+ side of the electrical device being powered*). The flow of current can be intense, with wiring turning red hot. Wires in the surrounding harness, plastic panels, even paint can blister and melt. That's why every circuit in your car is fused in some fashion–to cut off the flow of current as soon as practicable in the event of a short.

8. Relays

A relay is nothing more than a switch, either permitting current to flow or shutting off the flow, in response to an electromagnetic current. The "pins" – actually blades – on automotive relays are traditionally labeled using the numbers 85,86,87, 87a, and 30. The numbers 85 and 86 are always connected to either side of the electromagnet, the numbers 87 and 87a are opposing terminals, one of which is always an open circuit, the other is always closed, depending upon the way the relay is wired. The 30 terminal is the switch between those two and is always wired to the B+ terminal.[2] Two examples follow; wiring diagrams are shown below:

---

[2] More recently, those numbers have been replaced by 1,2,4,5, and 3, as shown on the relay in Fig.16-7, p. 420. Numbers 1 & 2 are now connected to either side of the electromagnet, the numbers 4 & 5 represent the opposing terminals, and 3 is the switch, always connected to the B+ terminal.

electric coil

contacts

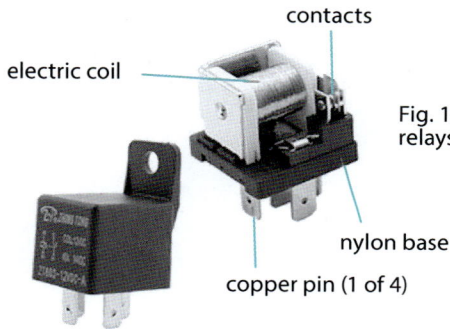

Fig. 16-5a & 16-5b. 4 and 5 pin relays

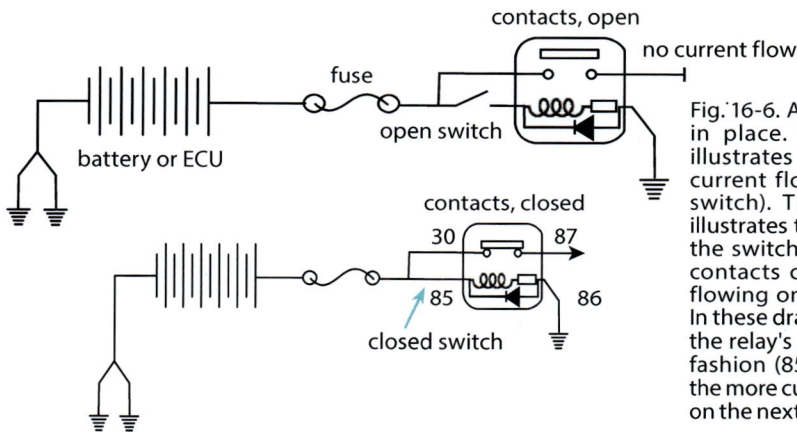

nylon base

copper pin (1 of 4)

contacts, open

no current flow

fuse

open switch

battery or ECU

contacts, closed

30   87

85   86

closed switch

Fig. 16-6. A 4-post relay, wired in place. The top diagram illustrates an open relay (no current flowing through the switch). The lower diagram illustrates the same relay with the switch closed, the relay's contacts closed and current flowing on to its destination. In these drawings, I've labelled the relay's pins in "old school" fashion (85/86/etc), I've used the more current nomenclature on the next page, Fig.16-7.. ©PT

16

The symbols on the side of each of the relays pictured in Fig. 16-6 are the copper coil windings (⌇⌇⌇), a resistor (⊏⊐), the path to ground, and doubling back, just past the resistor, is a secondary pathway back to the beginning of the loop. That odd looking triangle is representative of a diode (─▶⊢).

Diodes allow current to flow in only one direction – in the direction of the arrow. The purpose of all that hardware is to protect the ECU and other computers from high-voltage surges when the relay switches from one mode to the other. The wiring diagrams shown indicate just a battery and fuse driving the circuit, but in real life, that "battery" is much more likely to be the ECU commanding that some device be switched on or off. That loop allows current to continue to flow, gradually diminishing to nothing.

9. Testing Relays

You'll be using both an ohmmeter and a voltmeter. Remember that when testing ohms, the device being tested must be disconnected (removed) from the circuit.

- Four Pin Relays: The coil circuit, pins 85 and 86 {1&2} should show a resistance of roughly 60-100 Ω – this from the resistance of the coil plus the resistor.

Next,you'll need two jumper wires and a 12 volt battery source. You can use your car battery, but those jumpers need to be long enough that your testing is done away from the car. Never let the positive and negative leads come in contact with each other! See the cautionary note on p.410. A safer approach is to use a fully charged battery from your electric drill, saw, or lawnmower.

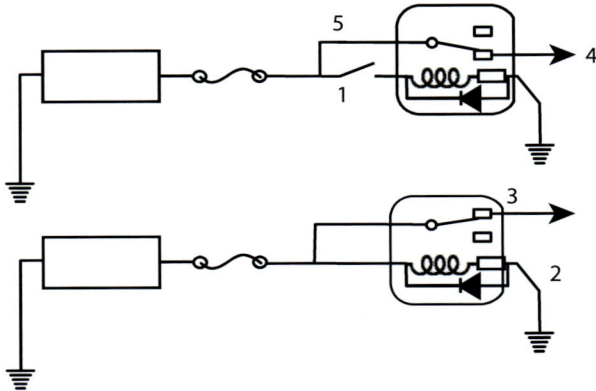

Fig.˙16-7. A 5-post relay, wired in place. The top diagram illustrates an open relay (current flowing through terminal 4). The lower diagram illustrates the same relay with the switch closed (current flowing through terminal 3).   ©PT

Apply current across terminals 85 and 86 {terminals 1&2}. You should hear the relay click. Disconnect it, it will click again, etc. If it doesn't click, you need a new relay. If it does click, attach your ohmmeter across terminals 30 and 87 {terminals 5 & 3}. You should see infinite resistance (pins 5 & 3 are not in contact, just like the top drawing in Fig.16-6). Leave the ohmmeter in place, and apply current once more to terminals 85 and 86. The relay should click and the ohmmeter should read zero resistance (pins 30 and 87 {5 & 3} are now joined across the relay's internal contacts).

Taking it one step further, use your testlight to verify that the terminal marked 30 {5} in the relay block shines brightly when grounded. Assuming it does, then the relay is good. If you run the power coming from terminal 30 {5} into the terminal labelled 87 {3} (within the fuse block), then the lights, or horn, or AC compressor clutch, whatever, should do what its supposed to do. All that leaves is the circuitry between the driver and that relay.

- Five Pin Relays: The coil circuit, pins 85 and 86 {1&2} should show a resistance of roughly 60-100 Ω – this from the resistance of the coil plus the resistor.

Unlike the four pin relays, which switch a single device on and off, the five pin relays are used in either/or situations – circuit breakers, high beams or low, fast or slow wipers, rear window defogger, etc. So the geometry of the terminlas 30, 87, and 87a {5,3, and 4} is different. In most cases, the terminal 87a {4} is connected to terminal 30 in its resting state (no current through the coils 85 and 86 {1 and 2}). When the relay is energized, terminal 30 switches from 87a to 87 (terminal 5 switches from pin 3 to 4 or vice versa, depending upon how it's wired.) You need the wiring diagram to know with any certainty.

So . . . Either terminal 30 is connected to 87a when the coil is at rest (no current flow) or to 87, *but not both*. Just like the 4 pin relays, the 5 pin relay should click when terminals 85 and 86 {1&2} are energized by a battery. But unlike the 4 pin relay, the switch is always on. Powering up the relay should do nothing more than flip the relay's contacts from 87a to 87, or vice-versa depending upon how the relay is wired. You need the wiring diagram to know with any certainty.

The last paragraph addressing the 4 pin relay is just as applicable here, as long as one-half of whatever that relay is supposed to be doing is doing exactly that with the relay at rest (no power). You need the wiring diagram to know with any certainty.

10. Testing Equipment

The most basic and the most commonly used tool is the 12 volt testlamp. To use it:

Connect the testlamp's alligator clip to ground (virtually any unpainted bolt head will do as long as it threads into steel), then touch its probe to any terminal or wire end. The testlamp will tell you unequivocally whether that point is powered: Its bulb will only light in the presence of 9–12 volts.[3] If that point is powered, current flows through the probe, through the bulb, and on to ground through the alligator clip.

Next, and just as important is the volt-ohmmeter (VOM). This must be a digital unit; and many have a lot of bells and whistles: auto-ranging, meaning you needn't worry about blowing up the meter by being on the wrong scale; audio signaling, also nice because you needn't look away from your work (some of the time), etc. Some of these devices can get pricy, neither of these features are necessary, but units that include both start at about $35.

First, the ohmmeter:

The VOM is switched to ohms, its two leads are used to test across a portion of the circuit being examined. A reading of Infinity means that the two terminals that you have chosen are not connected to one another (This could be totally expected, or it could mean an open circuit between your two leads.). More likely, you'll see a certain expected resistance (as defined by your shop manual).

> Anything you test with an ohmmeter must be disconnected from its circuit. There is a 9v battery inside the meter that supplies the current necessary for every test. Adding more voltage produces erroneous results and can potentially produce a blown meter.

Next: A voltmeter does not actually test voltage, rather, it will tell you the voltage differential between the two spots you're investigating. All testing with a voltmeter requires that the circuit under investigation be powered, typically by the battery–assuming that it's not dead.

---

3   9V will only barely illuminate the testlamp. For comparison's sake, connect the testlamp's ground to the car's battery ground, touch the probe to the B+ terminal. Another way of doing this: Perform the same test across a standard 9V battery.

- Attaching the black lead to ground and the red lead anywhere along a circuit will give you an absolute voltage reading (differential) between that point and body ground–typically 12 volts. The switch that controls that circuit must first be turned on.

- Attaching both leads within the circuit will probably not tell you much unless there's a problem. Most tests within the same circuit, particularly when taken without much in between, will typically present readings near zero, because there is no appreciable voltage drop between the points you have chosen. Resistors and coils will affect your readings somewhat, a diode will have no effect if it's pointing the right way and will result in no current passage if pointed the opposite way.

  But an open circuit between the two points you're measuring will provide you with a reading of 12v, because on the upstream side, you've got the battery, on the downstream side there is no current flowing and therefore, no voltage (V=IR). To do this right, you need a shop manual and accompanying wiring diagrams!)

- An ohmmeter reading across those same two points (with no voltage supplied to the circuit – the switch for that load first being turned off) will show Infinity.

The voltmeter is used on its DC (direct current) setting for all circumstances related to the automobile with one exception: All rotational sensors require that you switch over to the voltmeter's AC (alternating current) setting. Essentially, what that means is that diagnosing every wheel speed sensor requires the AC setting.

16

If you need to test a lot of alternators and starter motors, you should probably invest in a digital clamp meter, but these can be pricey and if you only encounter one or two problems a year, a battery load tester probably makes more sense.

11. Types of Fuses

There are three principal varieties of fusing, illustrated in Fig. 16-8:

- Your garden-variety fuses, those plastic spades or glass cylinders residing in the fuse panel. These panels are located either in the engine compartment or in the dash panel area, driver's side. Your owner's manual lists these fuses, their location, and most of the circuits they feed. The replacement of either type of fuse can be child's play, even for the technically inept. (Here's a hint: A fuse puller or needle-nosed pliers can help removal tremendously, as long as you don't accidentally short out one of the "hot" terminals within the fuse block by using an uninsulated pair of pliers improperly. Position the pliers with care so that an errant handle won't allow contact between both the terminal and the chassis concurrently.)

- Recycling circuit breakers are customarily used in circuits powering electric motors. Examples include heater-AC fan motors, radiator cooling fans, power windows, wiper motors, the AC clutch coil, etc. These circuits are more inclined to produce transient spikes of current that would blow an ordinary fuse. Since the circuit breaker can be reset, it provides a useful solution to this environment. Circuit breakers are usually hidden behind the kick and dash panels, I suppose to provide my trade with a steady stream of work. Even so, most owners' manuals give their location as well as the procedure for resetting them. Should you be bold enough to search them out, most of you will find that the task is quite simple. Should the breaker blow again, you'll need to start testing or unpluggine things to find out why.

- Fusible links: If something really ugly is in process on your car, such as a short circuit in the B+ wiring, or a battery jump-start that's being performed backwards (the positive terminal of one car mistakenly connected to the negative of the other, and vice versa), then that fusible link should fry like bacon, thereby preventing the car from going up in flames. Some cars have only one fusible link, most have two or three. These links can be formed as part of the positive terminal, as part of the wiring harness adjacent to the

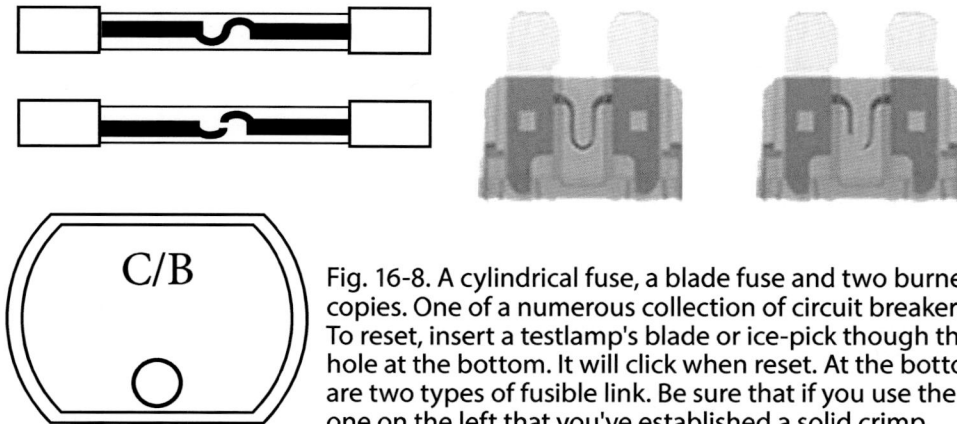

Fig. 16-8. A cylindrical fuse, a blade fuse and two burned copies. One of a numerous collection of circuit breakers. To reset, insert a testlamp's blade or ice-pick though the hole at the bottom. It will click when reset. At the bottom are two types of fusible link. Be sure that if you use the one on the left that you've established a solid crimp. D & E courtesy of Dorman and Bussman

16

battery, or they can be mounted into a fuse block nearby. Some are easy to replace, some are not (I'm thinking of those actually formed into the harness).

I've seen any number of "repairs" to the harness style of melted link, in which a piece of fusible wire is simply inserted between the battery's positive terminal and what's left of the harness, with A just taped into B. This can produce a union of very high resistance–read heat. Any time this type of fusible link repair is required, get a first-stringer involved so you won't have to buy yourself a new car!

Note that with all of the above fuse types, a short circuit in the harness causes the fuse to melt, creating an open circuit, thereby stopping the flow of current immediately. Fuses and fusible links melt in response to an excessively large flow of current, circuit breakers "trip" (switch off).

Note too that there is one wire on every car that is not fused; namely, the heavy cabling that joins the battery's positive terminal to the starter motor. The amount of current that flows down that cable when the starter is cranking is simply too great to permit fusing. Should this book embolden you to someday consider replacing your own starter, you cannot afford to forget that the B+ battery terminal must be disconnected before unbolting the starter cable at the starter. The same holds true for the alternator, but for a slightly different reason: Although its B+ terminal is fused, there's no sense in melting it down just to prove that you've cut one corner too many . . .

12. Inspection of Fuses

The fuse block receives its power from two sources; either:

- B+ wiring, straight from the battery: This line feeds circuits that receive power regardless of ignition switch position, such as the hazard warning and brake lights.
- The ignition switch–specifically its On and Accessory terminals–feed the rest.

There are three ways to inspect a fuse:

- Visually
- By replacement with a fuse that's known to be good
- With a test light

Of these methods, only the third will tell you whether the fuse is receiving power and whether or not that power is being transmitted across the fuse. Use of a testlamp will demonstrate that only a few fuses are powered with the ignition key in the Off position; the rest light up once the key has been turned to either the Accessory or On position.

A visual inspection won't necessarily tell you if a fuse is any good, because in most cases, fuse failure results from a hairline fracture. They can be extremely hard to see. This type of deterioration results from the repetitive heating and cooling that takes place over years of current flow. These variations in temperature distort the fusible metal and leave it brittle enough to crack, particularly in an environment so prone to vibration.

If you have an open circuit, a visual inspection is perfectly adequate: You can even see a sharp blue flash as a new fuse is installed and hear the "pop" as it blows. (This assumes that the relevant switch or relay is turned on so that current is ready to move into that circuit.)

Replacing the fuse in question (fuse A, circuit A) with one from a circuit that you know is working (circuit B) can eliminate the problem of missing a hard-to-see hairline fracture, but it won't tell you much more than that if it fails to fix the problem. Clearly, you will need to recheck your replacement fuse (fuse B) back in its original location (circuit B) should it fail to repair the problem in circuit A. If it will no longer power circuit B, then it suggests the presence of a short in circuit A. It could also mean that it, too, was an old fuse that was simply not up to the task of being disturbed. Whenever possible, use a new fuse *of equivalent amperage*; that way, you won't cloud your results.

A testlight is by far the best alternative. By connecting its alligator clip to ground–virtually any unpainted bolt head will do as long as it threads into steel–and then touching its probe to a terminal within the fuse block, the testlamp will tell you unequivocally whether that point is powered: Its bulb only lights in the presence of 9–12 volts.[4] If that point is powered, current will flow through the probe, through its bulb, and on to ground through the alligator clip. If the testlight fails to illuminate, try other terminals. If none cause it to light, you either have:

- A poor ground connection
- The ignition key has not been turned on
- Your battery is dead (nothing on the car will be working), or
- The testlight is defective. Investigate by attaching its ground clip to the battery's negative terminal, then touch its probe to the battery's positive terminal. The bulb should light strongly. If it does not, either the battery is stone-dead or the testlight is defective.

Meanwhile: If ithe fuse good, both of its terminals (both sides of the fuse) will light.

A blown fuse is indicated whenever one of its two terminals lights up while the other one won't. The terminal that lights is the powered side of the fuse panel (the battery or ignition switch side), the other is the fused side that leads on toward the circuit being protected.

13. Bulb Replacement

Exterior bulbs are typically accessed in one of two ways: either by removing the screws that secure an exterior lens cover, or by extracting the bulb's socket from its mounting panel inside the trunk (or fender). Most owners' manuals are fairly clear on these procedures.

If an exterior lens requires removal, a phillips-head screwdriver is usually needed. Don't misplace the screws! They can be very hard to replace because they come in strange lengths and are made of stainless steel to prevent their rusting into place.

---

4   See Footnote 3.

Hardware stores and auto supply houses are a good source for replacement bulbs. Since most cars use as many as four different kinds of exterior bulbs–not to mention the oddball sizes lurking behind the dash–you would be wise to take the originals with you on your shopping spree to avoid a second trip.

a = 1157

There are several different types of tail lamp bulbs that share the same base dimensions and overall size, but they are definitely not interchangeable. For example, the 1157 bulb is a dual-element bulb–one circuit being owned by your tail lights, the other by your brake lights, Fig. 16-9a. The 1156 is a single-element bulb, used in sockets that perform only one function such as the turn signals or back-up lights. The most common problems in the tail and brake light circuits involve the careless installation of the wrong bulb: An 1156 bulb will actually bridge the gap between the terminals in an 1157 socket, creating a direct short across these two circuits whenever either the tail lights or brake lights are energized. See Fig. 16-9b.

b = 1156

Note that the 1157 bulb installs nicely in one position, but it can be forced into the socket exactly 180 degrees (1/2 turn) out. The element within the bulb that is designed for the brake light circuit shines much more brightly than the tail light element so that brake light application will catch other drivers' attention. Jam it in backwards and you get the opposite effec

c = 168

Both of these bulbs are removed by first pushing them slightly downward into their socket, rotating counterclockwise, then pulling straight out. Examine the socket. Does its spring-loaded base move freely up and down? Is it corroded? If so, use an aerosol can of electrical contact cleaner and a bottle brush to clean it. If corrosion has really trashed the socket, it will require replacement. Very few manufacturers provide replacement sockets, opting instead for selling whole lamp assemblies at prices suited to the Ritz. What can you do? Either head for a junkyard or a friendly body shop–the smart ones save wiring harnesses out of selected total wrecks.

Bulb (c) is most commonly found in side-marker assemblies. These just pull straight out.

d = 3157

Fortunately, someone developed the 3156 and 3157 bulbs (d), which have largely supplanting the 1156 and 1157 bulbs. Their replacement is hard to screw up, but not impossible. Corrosion is still an issue, however.

Fig. 16-9
Courtesy of
Sylvania

The opaque lens covering the interior courtesy lamp usually pops right off with a small-bladed screwdriver. The bulb drops out by moving either one of its spring-loaded end terminals outward. These bulbs

16

come in a number of different sizes, most of which look the same, so have yours with you when you go to the store. Prior to reinstallation, turn off the light assembly–these bulbs can be blindingly bright–and HOT–with their cover removed.

Replacement of bulbs in the instrument cluster usually requires that at least a portion of the dash be disassembled. Be careful: The screws and clips that secure dash panels are well-concealed, and plastic loves to crack.

14. A Short in the Tail Lamp Circuit
– Troubleshooting a Short Circuit Using a Testlamp

Let's assume that you don't yet own a testlamp. And that your dash panel and tail light bulbs don't light. A quick check of your owner's manual reveals that the dash and tail lights share the same fuse. (That's true on most cars by the way.)

- Install a comparable wattage fuse in place of the tail light fuse with all of the lights turned off. If you have a rheostat to vary the intensity of the dash lights, then turn it off too, usually by rotating its knob counterclockwise.

- Now turn on your tail lights–just one click of the headlight switch. If your side marker lights and parking lights (tail lights) are on, then the fuse didn't blow. Turn up the dash panel lights using the rheostat. If they come on, verify that the side markers remained on. If all those bulbs are still working:

- Continue your investigation by turning on the headlights. (One more click of the headlamp switch) . . . If all is well even now, then either:

  – You had an old fuse and that's the end of it; or

  – You have an intermittent problem, probably with one of the sockets or with a failing bulb.

  – You have an intermittent problem that might require brake application, switching on your turn signals (first to one side, then to the other), going over a bump, etc. If so, you'll need to invest in a box or two of the correct amperage fuses, and be precise in your observations of what transpires just before the fuse blows next time.

It's conceivable that this circuit consists of more than just the tail lights and dash lights (but its unlikely). Consulting your owner's manual might provide sufficient information regarding any and all of the estuaries that branch off that line, but then again you might need the shop manual or a mechanic with that manual.

Let's take the other case: What if you install a new fuse in place of the tail lamp fuse, turn the headlight switch just one click (to the side marker, tail lamp position) and the lights don't come on? Your first step is to inspect that fuse: It's probably blown (See Fig. 16-8).

- Assuming that it is, then a testlamp becomes a minimum requirement, because now you are dealing with an open circuit–either before the fuse panel (battery side), but more probably, after it (within one of those bulbs

16

■

or in the wiring that leads to them). It's probable that a short within that circuit caused the fuse to blow. *That short circuit produced an open circuit.* The blown fuse protected the rest of the circuit from any further damage.

The testlamp is used to track the presence of voltage as you move farther out into the circuit. In most cases, a wiring diagram becomes handy, if not essential.

It's likely that the problem resides with one of the tail light or side marker bulbs, their sockets, or the wiring that leads to them, including the tail lamp relay (if one is on your wiring diagram . . . You'll find it on the battey side of the circuit in one of the fuse panels. If your dash panel lights do not employ a rheostat, then one of those bulbs could be the source of your problem. In all probability, the headlights are not involved: They typically have their own circuit, relay and fusing (Low and High).

Have any bulbs been replaced within the past month? If so, you could easily have the wrong bulb in some socket (see *§* 13, p.426). Otherwise, look for a broken lens that's taking on water, or body damage, or a lawnmower in the trunk (something heavy enough to have crushed or broken the rear wiring harness). Standing water in one of your trunk wells could have caused a short circuit as well. Most of these problems are fairly obvious.

For those unfortunate few whose search comes up empty, continue by examining the bulbs one at a time, their sockets as well. Is there corrosion down in a socket? If so, clean it out with a terminal (bottle) brush and electrical parts cleaner–and don't forget to pick up another ten fuses while you're out gathering supplies! Be sure to verify that every socket with one terminal at the bottom has a bulb with only one terminal corresponding to it, and that every socket with two terminals at the bottom has a corresponding two terminal bulb, Fig. 16-9, p. 426. These two-terminal sockets carry two circuits: one for the tail lights, one for the brakes. If someone were careless enough to install a one-terminal bulb, he would produce a direct short across these two circuits whenever either circuit is energized (The first time you apply the brakes, for example, or switch on the turn signals.)

These bulbs (1157's) reinstall nicely in one position, but can be forced into the socket exactly 180 degrees (1/2 turn) out. The element within the bulb that is meant for the brake light circuit shines much more brightly than the tail light element so that brake light application will catch other drivers' attention in the dark. Jam it in backwards and you get the opposite effect.

Armed with new fuses and your newfound familiarity with the car's circuitry, repeat the procedures for turning on lights as described above. Should the fuse blow again, try disconnecting a whole bank of lights before risking your next fuse. For example, all of your rear lights are probably on one harness that plugs in somewhere along a sidewall in the trunk. The side markers and parking lights located in the front of your car usually plug into the front harness individually, and must therefore be disconnected one by one.

Another spot to investigate is the tail lamp relay (where present). If it smells burned, replace it. Note that removing the relay prior to installing a new fuse does not indicate that the relay is bad (if the fuse no longer blows), because 90% of the circuit lies downstream from it: If you disconnect the relay, you disconnect the rest of the circuit as well–it's a meaningless test. Revisit *§*8, p. 418 and *§*9, p.419.

16

If you need to go any farther than already described, you would probably be better served by a skilled mechanic. This relates particularly to work behind the dash, to those with no familiarity with circuits other than what they've read here, and to those without a testlamp.

(A taillamp short is infinitely better than any short circiuit that can occur within the engine compartment, where a melted fusible link or circuit breaker that flipped has shut down a circuit. Those shorts can lead to fires if the link or breaker doesn't do its job.)

> Do not attempt to make any repairs involving the steering wheel or steering column without first reading *§* 20, p.435. Air bags are very dangerous if handled improperly.

15. The Importance of Being Grounded:
Checking the Taillamp Circuit for a Solid Connection to Ground

A good ground is just as important to a circuit as power. No circuit permits the flow of current unless it's a closed loop. On an automobile, that means having good electrical contact with the chassis on the unpowered (downstream, grounded) side of the electrical device.

During diagnostics, the possibility of a bad ground is often overlooked until all kinds of parts have already been swapped–always with costly consequence. If a shop ever suggests to you the need for some on-board computer or other expensive electronics, make sure to confirm that, indeed, they have already verified a solid ground connection.

A testlamp is used primarily to detect the presence of power–its availability to the circuit "downstream" from that point. But it does not necessarily indicate that a flow of current exists, even if the fuse for the taillamps is good. That's because there may not be a pathway for current to follow to ground–other than through the testlamp. In other words, there may be an open circuit downstream from the point you are testing, either a broken wire, a bad bulb, socket, or ground.

It's also important to understand that if the circuit that you're examining has a short circuit, then the testlamp will not light–you're testing a wire that has a "better" path to ground, one that offers no resistance whatever.

So . . . You can have a testlamp that lights (signifying that there's a problem further down the line) or you can have a lestlamp that doesn't light (signifying that there is either no power feeding the line or that the wire has a short circuit further down the line). What good is all that?

I would start with the wiring diagram. Which of the wires in that harness plug connected to the taillamps is supposed to be supplying power. Unplug the harness, then test that wire for power. Nothing there? Then you have an open circuit between the taillamp fuse and that lamp itself. Run the same test on the other taillamp circuit. You now have a tremendous amount more information than you did a minute ago.

16

You've already tested the bulbs and repaired/replaced any bulbs or sockets that needed attention (*§* 14, p.427).

Which wire(s) lead to ground. And where, exactly, is that ground? Both of those pieces of information are right there on the wiring diagram. Bypass the lights altogether; connect your testlamp between the two terminals—power feed to the plug and the line leading to ground. If no problem exists, then the testlamp should light. If not, then the problem resides either in that last stretch of wire to ground, the ground connection itself, or the wiring harness plug.

Remember the analogy of water falling down a mountain? The bulk of the water flows down the path of least resistance. What that means to a testlamp is that if you touch the probe to a point that has an easier path to ground, the test lamp simply won't light. In practical terms, if you apply your probe to a terminal that you expect should light, and yet you get no illumination, unplug the terminal and test again on the battery-side of the now-open terminal.

Assuming that your alligator clip is connected to ground, you will never detect the presence of current on the ground (downstream) side of an electrical gizmo. That's because the testlamp "sees" no voltage difference, both the probe and clip are occupying the same level electrically—ground. This observation—that a testlamp will not light on the chassis side of things—is why we mechanics view the chassis as the electrical end of the line. But if you move your probe to the powered side of the device (upstream from the device being tested, the positive side of the circuit, and unplugged at that point), it will encounter the undiminished force—voltage—capable of pushing those electrons through the bulb in your testlamp.

Let's review how to look for power:

- Connect the testlamp's alligator clip to ground (virtually any unpainted bolt head will do as long as it threads into steel), then touch its probe to any terminal or wire end that has been unplugged from its circuit "downstream" of that point. The testlamp will tell you unequivocally whether that point is powered: Its bulb will only light in the presence of 9–12 volts.[5] If that point is powered, current flows through the probe, through the bulb, and on to ground through the alligator clip.

- Remember that most circuits receive their power through the ignition switch; therefore they can't illuminate a testlamp without the ignition switch in the On or Accessory position. The engine does not need to be running.

- To determine whether your circuit is closed and to verify a good ground connection, attach your *alligator clip* to any known source of power (any terminal that illuminates your testlamp when connected). Touch the probe to the wire that you suspect runs to ground. If it does, the testlamp will light—assuming good connections.

> Do not attempt to make any repairs involving the steering wheel or steering column without first reading *§* 20, p.436. Air bags are very dangerous if handled improperly.

---

5   See Footnote 3.

16

16. Turn-Signal Circuit

Turn signals usually alert the driver to a bad bulb either by flashing at a preposterous rate or by not flashing at all. The problem usually resides with the bulbs up front–home to cracked lenses that love to fill with water. Corrosion within these sockets is common; see §14, p.428 for remedies. Unfortunately, since these sockets are usually formed as part of the lamp assembly, replacement of the socket alone is a low-percentage shot unless you have done some homework first. If both the left and right turn signals are exhibiting identical symptoms, suspect the turn signal flasher. But first check your bulbs, at least visually. If the entire system is dead, first try turning on the hazard warning lights, check its fuse as well.

The hazard warning lights employs the same bulbs as the turn signal circuit–as well as the same flasher–but since all four bulbs operate simultaneously, the flasher is typically a five-pronged unit to accomodate its increased demand for current.

17. Headlamps

> Do not attempt to make any repairs involving the steering wheel or steering column without first reading §20, p.435. Air bags are very dangerous if handled improperly.

There are two basic types: the halogen lamp insert Fig. 16-10A and the old school sealed-beams of yesteryear, Fig. 16-10B & C, the latter two modernized with halogen innards.

❑ The halogen insert is (usually) a whole lot easier to deal with. These bulbs are accessed from behind the headlamp lens. On certain cars, this requires that the battery be repositioned to allow clearance. In most cases, all that's required is removal of the battery strap; leaving the battery's electrical connections undisturbed.

Begin by unplugging the lamp's electrical connection–usually a tab must be depressed before the harness end will move. If it seems frozen, a shot of CRC 5-56 or WD-40 will usually free it. The halogen inserts are all held in place by a clamp. There are a number of clamp types; your owners' manual should tell you exactly how to release them. If it doesn't, consult your shop manual. (Most come free by rotating a collar counterclockwise.)

The bulb pulls straight out. This may be easier said than done, in that the bulb has an o-ring to seal it against the headlamp housing. It may feel unwilling. If so, try your lubricant. Once it's out, observe whether you broke off any of the central tabs on the retaining collar. If so, it should be replaced (a dealership item). During installation, avoid touching the glass portion of the bulb with your fingertips–oils supposedly shorten their lifespan. Installation is performed in exactly the same way as removal.

For a fairly sizeable piece of glass, the lens portion of these headlamps is clearly located in the wrong place: Living right out front and longevity are two ideas that just don't go together. If yours should suffer a "bullet" hole, you can either spend $100 to $150 for the lens assembly, or you can patch the hole before the lens starts filling with water (its interior should be bone dry). Use a small dab of clear silicone adhesive, available at any hardware store. If you don't find out about this problem

16

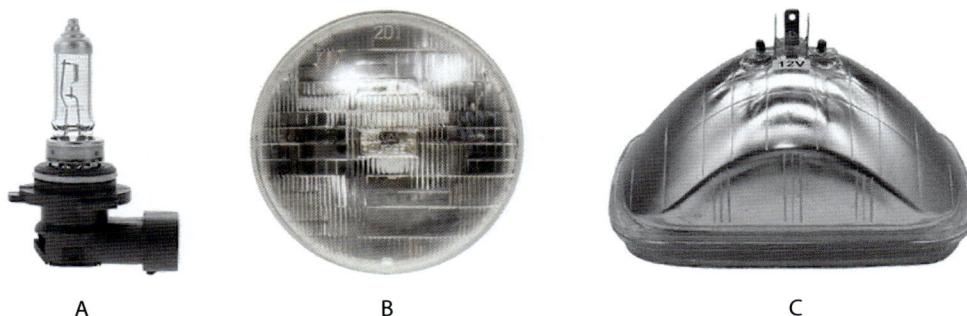

A   B   C

Fig. 16-10. High and low beam headlamps (L&R) have three terminals (high, low, ground). The headlamp in the middle is a low beam only (two terminals). Courtesy of Philips Electronics

until after the first bulb has blown, you can evaporate the water using a hairdryer applied through the bulb's mounting hole in the rear of the assembly–assuming that you've got all day. Or you can drill a drain hole in the bottom of the lens, let it dry, then seal both holes. If you do use a hairdryer, start with the lower heat settings until you determine how much heat that plastic is willing to take.

If you use a drill to cut a new hole through the glass, start off at slow RPMs and build from there. Oiling the drill bit helps to cool it, thereby reducing the chances of cracking the glass. Naturally, if the bottom of that plastic housing is accessible, forget about drilling through the glass–the plastic is much softer.

16

Every headlamp bulb–both halogen and sealed beam–has two or three adjusters. All face rearward, into the engine compartment. These should not be touched. They have nothing to do with getting the bulb in or out, and once you disturb them you need to realign the headlamps.

❑ All of the sealed-beam bulbs (all are halogen-lit these days) are approached from the front. Some require that the headlight surround (the decorative plastic surrounding the headlamp) be removed. This can involve having to remove the central grille as well.

The round headlamps typically have three retaining screws that hold them; these are backed out four to five revolutions each–just enough to permit the outer retaining ring enough freedom to rotate counterclockwise about five degrees so that the ring can be pulled (horizontally) straight off. The bulb then comes out, its electrical plug pulls straight off the back. The purpose of not removing the retaining screws entirely is that because they are made of stainless steel (to eliminate rust); they are consequently nonmagnetic and live for the opportunity to fall off the tip of your screwdriver into some unretrievable crevice. If you encounter a stubborn screw, hit it with CRC 5-56 before you strip out its head.

Any sealed-beam bulb can be replaced with a halogen sealed-beam replacement, but both sides should be installed toogether (in pairs).

Observe the back of the bulb. High/low combination bulbs have three terminals, the high-only beams carry just two.

❑ Square bulbs have four screws securing them, and unfortunately, in most cases all four of these have to be removed in order to free the bulb. Try a drop of *Stick-Um* on the tip of your screwdriver to help latch on.

Every headlamp bulb–both halogen and sealed beam–has two or three adjusters. These should not be touched. They have nothing to do with getting the bulb in or out, and once you disturb them you need to realign the headlamps.

18. The Horn Circuit

Horns are usually located ahead of the radiator, below and behind the headlamps. A horn that doesn't sound like it used to can often benefit from a sharp smack (to clear any loose rust). Another possibility: Make sure that the horn itself is not lying up against the sheet metal to which it's bolted. Two points:

- Cars typically have two horns of differing pitch: one low, one high.

- Most of you will see two terminals on each horn, one carrying 12 volts of power, the other terminal heads to ground. Older cars have horns with just one terminal–12volts of power–ground being supplied by the horn's mounting flange.

Most horns have power supplied to them at all times, but they can't honk until you feel sufficiently enraged to provide them their pathway to ground: When you depress the horn button (or horn field in the center of the steering wheel), the relay's 86 terminal should "see" ground–closing the ground side of the circuit through the steering column itself. Possibilities for failure include:

- A blown horn fuse–test it following the procedures in *§*12, p.424.

- A melted horn relay–test it following the procedures in *§*9, p.419.

- A worn-out horn pin: A spring-loaded pin centrally located behind the steering wheel, usually incorporated into the turn signal/combination switch. It sticks straight out toward the central portion of the steering wheel. This pin rides against the horn ring –a ring of brass built into the back of the steering wheel–and provides the point of contact between the horns and horn button regardless of steering wheel position. This problem is not likely on cars with less than 60,000 miles.

> Do not attempt to make any repairs involving the steering wheel or steering column without first reading *§*20, p.435. Air bags are very dangerous if handled improperly.

- A worn-out horn ring, or a horn-ring in need of lubrication: This diagnosis is virtually a locked bet if the horn works in certain positions of the wheel but not in others. A scratchy sound when turning the wheel, especially when the car is cold, indicates insufficient lubrication. Leaving it unattended soon wears out the horn pin. A little dab of grease, judiciously placed around the horn ring is all that's required (but the steering wheel has to come off first).

16

- A defective horn: The above four problems will knock out both horns; what follows concerns a single horn failing.

First verify that the horn is getting power. You'll need a testlamp, but before you can test for power you need to disconnect the horn's positive terminal plug. (Remember that whenever a line is to be tested for power, any wire connected to it could be a pathway to ground. That possibility would negate any and all tests of power supply–at least with a testlamp. So the power lead must be disconnected before it can be tested.)

Take note of the hot wire's color. No hot wire? Then you have a blown fuse or a defective relay. Another possibility: a meltdown within the engine's fuse block.

Use a jumper to connect the hot wire to the positive terminal of the horn; connect a second jumper to the horn's other terminal. Touch the second jumper's free end to ground. The horn should honk loudly. If it doesn't, the horn needs to be replaced. If it does, then the horn's ground circuit needs to be traced (your shop manual) or a secondary ground circuit needs to be installed.

But first, check the second horn.

The following are several sketches of how horns can be wired.

16

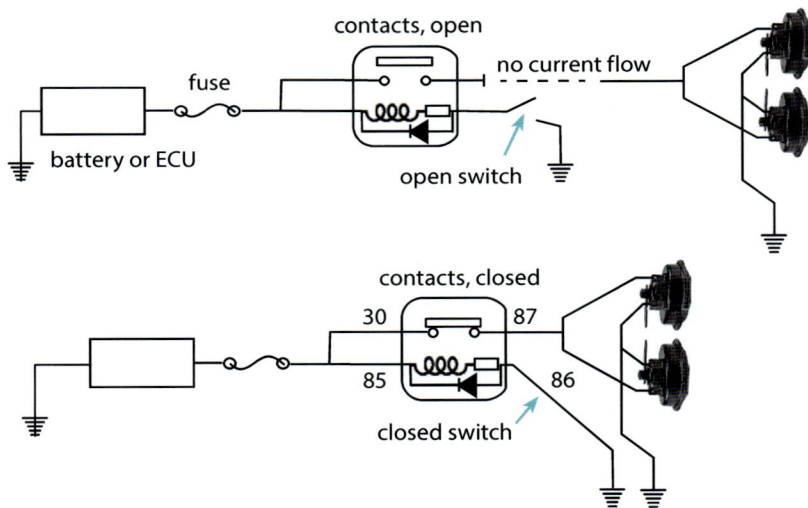

Fig, 16-11. A four-pin relay driving two horns.   ©PT

86

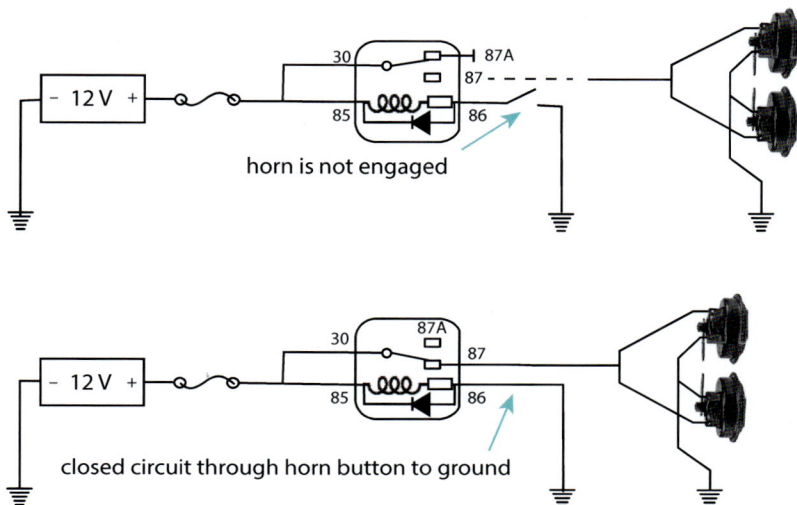

horn is not engaged

closed circuit through horn button to ground

Fig. 16-12. A five-pin relay driving two horns   ©PT

19. Testing the Heater–AC Fan Motor

The following test assumes a two-terminal fan motor. You will need a testlight to locate an alternate source of power (§10, p.421) and two jumper wires: one to provide power, the other to provide an unquestioned path to ground. For the "hot" lead, use the fused side of any comparable fuse in the driver's side fuse panel.

Disconnect the motor from its wiring harness plug. Connect both of your jumpers to the motor first, taking care that they are not contacting one another in any metallic way.

Next, take the free end of either jumper and secure it to an unpainted body bolt–your ground connection. It doesn't matter which terminal you choose to ground, fan motors turn just as well in either direction. (Of course, they don't blow as well in reverse because the fan blades were sculpted to push air in one direction only. That's why the wiring harness' two-terminal plug can only plug in one way.) Finally, take the free end of your remaining jumper and touch it ever so gingerly to your source of power.

> If you have a short internal to the motor, a strong spark will leap the gap as you bring the jumper near. Don't complete the connection of the circuit; doing so could turn those jumpers red hot within moments. Before condemning the motor, verify that you have not created the short yourself by somehow bridging the motor's terminals.

> The jumper that you attach to your power source cannot be allowed to touch any metal other than one of the fan motor's terminals.

A hot spark across the gap (arcing) indicates the need for a new fan motor. If you see no arcing, go on to connect the two. If the motor does not respond, then clearly it's no good. If the motor spins freely and noiselessly, your problem lies within the control circuit: either in its power supply, the fan relay, the fan speed switch, the fan resistor, or some problem within the harness itself. Some of you are quickly getting in over your head; consult a mechanic with good credentials in electrical diagnosis. Or pull out your shop manual and get to work!

Replacement of the heater fan motor is no picnic, requiring contorted positions, standing on your head for extended periods of time. I'd recommend replacing both the motor and fan cage at the same time. (Plastic cages warp rather easily. They usually require special order.) Just replacing the motor and its cage will not necessarily cure all ills. A failing motor usually burns out one or more of the fan speed circuits by cooking portions of the heater fan resistor. To inspect the resistor, you first have to find it: It is always located in the passageway leading away from the fan, this to assure that it has cool air blowing across it whenever current is flowing in the circuit. Fan resistor coils grow very brittle with age.

Fan motors often burn out as the direct result of a plugged drain hose. See Gurgling Sounds, p.393.

20. Working Around Airbags

The most likely airbags you'll have to contend with are in the driver's steering wheel or in one of the door panels. In all cases you should consult your shop manual first. My procedure in all cases is to get the airbag disconnected as soon as possible. But if you begin by disconnecting the battery's positive terminal, you'll have a whole lot less to worry about. Unfortunately, unless you own a scanner that will store them for you, you'll lose you favorite stations and possibly a whole lot more (seat positions come to mind). See Caution, p.410.

As for the steering wheel, you first need to remove the center cap or horn ring. You'll find a locking nut holding the wheel in place. To avoid smashing yourself in the face with the wheel as it comes free, just unloosen the central locknut four or five turns, then work the wheel free. Once loosened, you're in control. There may be lots of electrical connections that need to be disconnected; start with the airbag.

Door panels can be difficult. Follow your manual step-by-step. You'll still break one or two retaining clips – it's in the nature of door panels.

21. Windshield Wipers

In winter weather, always remember to turn off your wipers before you turn off the ignition. Following a drive in snow or freezing rain, the wiper blades will quickly freeze to the glass–particularly if you've been running with the defroster turned on. If the wiper switch is left on when you shut down the car, the wiper motor will want to move as soon as the ignition key is turned into either the Accessory or On position. This holds true even if you remember to turn the wiper switch off before turning the key–the wipers still need to cycle themselves into their parked position at the bottom of the glass.

16

What can happen? Usually the motor will exert sufficient force to bend the wiper blades, thereby changing their blade angle. The result is noise–either a scrubbing or chattering across the glass–plus a loss in wiping action. Repair should be left until the blade is no longer frozen; a judicious tweak with hand or pliers (wipers off) can bring the blade back to its original perpendicular orientation. You want the rubber insert to flip, then flop, first in one direction, then the other, as the blade changes direction, §1, p.443. Badly bent blades require replacement.

The wiper motor can also refuse to stand motionless. If you're lucky, it'll just pop out of its union with the wiper link (Fig. 6-6, p. 99). Repairs aren't too bad–assuming an accessible wiper motor–A is reconnected to B and all is well except that both pieces will be much more likely to do it again. But if you break the plastic cup that secures link to motor, you'll be staring at a $300 bill.

Most factory wiper blades have replaceable rubber inserts. Most people are satisfied with an annual changeover, the finicky need a fix perhaps every six months. Beware of the gas station/lube joint jockeys! These shops do not stock factory inserts, most stock junk, and by the time you get your car back, your factory blades are long since in the trash. If you need wipers, take the time to call your dealer to inquire about their refills, because once you have the junk, factory refills won't fit

What can I tell you about washer solvent? Never use plain water, particularly in winter: It'll freeze and might ruin your washer motor. Don't use antifreeze either, that goes in the radiator! Washer solvent comes both premixed and in concentrate. In proper dilution, it won't freeze.

For more on wipers, see §1, p.443.

22. Rear Window Defogger

The rear window defogger is nothing more than a series of high-resistance wires embedded in the glass. (Resistance to current flow creates heat.) Run your finger (not your nail) across the inside surface of the rear glass. If you can feel the bumps, you can break the wires. Avoid stacking things up against the glass that could rub or scratch these wires. If you do need to repair a line or two, 3M makes a repair kit.

If none of the lines are functioning, you'll probably need to replace the defog fuse or recycle its circuit breaker–consult your owner's manual for location. If that's not the problem, make sure that your connections at the glass are secure, preferably using a test light. Do not tug at the terminals that adhere to the glass, they pull off rather easily and don't reattach. If you're using a testlight, lightly touch the area surrounding the terminals. Don't scratch through those electrically conductive end panels, and never touch the wires themselves–your probe can easily break through a line. Note that one terminal (left or right) should light your testlamp and that the other one (right or left) won't (because it's the equivalent of chassis ground). If you have no power to either terminal, the possibilities include:

- a defective defog fuse
- a melted defog relay or circuit breaker

- a defective defog switch

- a broken wire (quite common on wagons and liftbacks, always in the vicinity of the hinge that carries the wiring harness bundle out into the tailgate door).

If you have power and the defogger doesn't work, then you have a bad ground or the defogger itself is too old to be effective. These wires crystallize over time, yielding a zillion open circuits across the glass. This process takes ten years or more. The aftermarket defoggers that I've encountered aren't worth the box they're sold in. If you intend to keep your car, get two or three quotes on rear glass replacement.

23. Fuse Panels

It is perfectly safe to test individual fuses and relays, or to remove all of the relays for inspection. This enables the ability to look for a meltdown of the panel's surface directly beneath any one of the relays. However, I wouldn't pull more than one ot two at a time, this to ensure that the relays gets reinstalled where they belong. Not all relays that melt do so on the visible side of the box; in fact, many of these meltdowns occur within the harness plug itself and that requires removal of the fuse panel.

These "black boxes" must be handled with the same respect afforded the alternator or starter motor: The battery's positive terminal must be disconnected before you begin disassembly, otherwise you run the risk of a serious shock, fire or both. Unfortunately, unless you own a scanner that will store them for you, you'll lose you favorite stations and possibly a whole lot more (seat positions come to mind). See Caution, p.410 Once that's done, unbolting either of the fuse panels and removing its back cover from the upper side of the fuse panel is safe.

Looking for a plug that melted down is among the very last places to look for a problem within a circuit. A defective relay or blown fuse is far more likely, as is a short or open circuit.

24. Music

If you have a radio that's not working properly, don't be too hasty about its replacement: Poor sound quality is often due to defective speakers. Imagine being asked to vibrate both at low and high frequencies, often at high volume, in temperatures both below freezing and above 120°F (a typical dashboard temperature for cars sitting in the summer sun). Humming, static, booming–all can be due to one or more defective speakers. By using your balance (and fader) knobs to isolate each speaker in turn, you can easily detect a bad one. If you do need speakers, remember that factory speakers fit!

If you have an older car with a bum radio, it should be obvious that an electronics store will burn circles around any price the dealer could quote. But you have to be careful: There are some great auto radios out there, but there are also less than ideal systems that none of us have ever heard of. So spend some time shopping–and listening–before you settle down to buy. Unless you're very handy, the store you buy from should be in the installation business as well, so that if the thing doesn't work right, there will be just one company responsible. There's an added benefit to this: You get to hear one price before you buy.

25. Installation of Aftermarket Equipment

If you don't already know what I'm about to say, then you've been dozing in the back row! There are very few add-ons that will actually enhance the value of your car. Every time you add some gadget, you open your car up to a chopped wire here or a loose connection there. Have you ever walked around back of the cellular phone store or auto electronics shop to take a look at their shop? Count the number of people over 22 years of age. Now I'm not prejudiced against young people–Hell, I used to be one! . . . and I've employed and trained quite a few–but I can assure you that seasoning and experience are not their strong points. Their work will probably be sufficient to get you back out on the road, delighted with your new gadget–quite possibly for a period of years; but when corrosion or rust or loose wiring comes back to haunt you, most of you will have long since forgotten that your stereo system or anti-theft device was an aftermarket add-on. For the undaunted among you, a few words of advice.

Fig.16-13. An inline fuse holder, suitable for radio
appllications, glass cylinder fuse.

- All equipment must be fused. Any power pickup off the fuse panel must employ a fuse Any work that involves simply jamming a wire underneath the blade or endcap of a fuse should be rejected.
- This installation cannot involve any modification to a circuit: Should this installation involve splicing wire into any portion of the automobile's wiring harness, be sure that it's visible.
- All connections to the harness must either be:
  - Located at a preexisting harness plug where terminal ends can be used.
  - Soldered into place, using a heat sink whenever necessary to protect delicate electronics.
  - In either case, all connections must be thoroughly sealed.

16

 Two thoughts come immediately to mind: First, you can rest confident in the knowledge that using this addendum will place you in a class somewhere between lawyer and slug. Second, you'll still have to check their work.

Verify that connections into the fuse panel are actually fused. They could accomplish this in two different ways:

- The wire that supplies power to virtually every radio comes equipped with an in-line fuse holder already built in. The power lead to every radio that is so equipped can be installed on either side of any fuse in the fuse panel. (Naturally, if the accessory has a memory chip, it needs to be wired into a circuit that is powered continuously–such as the tail lights.)
- If no fuse holder is evident, then the radio must be connected to the fused side of one of your fuses. The only way to tell which side of a fuse is protected is by using a testlight. The fuse to which the accessory will be attached is temporarily removed. The testlight's alligator clip is connected to ground (virtually any unpainted bolt head will do), the probe is then touched to the

> two terminals to which that fuse was connected. The terminal that lights is the powered side, the terminal that does not light is the protected, or fused side. Your accessory should be connected to the protected side.

As for undoing electrical tape to verify soldered connections, I personally wouldn't waste my time unless I smelled a rat. However, it is extremely important for you to know what circuits have been altered, and where. Write this information down and put it in your glovebox as insurance against the day that you need it. And from that time on, any time you have an electrical problem that doesn't fix with a fuse or a bulb, raise the issue when you first talk to the repair shop, and mention it again if a fix remains elusive.

26. DIY –

26A. Building a Jumper Cable –

There are any number of reasons why you might need a jumper, all of which begin with the need to replace a portion of wire with a section that you know is sound. I built my first to elimate the ignition/starter circuit from my second car (a 1958 Peugeot 503) to verify whether it needed a new starter. (It did.) Starters were a whole lot easier to access in those days – as were batteries.

Start with a 5' long piece of 16 gauge(+/-) automotive wire. It typically comes as a single wire, not twisted like household wiring. Check with a nearby auto supply store. Scavenge if you can. Use a piece of household (twisted) wire if you need to. It all looks the same to an electron. Make sure that your cable is long enough to reach both the starter motor and the majority of the engine's electrical components, but don't go overboard with length; you never want to end up with that cable getting caught up in a moving drive belt.

Inspect the terminals on your car. There are many sorts; you just want to duplicate one or two kinds. Take a look at the terminal on your starter (see the green arrow in Fig. 16-3, p. 412). At least one end should employ a fitting that would work on it. There are tools specifically made to crimp electrical connections, but a pair of pliers is perfectly suitable for an every-once-in-a-while task.

There. You now own your first jumper, which can be used to test your own dead starter some day.

26B. How to Use a Testlamp –

A detailed discussion spans two sections: §14, p. 427 and §15, p. 429.

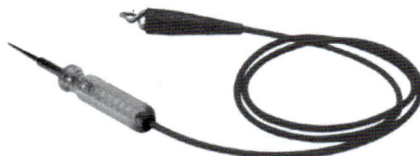

26C. Using a Micrometer –

There are any number of reasons to use a micrometer rather than a testlamp, the primary one being that a micrometer can read resistance within a relay or a circuit. It can also distingush the state of a battery's charge or the output of an alternator. It's also very handy when digging into the circuitry of an ECU. Good ones come with a temperature probe, essential for cooling system work.

But I've limited myself as much as possible in this chapter to the testlamp because its use is so instructive in teaching. For more on the micrometer, consult its owners' manual.

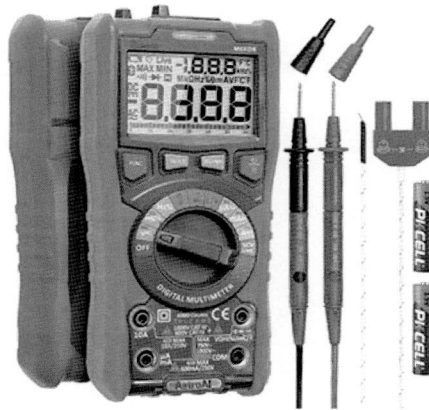

AstroAI 6000 multimeter with temperature probe

16

16

Chapter 17. Some Loose Ends: Wipers, Exhaust Work, Vibrations, Water Leaks, and Forgotten Photos

17

Notes

Some Loose Ends

1. Windshield Wipers

Wiper blade design is as varied as patent law permits, and yet all share one of two basic designs:

- The two-piece unit with replaceable rubber insert (@), Fig. 17–1A
- The one-piece unit in which metal (or plastic) blade and rubber insert are replaced as one, Fig. 17–1B

Fig. 17-1A (L) and Fig. 17-1B (R)
Two basic types of wiper blades

Fig. 17-1A Courtesy of Subaru
Fig. 17-1B Courtesy of Trico

A two-piece blade can be left on the car until its pivot grows sloppy or until the blade becomes distorted or bent by being frozen to the glass. Inserts can be replaced whenever they start to streak or smear. The lifespan of a one-piece blade is limited by the condition of its rubber edges, the first part of the assembly to wear.

Wiper blades attach to the wiper arm in a variety of ways, from clip-ons to back-flips to little bolts. The structure of the blade and the design of the insert are even more imaginative, but through all of these variations there is one underlying truth: The factory has engineered its blade to fit a specific piece of glass. The curvature of the glass and the geometry of the wiper arm define how much tension is required in the blade. Without proper tension, the insert won't flip-flop as the blade changes direction (Fig. 17-2A and Fig. 17-2B) resulting in a scrubbing sound as the insert chatters across the glass.

17

Fig. 17-2A & Fig.17-2B. Wiper blades in motion. Fig.17-2A is going up the glass (insert flipped downward), Fig. 17-2B is going down the glass (insert flipped up).

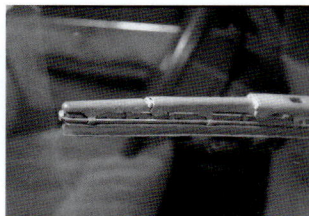

Most shops don't stock factory wiper refills; there are just too many of them. Instead, they rely on aftermarket inserts and/or blades that can attach to a number of cars (short of), but unfortunately, fit none of them as well as possible. If your car arrived at the shop with a decent set of original-equipment blades that were designed for replaceable inserts and it's now equipped with something else, see the service manager before you leave. Dragging yours out of the trash is only possible for a few hours, usually until the close of business. Having to replace your new blades will cost every bit of $30 when you decide that you want to see clearly again. Naturally, if your car arrived with aftermarket blades, then a generic refill will fit as well as any, but if you are seeking clear glass and quiet wiping during a light drizzle, you'll do best with the original-equipment blade and its correct, factory-issue wiper insert, available through your dealer and some independent specialists. While you're there, ask for a demonstration of proper installation. Replacement wiper inserts and replacement blades are clean and shiny, with crisp rubber edges where they meet the glass. The edges should flip direction every time the wiper blade changes its direction. If they don't, they'll chatter their way across the glass.

Your washer nozzles should be properly aimed, targeting the approximate center of each blade's path. New inserts should clear the glass without streaking in one or two swipes (assuming that no bugs are glued to the glass). Check that your blades do not smack against either the cowl or pillar post as they travel across wet glass on high speed. If one does, its wiper arm needs realignment at the wiper pivot. This is the type of issue that an owner should notice before dropping off the car. The observation should be included in your note to the shop, permitting sufficient time during the course of the day for your mechanic to correct the problem, instead of being under your watchful gaze when the both of you should be on your way home!

Wipers need helping hands. Dirt gets trapped along the wiper's rubber edges. Glass cleaner soaked into a paper towel works wonders. As for bugs, use a single-edged razor blade applied to the glass at a thirty-degree angle.

2. Drive Belts

17

Drive belts come in all shapes and sizes; to generalize about them is to invite misinformation. But the following statements are universally true:

- Belts that squeal first thing in the morning usually just need to be tensioned (tightened), but if they're oil-soaked they should be replaced.

- Examine the V-shaped surface(s) that ride(s) through the pulleys, looking for cracks. A zillion shallow cracks is normal and not sufficient cause for replacement. But two deep cracks located near one another are just a step away from spitting out that chunk. Clearly, a badly cracked belt should be replaced. See Fig. 17-25, p.461.

- The serpentine configurations (Fig. 17–3) wipe out belts twice as fast as the "old-school" triangular arrangements because the belt must double back on itself. Traditional belt configurations permit life spans of 60,000 to 90,000 miles under typical conditions; the "serpents" start losing chunks of rubber in as little as 35,000 miles.

- Engine heat and friction dry out belts over time, reducing their pliability and glazing their working surfaces. The result is a noisy belt – they start to sing. Noisy belts don't necessarily require changing even if they have 10,000 teeny cracks – except to soothe an owner's fraying mind.

- The condition of a belt through 90% of its span cannot be generalized to speak for the remaining 10% – a thorough inspection includes the belt's entire length. To see those sections that are now lying in contact with a pulley, use the ignition key to "bump" the engine about thirty degrees. Dab a spot of correction fluid (*Liquid Paper, White-Out*) or paint onto the backside of the belt to provide a point of reference. Hold the ignition key in the Start position for no more than the blink of an eye, otherwise your mark will be out of view, halfway round the track.

- Drive belt tensions are checked with the engine off. (Duh!) Locate the belt's longest accessible span. At its center, use thumb and index finger to push and pull on it within the plane through which it travels (green arrows, Fig.17-3). Expect about a 1/4 inch of play; more than that should be addressed by retensioning the belt. Really loose belts produce a screeching sound when revving the engine, particularly while it's still cold. If the belt is soaked with oil or coolant, at minimum it will have to be replaced.

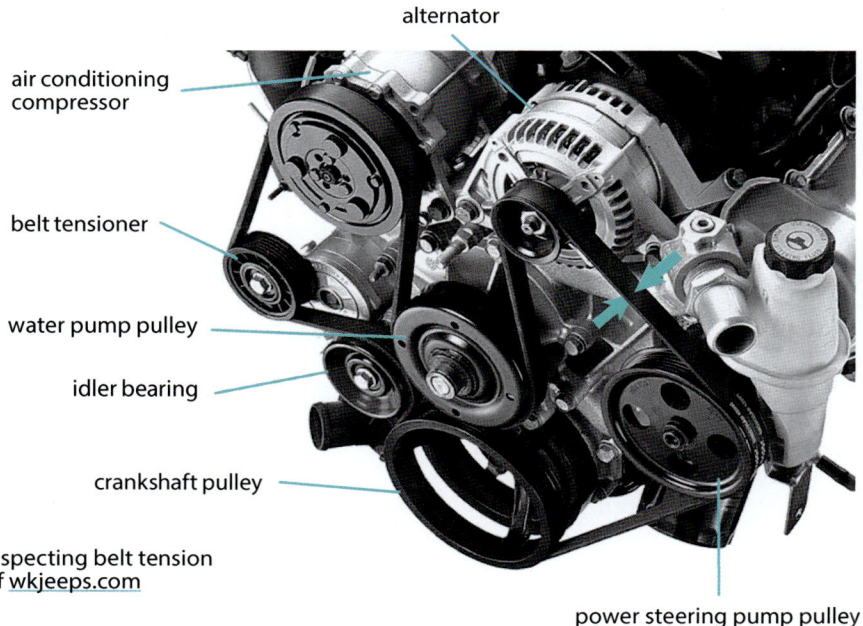

air conditioning compressor

alternator

belt tensioner

water pump pulley

idler bearing

crankshaft pulley

power steering pump pulley

Fig. 17-3. Inspecting belt tension
Courtesy of wkjeeps.com

17

- If you are going to replace one drive belt, have them all replaced. Why pay for labor twice, completing in two steps what should have been considered from the outset as the rest of the job? This is not to say that the need for one drive belt should escalate into timing belt replacement – they're different animals. However, the reverse is true: Timing belt replacement provides the perfect opportunity for drive belt replacement, because every last one of them has to be removed to access the timing belt. In other words, except for the cost of the parts, drive belt replacement should not cost a cent in labor charges when performed in conjunction with timing belt replacement.

- Timing belts don't squeal because they can't slip. Their notched teeth are matched up to corresponding grooves cast in their drive pulleys. When a timing belt starts losing its teeth, the engine loses power, then quits, Fig. 17-4.

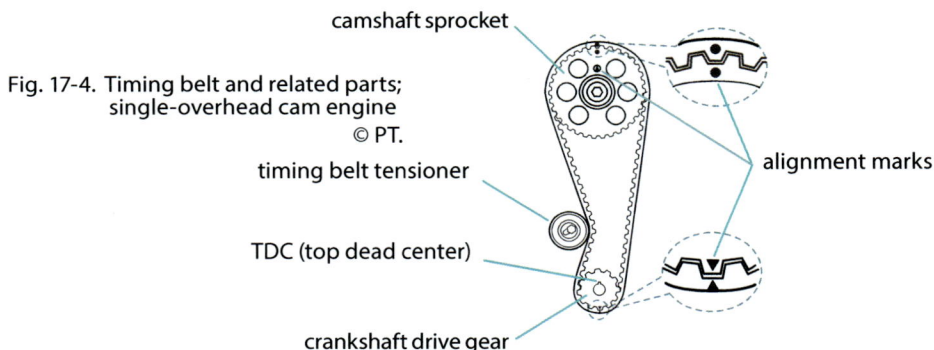

Fig. 17-4. Timing belt and related parts; single-overhead cam engine
© PT.

camshaft sprocket

alignment marks

timing belt tensioner

TDC (top dead center)

crankshaft drive gear

- **Drive Belt Tensioning:** Most drive belt tensioners look similar to the one shown in Fig. 17-3. You'll need a six-point socket and breaker bar or a six-point wrench to loosen its central lock-nut, and a long screwdriver or similar tool to apply sufficient pressure to tighten the belt. With tension applied to maintain a tight belt, the central lock-nut must then be retensioned with your other hand.

Other tensioners employ a long bolt to provide tension to the belt, but they too require that the central locknut be loosened first, then tightened following adjustment.

Still others employ a hydraulic tensioner; these seldom need attention.

17

3. The Exhaust System

A typical exhaust system includes the exhaust manifold, front exhaust pipe, catalytic converter, midpipe, rear pipe, muffler, and tailpipe, Fig. 17-5 and Fig. 17-6. Clearly, there are variations:

- The rear pipe, muffler, and tailpipe often come as a single-piece unit (Fig.17-5)

- Many (most?) systems include a resonator, too (Fig.17-5).

- A long rear pipe can substitute for a nonexistent midpipe

- The catalytic converter can be located at the exhaust manifold's lower end

- Air injection, O2 sensors and other piping can thread into the exhaust manifold, front pipe, or catalytic converter.

- Exhaust manifold shown in Fig. 17-7.

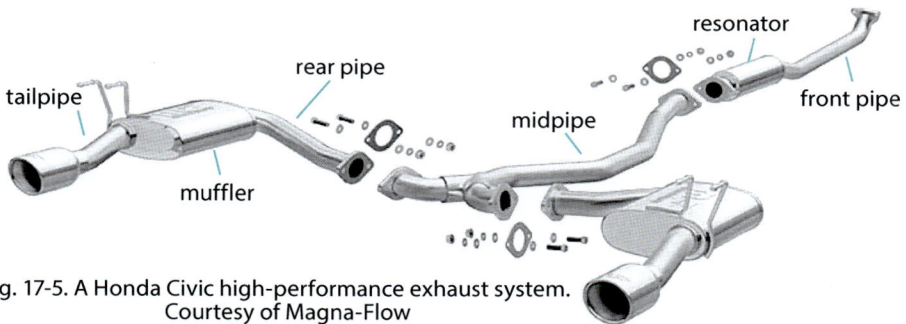

Fig. 17-5. A Honda Civic high-performance exhaust system. Courtesy of Magna-Flow

17

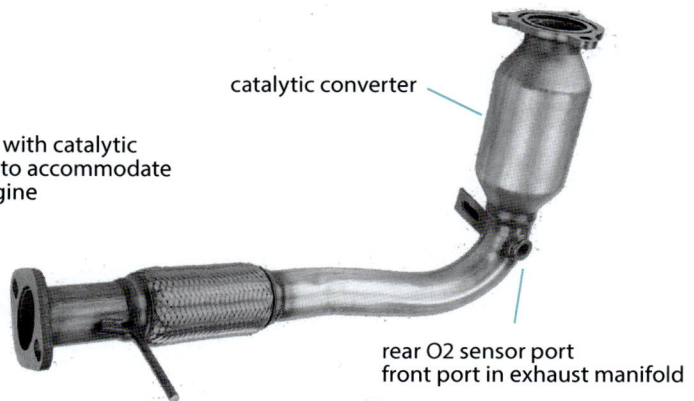

Fig. 17-6: Front exhaust pipe with catalytic converter & flexible mesh to accommodate a transverse-mounted engine

catalytic converter

rear O2 sensor port
front port in exhaust manifold

Most front-wheel drive cars are equipped with a transverse-mounted engine, Fig. 17–7. For the engine to express its torque, the front exhaust pipe has to be both compressible and elastic. Most factories use woven wire and a bellows design (Fig.17-6) to eliminate the exhaust system's being continuously shoved forward and backward under acceleration and deceleration.

Fig. 17-7. The transverse-mounted engine in a late-model Honda Civic. Note the #1 O2 sensor mounted in the top of the exhaust manifold.

To reduce vibrations, the exhaust system is suspended by rubber cushions beneath the body. A series of support hangers – one on the body, a corresponding one on the pipe–are juxtaposed one atop the other, joined by rubber cushions, Fig. 17–8.

Fig. 17-8. A support cushion, one of four for the mufflers shown in Fig. 17-5. The far end is supported by the exhaust manifold. There must be other hangers, not shown.

17

Individual exhaust pipes are joined to each other using flanges or slipover couplings Fig. 17-9 and Fig. 17-11 The flange approach is less likely to leak, but because each flange must be welded to its pipe at a specific angle, this approach requires more precise manufacturing, severely limits the applicability of every piece, and consequently commands a higher retail price. The slipover joint permits much greater freedom in aligning two adjacent pipes in that one can be rotated relative to the other. While this provides added flexibility during the occasional problematic installation, this same freedom gives "Butch" the opportunity to take a simple job and screw it up – positioning an exhaust pipe hard up against the chassis where it can send engine vibrations throughout the truck. If you still have the original-equipment exhaust system on your car, look at it before it's replaced. If it's flanged, you might find that an aftermarket replacement is less than satisfactory. There are only three reasons to replace an exhaust system component:

- Exhaust leakage
- A vibration in the body of the car produced by contact between pipe and body
- Internal blockage and consequent power loss. See § 9H, p.244

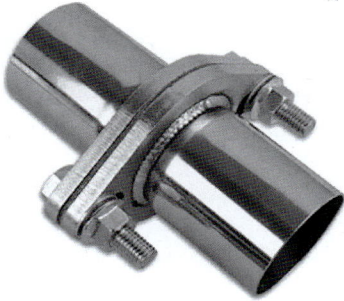

Fig. 17-9. This is a replacement flange for a perfectly straight piece of pipe.

Most flanges that you see will not be straight, the pipes themselves being held in place by clamps like those shown in Fig. 17-10.

Note the stainless steel hardware, essential if you ever want to get the two pieces apart (without a torch).

Fig. 17-10. This type of exhaust clamp is used to secure pipe to pipe. All of the pieces shown are made of stainless steel to prevent rust. It wasn't always that way!

x2

Fig. 17-11. These unions are more commonly seen in truck applications. If not stainless steel, they are nearly impossible to remove without a torch. Courtesy of Walker Exhaust Systems

3A. Exhaust Leaks

Exhaust leaks always make noise; the bigger the leak, the more obvious it will be. Tracking its source is simple: Lie down alongside your car while you have a friend rev the engine (in Neutral or Park) up to 4,000 RPM and immediately back to idle, up to four grand and back, repeating the cycle while you home in on the noise.

Another approach is to have your assistant wad up a rag to plug the tailpipe outlet (engine idling in Neutral or Park). A leak will hiss or burble as escaping gases rush through the opening.

> Exhaust gases are extremely hot. Your rag must be large enough to seal the tailpipe opening while still protecting your hand. Do not rev the engine while the exhaust system is plugged: The force of the escaping gas might first surprise, then burn you. Do not plug an exhaust system for more than thirty seconds at a time because excessive back pressure can stall the motor.

17

Note that virtually every slipover coupling will leak to some extent when the tailpipe is plugged. Further, most mufflers have a small drain hole at the bottom of their forward or rearward face. Obviously, this hole will "leak" if the tailpipe is plugged. Consider it normal.

Visually, exhaust leaks take two forms: one or more holes ranging in size from a pinhole to a fist, or a blackened, sooty trail exiting a rusted seam or cracked flange.

> If you must feel for the leak, remember that exhaust gas temperatures rise as you move nearer the engine. The exhaust manifold's internal temperature approaches 1,500 degrees; skin literally melts at that temperature.

> Flakes of rust drop from every exhaust system – as if on cue – once your eyes are centered directly beneath. A speck of rust can be excruciating in an eye, slicing its way right on in, lodging there to scrape against your eyelid every time you blink. Flooding the eyeball with eyewash or water can dislodge the less tenacious flakes, the remainder demand the skills (and tools) of an opthamologist. Don't let that eye fester till midnight, because the wee small hours are by far the worst! Safety goggles are always sound practice when crawling beneath an exhaust system. Glasses offer the illusion of safety with none of the benefits.

> The rearmost hanger—the one supporting the tail end of the muffler—has to be present. Without it the entire system will bounce until something cracks, an expensive lesson in the consequences of neglect.

3B. Exhaust Vibrations

Exhaust vibrations come in three flavors: Those caused by rust, those that result from a misaligned exhaust support cushion and those produced by a pipe that's lying in contact with the body,. The first of these is obvious. The second can result from:

- An under-chassis collision – driving over a cinder-block for example, or backing into a steeply-rising bank. An exhaust system that has been damaged (bent) by running it overtop of an unyielding object will show a misalignment between (formerly) corresponding hangers – the one on the body, the other on the exhaust pipe. Every exhaust support cushion (Fig. 17-8) should be oriented vertically between these two. Any cushion positioned at an odd angle should be investigated further, particularly if its two hangers no longer line up.
- A botched job of installing aftermarket exhaust components
- A pipe that has warped with the heat, upwards into the body. Vibrations of this sort can be intermittent; the offending pipe might only be in contact when it's cold or when its hot.

17

Some of these misalignments can be fixed by softening one or two areas of the pipe with a torch, then forcibly bending it away from the body. This should never be performed on a pipe that has lost cross-sectional area because of a collision, because reduced exhaust gas flow can do significant damage to an engine.

Individual exhaust pipes are joined to one another using flanges or slipover couplings, Fig. 17-9 and 17-11. The flange approach is less likely to leak, but because each flange must be welded to its pipe at a specific angle, this approach requires more precise manufacturing, severely limits the applicability of every piece, and consequently commands a higher retail price. The slipover joint permits much greater freedom in aligning two adjacent pipes in that one can be rotated relative to the other. While this provides added flexibility during the occasional problematic installation, this same freedom gives "Butch" the opportunity to take a simple job and screw it up—positioning an exhaust pipe hard up against the chassis where it can send engine vibrations throughout the car. If you still have the original-equipment exhaust system on your car, look at it before it's replaced. If it's flanged, you might find that an aftermarket replacement is less than satisfactory.

3C. Vibrational Noises Caused by Rust

❑ Rusted Baffles Inside the Muffler

These can buzz or rattle when the engine is revved. They are easily verified by banging on the muffler's case with a rubber hammer. A good muffler is quiet; a jingling sound gives the bad ones away. Rusted baffling is not sufficient cause for muffler replacement unless the noise is getting to you or you have power loss associated with it. Occasionally, a baffle will drop into the exhaust stream, blocking its flow. In that case, the muffler must be replaced because of the engine's power loss.

❑ Heat Shields

There are two different sorts of heat shields: The first type can be purchased separately from the pipe (and car body) it protects, the second is part of the pipe itself and is sold that way. The former I'll call body shields, the latter, heat shields.

Heat shields usually surround the front exhaust pipe, the catalytic converter, and the exhaust manifold. The larger body shields can be replaced with factory components or + that is typically sold in small rolls. As either pipe or body shield rusts and chunks fall away, the offending piece will buzz, usually at a specific RPM. Track the location of the noise by having your friend again rev the engine, holding the throttle at whatever engine speed creates the noise, engine in Park or Neutral. Confirm your suspicions by placing a block of wood (not your hand!) hard up against the shield to dampen the vibrations.

Every heat shield has its function; not one was engineered into your car without a good reason. That being said, to replace the offending shield usually requires replacement of the entire factory pipe (most aftermarket pipes have no shields)–an expensive proposition. Quite often, the only practical solution is to cut off the shield with a cutting torch or die-grinder.

If the shield is there just to deflect heat away from the passenger compartment floorboard, try living without the shield for a week or two–cutting it off is much cheaper than replacing the pipe–or consider investing in a roll of exhaust insulation. Chances are there will be no appreciable downside to body shild removal beyond some additional heat to the floorboard. Be sure to seek your mechanic's advice regarding any potential problems that might surface if it's re moved.

> If the heat shield is adjacent to and protecting a hydraulic component of the steering, some critical engine component, or the gas tank, it'll have to stay. Or it can be replaced with a new factory shield or an aftermarket substitute.
>
> Every original-equipment catalytic converter has a downward-facing heat shield to reduce the chance of fire when parking over a pile of leaves. If yours is coming off you'll need to be particularly cautious when parking, particularly during the fall.

❏ Rusted Exhaust Hangers

If only one hanger (body or pipe) out of five has rusted sufficiently to create noise, you should be able to get away without it until the pipe that it supports rusts away. Yanking off (possibly cutting off or torching off) a rusted hanger takes a matter of minutes. Replacement of that same hanger when the pipe is being replaced shouldn't be much more than the cost of the part, but in virtually every case it will have to be special-ordered.

3D. Replacement of Exhaust Components

Automobile manufacturers spend huge sums in the pursuit of quality in their parts. The muffler on a Taurus was engineered specifically to fit that car; it has a specified amount of back pressure, a specific amount of resonance. Ford builds them to very tight tolerances and they fit well. Only a few aftermarket exhaust manufacturers produce components that are even comparable. Those that do usually supply the factories as well as the aftermarket.

17

The national chains use their own brands and suppliers: If the muffler for a Chevy Malibu has roughly the same dimensions as that for a certain Toyota, then simply using a different inlet or outlet pipe enables it to fit both. That very same inlet pipe might be used on an old Nissan, providing a veritable smorgasbord of mix and match combinations! But what about the hangers that suspend the muffler beneath the car—are they the same on both models? No . . . What about the positioning of those bends in the pipe? If things don't line up exactly, you'll get a vibration whenever adults ride in the back seat. And what about those mufflers that start out their lives suspended by only one hanger? (More the rule than the exception.) When that inlet pipe rusts through, the front of the muffler drops to the pavement and for a few brief moments beats itself against body and ground, until it breaks free to take up residence in the middle of the freeway. You've heard the saying: You get what you pay for. In the wonderful world of exhaust repair, there's a twist to that maxim: Sometimes you don't even get that!

❑ **Front Pipe Replacement**

On occasion, replacement of the front pipe can be an ugly affair. The studs that secure the front pipe flange to the exhaust manifold are subjected to wide temperature variations, over and over again. They can become brittle, then snap off or strip when a tool is applied. While some studs can be repaired on the car, many require exhaust manifold removal to provide sufficient access.

Every other joint in the exhaust system runs at cooler temperatures and employs either disposable nut-and-bolt sets or u-clamps, Fig. 17-9 – Fig. 17-11, p 449. The worst development with any one of these joints is the discovery of rust—invisible until the rotted piece is actually removed – which condemns the adjacent piece to the dumpster as well.

Exhaust system work is best left for the professional shop. A torch set and the occasional special service tool are a must in most cases.

3E. Inspecting Exhaust Work

There should be no appreciable amount of noise coming from the system, either at idle or when the tailpipe is momentarily plugged, §3A, p.449. There should be no new sounds (buzzing or vibrational stuff) when the engine is revved. The same should hold true on the road test. Take a bumpy road for part of your test, include a speed bump or two. If you have a new midpipe or muffler, put two passengers in the back seat (because a poorly-aligned pipe can clunk or rumble with added weight in the rear).

Check for the presence of exhaust pipe hangers and inspect their positioning. Each body hanger should have a corresponding exhaust pipe hanger situated directly below it; the two should be joined by rubber cushions, Fig. 17-8. If you're missing a few hangers, you now own a piece of junk. If you're missing a few hangers and those that are present don't line up with their counterparts, then you now own a misaligned piece of junk. Take it back!

3F. Oxygen Sensors

The oxygen sensors (Fig. 17-12) bolt into the side of the exhaust manifold or front pipe and provide information about exhaust gas composition to the emission control (EC) or electronic fuel injection (EFI) computer §3D, p.196. Its signal enables the computer to adjust the fuel mixture as needed to assure a clean-burning engine. Engines that are burning too rich receive a leaner mixture, and vice versa.

17

Fig. 17-12. A late model O2 sensor. The wiring plug for O2 sensors can range from 1 wire to 4. Courtesy of Denso

A defective oxygen (O_2) sensor can cause terrible emissions readings and a poorly running engine, but will seldom produce an engine that won't run. Every trouble code system on every ECU or EFI computer has some means to identify a defective O_2 sensor; unfortunately, not every defective sensor will trigger that code.

3G. Catalytic Converter

The catalytic converter is an expensive little unit that looks kind of like a small muffler. Most are placed directly behind the front pipe, in the front pipe (Fig. 17–6, p. 447), or at the base of the exhaust manifold. Its purpose is to convert unburned hydrocarbons to carbon dioxide and water. Don't ask me how it does that, there's magic going on in there. The exhaust gases pass over a catalyst (either palladium or platinum—that's why they're so expensive) and shazam! . . . Cleaner air!

The catalytic converter should last the life of the car unless:

- Its case rusts through
- Somebody has the bad taste to pump leaded fuel into the tank, or
- If the engine runs for an extended period on an excessively rich (or overly lean) fuel mixture. Either one of these will cause the catastrophic converter's internal temperature to climb the sky as it attempts to deal with all those hydrocarbons on the loose (or lack thereof). The converter will finally suffer a meltdown–literally–and consequent loss of power due to internal blockage, §9 H, p. 244. The only way to know for sure is to visually inspect its interior, either by completely removing the converter or by unbolting an adjacent pipe to take a peek.

4. Vibrations

4A. Vibrations That Occur While the Car Is in Motion

The most likely possibilities are:

- One or more out-of-round tires, or a ply separation, §1 through §4, pp. 317-326
- One or more bent wheels, §1 through §4, pp. 317-326
- Brake pulsation: Ordinarily, this occurs only while the brakes are being applied (§5, p. 266)
- Frozen u-joint on the driveshaft (rear-wheel drive cars), §5, p. 373
- Bent driveshaft: A bent driveshaft is easy to see and even easier to remember.
- Bent axle or axle flange: almost always the result of a collision.
- Worn CV joint (felt or heard only when cornering), §4, p. 372

4B. Vibrations With the Car Sitting Motionless at Idle

If your car has an automatic transmission, see if the vibration goes away when the transmission is put into Neutral or Park, then returns as soon as you put the transmission back into Drive. If so, it could simply be an idle speed that's been set too low: An automatic transmission drops engine idle speed by about 200 RPM whenever it's put in gear. First check that the idle speed is set to spec. If so, try elevating the idle speed by as much as 100 RPM over factory spec to see if that helps. If not, get the engine/transmission mounts inspected—one is probably collapsed or torn.

Note: I've used the term motor mount here to include the transmission mount(s) as well. On a rear-wheel drive car, it's the transmission mount that tends to tear, the passenger's side motor mount that likes to collapse. On a transverse-mounted–front-wheel drive configuration, the mount closest to the firewall is the one most likely to fail. It is located centrally between the two passengers at the same height as their feet.

If the vibration is essentially the same whether you are in gear or out, then the problem can usually be attributed to a rough-idling engine, §6A, p.236.

If your car is equipped with a manual transmission, there's no simple way to differentiate between a collapsed engine mount and a problem with the engine itself. In these cases, a trained set of eyes are your best tool.

Two other, less likely possibilities require a recent collision: Ask the body shop to reinspect for a bent motor mount bracket or adjacent frame component, or for an exhaust system component that's been shoved into contact with the body.

4C. Vibrations That Occur When the Engine Is Accelerated

There are two likely possibilities:

- Engine misfire caused by a failing ignition component, §6, p. 236 and §8, p.241
- Exhaust pipe or exhaust hanger contacting the body, §3B-3D, pp.450-452.

For other possibilities, see Tables 3–1 through 3–3, pp. 56-63

5. Water Leaks

5A. Into the Passenger Compartment

If you find water inside the car, first establish which quadrant is getting wet—passenger's front, driver's rear, etc. If only the front mats are involved, make sure that what you've found is water, because antifreeze can leak from the heater core or its piping, soaking the carpeting ahead of the front seats. How can you tell? Antifreeze feels slippery and smells sweet. Depending upon the location of the heater core, either one or both front quadrants can be soaked. If you do find antifreeze, get the leak repaired pronto—it's a coolant leak, after all. Furthermore, antifreeze will ruin a carpet in no time. While water eventually evaporates, antifreeze lingers on. You'll need a good mechanical shop for this job, most heater cores are buried deep in the dash. Heater core work can be very expensive, even when no plastic dash components end up broken along the way.

If the wetness is limited to the carpets ahead of the front seats and antifreeze is not the issue, the possibilities include a leaking windshield, door weatherstrip, door glass weatherstrip, or a blocked sunroof drain. A leaking sunroof is easy to diagnose because the water almost

17

always finds a way to drip right into your lap. If it's a factory-installed unit, compressed air applied to its drains (2 or 4) is usually all that's required. Aftermarket sunroofs typically have no drains, the leakage is due to an ill-fitting or torn weatherstrip.

If the front carpeting is getting wet but the seats are not, suspect the windshield. On cars possessing recessed wipers, first try clearing the cowl area at the base of the windshield of any accumulation of leaves. If this doesn't solve the problem, have the windshield removed by a glass shop for resealing.

Wet seats can also be caused by a door window that's not sealing well.

- If your doors have an enclosed track for the glass to run in, Fig. 17-13, then leakage past either the door weatherstrip or the window surround becomes the issue. Look for a tear or a misaligned door.
- If your door glasses are not enclosed in a frame, Fig. 17-14, then look for a poor fit between door glass and roof line. A careful visual inspection is usually all that's necessary to locate the gap. Find a good glass shop or body shop to realign the glass.
- On older cars, a new window regulator, Fig. 17-15 (the crank assembly that moves the glass up and down–motorized or manual) might be required due to wear (sloppiness) in its travel. Another possibility is a torn weatherstrip.
- Older cars can suffer a weakened door hinge: Open the door approximately 45 – 60°, grab its outer edge and lift. Slop in the hinge will be immediately apparent as the door rocks upward while the car remains motionless.

door weatherstrip

window surround

17

Fig. 17-13, Fig. 17-14. Door windows with and without an enclosing track..

Fig. 17-15. A door window regulator, powered, for Toyota Tacoma, Courtesy of 1A Auto

Wet carpeting breeds mold and mildew. To rid the carpeting of water, first remove any floor mats. Leave them out for at least a month: There's a lot of padding under most carpets, used for soundproofing and insulation. It acts like a sponge and takes considerably longer to dry than the surface carpeting. Sop up what you can with bath (dog) towels. Leave the doors open as often as possible to foster air circulation. (Opening all the windows just won't ventilate the carpet adequately.) While driving, turn the interior fan on full blast, directing it toward the floor. If you have air conditioning, use it with the fresh-recirc button/lever set on recirc. The heat will act to evaporate the moisture, the AC will act to dehumidify the interior. When things start to improve, buy a few boxes of dessicant (available in hardware stores; it pulls moisture right out of the air), and don't forget to apply mildew spray liberally and routinely.

5B. Windows That Roll Down Cockeyed, Windows That Bind

In all probability, you need a new window regulator, Fig. 17-15. The rare exception can be saved with lubrication and proper alignment, but to lubricate most of them adequately, the regulator must first be removed from the door. Why pay two labor charges? Get a new one on the first go-round unless you find yours is loose.

The motors that drive electric windows have nothing to do with the alignment of the glass, they deal strictly with moving the regulator up and down. The only reason to replace the motor is its failure, usually due to a dead short or open circuit within the motor's windings. Another likely cause, particularly with the driver's door (because it gets opened and closed more than any other) is a broken wire in the wiring harness connecting the door to the chassis. There'll be a bunch of wires inside a rubber boot between the door and chassis.

5C. Into the Trunk

If your trunk wells are filling with water, you have a problem with the lenses, a body seam, a torn trunk weatherstrip, or a trunk latch in need of adjustment. Small amounts of water can be mopped up with old towels, a small pond is best addressed by removing the rubber drain plug(s) located at the bottom of each well. To uncover the source of the leak, ask a friend to hose down the closed trunk with a gentle stream of water while you lie curled inside, comforted by your trusty flashlight. (Claustrophobics need not apply!) Run water around the entire perimeter of the trunk lid, as well as directing it down onto the tail lamp lens assemblies. Cracked lenses require replacement. Unbroken lenses can be resealed by first removing them, then using a strip of windshield sealant in place of the deteriorated gasket. Leakage past a body seam can be addressed with exterior grade weatherproofing caulk.

5D. As a Result of Body Damage

When your car was assembled, all welded seams were first treated to a bead of body caulk. Caulk can crack or split under the force of impact, yielding a nasty little leak that can be very hard to track down. A good body shop can be invaluable because they know how these seams fit together. They'll start with a hose or shower head, and finish with a healthy dose of patience.

A windshield can be the culprit as well. Windshields are often removed as part of body work; they can spring a leak whether they've been removed or not, just like a body seam.

17

6 . Graphics That I Forgot to Add, and Some Text

Once more: For an engine to start, you need good compression, the right combination of fuel and air, and a properly timed spark. We'll assume good compression.

The amount of air supplied is essentially controlled by the opening of the intake valves–how far they open and for how long.

Over the years, there have only been two mechanisms to control fuel flow: the carburetor (1890's through the early 1990's), and fuel injectors. The carburetor is strictly a mechanical beast, controlled in part by temperature (cold vs. hot) and vacuum switching valves, plus the obvious control supplied by the driver's right foot. The fuel injectors, typically one per cylinder, are controlled by the electronic control unit (ECU) which takes its cues (as with the carburetor) from the driver.

Fig. 17-16. A Toyota carburetor from about 1987.

Fig. 17-17. An EFI injector from a 1996 Corolla
Courtesy of Toyota

Spark generation is a far more complicated topic. To time it properly, designers have employed magnetos, points and condenser (including distributor rotation to set the distributor's timing), and throughout most of that history, the coil used to "generate" the spark has been, for the most part, a single unit mounted on an interior fender or housed within the distributor.

My own experience has shown that whenever an engineer decided to put an igniter or coil inside a distributor, that the temperatures involved produced relatively rapid parts failures.

We have finally reached the point where each cylinder employs its own coil.

With all that said, a photographic portrayel of at least some of that history:

.

Closed Points   Open Points

camshaft lobe

distributor shaft

points

points gap

pivot pin

rubbing block

pivot arm

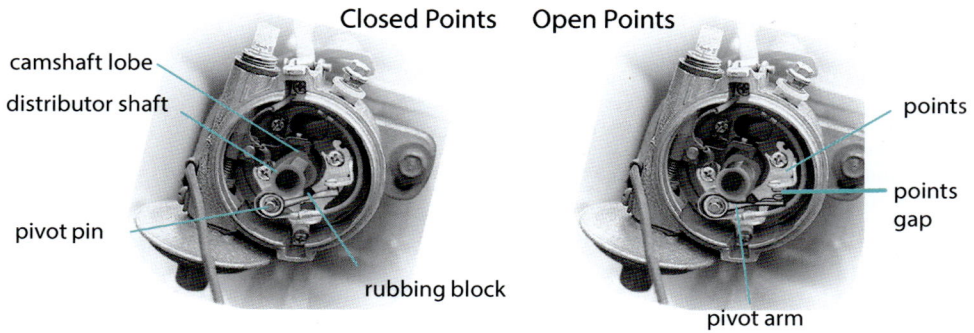

Fig. 17 - 18. A distributor from the late 1970's. Condensers are long gone. The points have been lightened to set them off from the distributor's base plate. The distributor shaft spins in sync with the engine's camshaft. The points' pivot arm oscillates in and out, tracking the distributor cam's lobes (high points) and flat spots with its rubbing block. This continuous opening and closing of the points triggers spark generation.

Fig. 17 - 19 An ignition coil (the cylinder) with igniter attached (the rectangular box on top). No more points; the igniter replaced them. This type of igniter (away from the distributor) seldom failed. The distributor was still very much present; it had to "distribute" the spark to the individual spark plugs..

fuel injector (1 of 4) attached to the fuel rail on its one end (highlighted in green) and inserted into the intake manifold on its other.

distributor and wire set

17

Fig. 17 -20. Ten years later, the distributor has grown, now housing both its own coil and igniter. The wire set is still present for another few years, but now the engine is fuel-injected.

Fig. 17-21. A modern day coil that sits overtop of its spark plug, within the spark plug well. There is one coil per cylinder, the ECU controls its firing relative to TDC (top dead center) as determined by the crankshaft position sensor. Courtesy of DENSO Automotive Parts

Fig. 17-22. A direct fuel injector (DFI), found on most contemporary engines. It injects fuel directly into the combustion chamber to which it's attached. Courtesy of General Motors

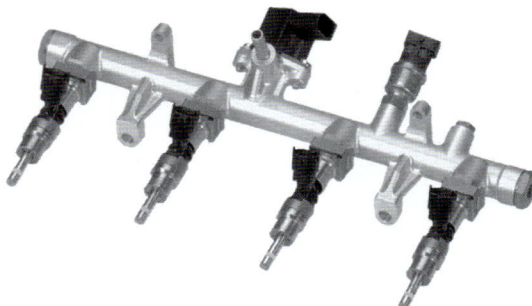

Fig. 17-23. The fuel rail for a direct fuel injector (DFI) system. Courtesy of General Motors

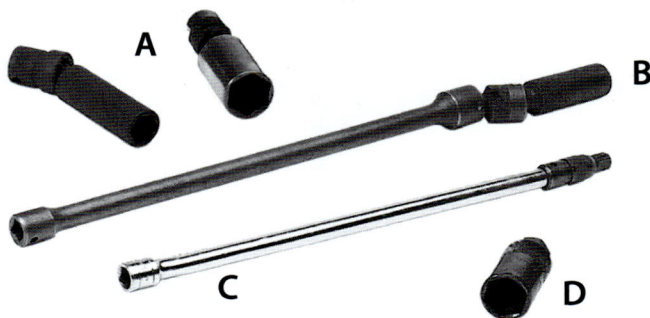

17

Fig. 17-24. Spark plug sockets. A. Sockets with swivels, large and small sizes. The smaller socket fits most spark plugs these days. B. Useful only for hard to reach plugs; this tool should never be used with an air-gun because of its speed. C. A locking extension, with straight-shot access; the locking mechanism prevents losing a socket down in a hole, as in Fig. 17-20. D. A large socket with no swivel.

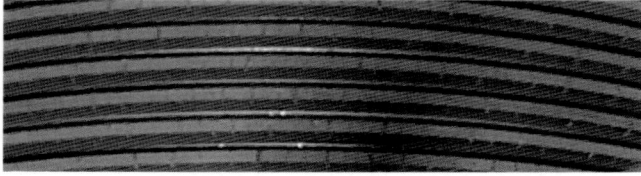

Fig.17-25. I wouldn't lose sleep over this belt, as long as it looks the same throughout, but I would definitely change it when I could afford to.

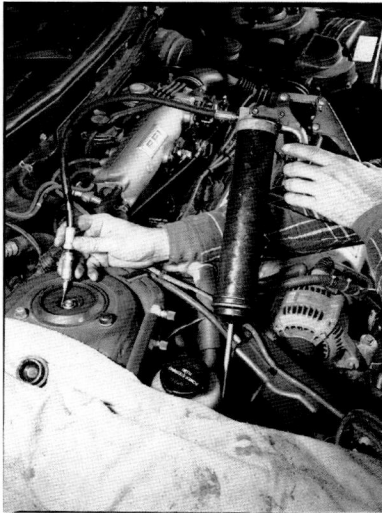

Fig.17-26. Lubing the bearing at the top of the strut tower. Auto-suppliers sell grease pens.

17

17

Chapter 18. Buying a Used Car

17

Notes

Buying a Used Car

1. Introduction

A good quality, low-mileage used car can last essentially the same length of time as a new one and can hold up just as well, but will cost several thousand dollars less. Whenever possible, buy a young car, preferably no more than four years old. Stick to those models that stay in production year after year, cars that are well represented on the street. Avoid the first year or two of any product line as well as the first year or two following a major styling change. Don't waste your time on the fringes–cars that are no longer in production, the odd-ball import, the "classic" old heap. And remember that flash and pizazz do not make a good car—engineering does.

If you are unsure of what to buy, begin with the *Consumer Reports Annual Auto Issue*. Available online for the price of an anual subscription or at your local library in their reference section, each years' edition contains a listing of their preferences in used cars. If you already know what you're looking for, you should still consult *Consumer Reports* to verify that the model and year you're leaning toward is considered a sound automobile. This publication is an extremely valuable resource on cars that are three or more years old. (The newer cars just haven't been out long enough for sufficient data to roll in.)

A strong case can be made for gauging the quality of a particular manufacturer by their track record over the past six to eight years, particularly when these records are compared with the offerings of other manufacturers. Different models from the same company can show wide swings in reliability, depending in large part on the age of the design, the number of units manufactured, and the assembly plant in which it was built.

The *NADA Used Car Guide*[1] (commonly referred to as the Blue Book, which for some inscrutable reason is actually orange and black) is a monthly publication containing the current average price for all cars manufactured during the past eight years. (They produce an older cars book as well; the book you're holding has an order form in it.) This information is also avalable online via *KelleysBlueBook.com* or *Edmunds.com*. Any one of these resources is an excellent way to get a feel for what things cost—a reality check of sorts. Every car is valued in three ways—wholesale, bank loan, and retail. The newer the car and the finer its condition, the more likely it is that you will be buying at retail. Conversely, the rougher looking cars will probably be going for wholesale or close to it. If you end up buying from a new or used car dealer, expect to see retail regardless of the car's condition, particularly if a warranty is required by law. In any case, remember that the bank will typically give you no more than the loan value, regardless of what you pay for it.

All three sources are very specific: A particular model might be listed on six or eight separate lines to distinguish between the different option packages that were available at the time of its new car marketing. You often see a dramatic swing in price. Furthermore, other options such as air conditioning, electric windows, or sunroof are listed at the end of each section and must be added onto the base figures where applicable. Finally, there

---

1 National Automotive Dealers Association, Costa Mesa, CA.

are low-mileage and high-mileage tables at the front and back of each copy. Photostat or print those pages that are relevant to your search, including the options list and the high- and low-mileage tables—you never know what awaits you out there!

The National Highway Traffic Safety Administration has a toll-free number for those interested in the recall and safety defect history of a particular model. This call could be of immeasurable value to those warming to the wrong car! The number is 1-800-424-9393.

2. The Cardinal Rules of Used Car Purchasing

❑ First and foremost: Don't rush this process! You are about to embark on a major purchase, one that you have to live with for a number of years. If you need to have a car fast, consider renting or borrowing a car to provide you more flexibility.

❑ Regardless of how many cars you decide to look at and regardless of price, seller's needs or your needs, the most important thing that you can do to avoid getting burned is to get the car to a competent mechanic who is familiar with this particular type of automobile. Where should you go for this? Either to an independent shop specializing in that make or to a dealership.

❑ Most owners are not willing to entrust their car to a total stranger without some assurance that (1) you're serious and that (2) you'll be back! A $500 check as deposit along with a note signed by both parties stating that you will purchase this car contingent upon the approval of your mechanic is sufficient in most cases, assuming the owner has nothing to hide. If the owner is unwilling to allow an independent checkover, I'd hit the road no matter how convincing the sales pitch. Assuming you've found a good and thorough mechanic, the checkover provides a professional appraisal that could persuade you to buy the car or save you from potential disaster. If your mechanic condemns the car, the worst that can happen is that you might have to stop payment on your check.

❑ Never go looking for a car in the dark. Mechanics make most of their diagnoses through direct observation; don't limit your own powers of observation by denying yourself the advantage of your eyesight!

❑ Find out on which days the local newspapers publish their best deals on classified advertising. These are usually three-day ads, Friday through Sunday. So take Friday off and begin reading at 6:00 a.m. so that you can be the first person to call on the good ones. Having an appointment already set up with the repair shop can make things flow so much more smoothly!

❑ Don't discuss price with the owner until your mechanic has completed his inspection. After all, the price should really depend on what's right or wrong with the car. If you and the seller have already gone through the process of bargaining, then the value of that inspection is reduced to little more than a recommendation to buy or keep looking. However, if you leave the issue of price alone until after the car has had a thorough going-over, then you've accomplished several objectives: You now have either a recommendation to buy or bail out, you have a list of problems with a dollar amount attached—a list that can have a tremendous impact on the outcome of your negotiations—and the car's owner has had a few hours to worry about the whole situation.

18

3. How To Examine a Used Car

- ❑ Carry a small notepad with you.

- ❑ Start your inspection by looking over the body. Although a minor fender-bender should be no reason to condemn a car, mismatched paint might swing the value of the car by a thousand dollars or more. Furthermore, if the car was worked over in a butcher shop, maybe you should spend your time and inspection costs on a more desirable candidate. Even if lousy body work doesn't bother you, it may still tell you something about the owner's other escapades with the repair community.

  Any car that's been in a collision will provide evidence to the careful observer. Slowly walk around the car looking for panels of a slightly different hue, or for buff marks (circular swirls) in one or more panels that you don't see in the others. Sight down each side looking for any irregularities—unevenness, waviness, imperfect body lines—that you don't see on the other side. Open each door, the hood, and the trunk to look for taping lines that ordinarily would be concealed.[2] These lines can be very obvious even to a novice, so take the time to look and do it in good lighting. Naturally, if the car has been repaired by craftsmen, visual clues may be quite subtle, in which case you probably won't need to worry about the fact that it was hit —unless it was a major collision. BIG HITS are quite noticeable from beneath the car once your mechanic has the car on his lift. Most involve frame or suspension damage that require new parts, virtually all of which were originally tagged with adhesive part-identification labels. Believe me, body men don't spend their time peeling off labels from parts that are not visible to the customer!

  Look for rust, too. An occasional nick is no big deal, but if the rust has metastasized—bubbling up the paint here and there—resume your search elsewhere because you're only seeing the tip of the iceberg.

  Are all the parking lamp, side marker, and tail lamp lenses intact? How do the tires look?

- ❑ Next, act a bit stupid, asking the owner to show you where the engine's oil dipstick is located. If he doesn't know, it's not a good sign! Is the oil clean and full? (Oil appears somewhat darker when cold.) To get an accurate oil level reading, you'll recheck this level again once the car has been warmed to operating temperature. (I'll remind you.) Now wipe off the stick and inspect it for blackened deposits baked on above the Low–Full grid, Fig. 18–1. These deposits indicate an engine that has been run low on oil; what oil was left got so hot that residues literally baked onto all interior engine surfaces. If you find this condition, walk away—the owner is dumping the car because it burns too much oil.

18

Fig. 18-1. A dipstick that shows its engine was repeatedly run low on oil.

---

2 When a car is repainted, large sheets of paper are used to cover areas that are not being redone. Masking tape is used to secure the paper's edge. Whenever possible, the line between old and new paint is hidden behind a door along its pillar post (the vertical portions of the car that the door closes against), or under the hood, the trunk lid, etc.

❑ Open the oil filler cap located on top of the valve cover, turn it upside down and inspect its interior. What you would like to see is a nice, clean surface; the interior of the engine will correlate strongly. Any engine that has run for extended periods on dirty oil or low oil levels will have stalactites of sludge built up on the inside surface of the cap, Fig. 18-2.

If you encounter a grayish white-to-chocolate milk-colored scum inside the oil cap, bless the day you read this section and leave without delay. The owner is attempting to unload an engine with a blown head gasket, or worse.

❑ If the engine is stone-cold, open the radiator cap to verify that the cooling system is full. If it's low, make sure that your mechanic knows about it! Check the level within the coolant overflow bottle itself (Not relevant on engines with a surge tank, Fig. 9-2, p. 149).

❑ Before you spend any more time kicking tires, drive the car to see– among other things–if you like it.

❑ Check to see that all of the dash panel warning lights come on when the ignition key is turned to the On (not Start) position—oil pressure, coolant temperature, charge warning lamp, engine warning light, air bag warning light, door ajar light, etc.

❑ How is the seating? Does the car have a rear-view mirror on the passenger's side to match the one on the driver's side? How is the vision to the rear and to the side?

Hopefully the car will be cold: A warm engine can mask cold-starting problems; a warm automatic transmission can hide slipping problems that are only evident when the car is first started cold.

❑ If the car has an automatic transmission, it should definitely not clunk when put in gear—either cold or hot. When you give it gas, the car should begin to move immediately, not sit there idly slipping. Check reverse as well, both for slipping and clunking. The transmission should shift smoothly both upward and downward through the gears. Count your forward gears from a stop. (Modern automatics have four to six forward gears.) You may have to verify that online before you go looking.

❑ On a manual transmission, check the clutch for slippage following the procedures in §2D, p. 366.

❑ Pay attention to the car's performance and to any noises that you might hear—cold, hot, when turning, when braking. Take notes!

❑ Does the car steer straight? Stop straight? Does the steering seem vague or inconsistent, as if there were some looseness in the linkages? Get it up to highway speed: Does the steering wheel shimmy (vibrate)? It's a bit early to start bothering with diagnosing pulling or pusation problems on somebody else's car, but if you can't help youself, see §2 through §6, pp. 259-270,. and §3, p. 320.

❑ Hit the brakes firmly on a downhill stretch. (Don't slam 'em on!) Does the brake pedal pulsate under your foot? Does the car pull to the right or left? Speed up, then hit 'em again. Did the pedal seem to fade?

❑ On a fully warmed engine, stomp on the accelerator while moving up a hill. (Pedal to the metal!) Are there any clouds of smoke behind you? If so, it's time to move keep looking.

18

❑ What about the temperature gauge? Does it work? Where is it reading? What about the other gauges?

❑ Check the switches, the lights, the wipers, the air conditioning: Are they all working?

This list covers a lot of ground, so take your time on this road test. After all, if you buy, you'll be living with the decision for a long time!

❑ If you really like this particular car, ask to see any and all service records.[3] Pay particular attention to the latest bills; any hints as to why it's up for sale? Check the dates on the repair orders. Do they indicate routine servicing or just damage control?

❑ While you're inspecting this stuff, leave the engine running at idle (in Neutral or Park) for at least five to ten minutes. If equipped with an automatic transmission, check its fluid level (if relevant), §1A,p.357. Now blip the throttle—quickly revving the engine to 4,000 RPM (in Neutral or Park) before quickly releasing it—while watching for a plume of grayish smoke to exit the tailpipe. Any smoke at all indicates an oil burner, a billowing cloud means that the car won't pass emissions testing. Sorry, keep looking.

❑ Immediately after turning off the engine, recheck the oil level, §1E,p.97. Over a quart low? You're looking at a neglectful owner or an engine that's consuming some oil. About a quart overfilled? You're looking at an idiot or at an engine that was recently extremely low on oil, a situation that led the owner to overreact. (Possibly both!)

❑ Take the time to finish up the incidentals. These items are usually glossed over, even by thorough mechanics, but they are essential and do cost money to repair. They can also be used as points of negotiation.

❑ Do all the windows roll up and down smoothly?

❑ How about the door locks?

❑ The seatbelts?

❑ How's the sound system?

❑ Is there a decent spare tire and complete jack assembly on board? What about the trunk wells: Are they dry, or is there a leak somewhere?

❑ What about the carpeting? Is it wet or mildewed? If so, you now know why this guy is dumping the car: Allergies to mildew can be severe. Bail out and consider yourself lucky.

❑ Still moving along swimmingly? Then establish the ground rules for the mechanical inspection and if necessary, write a check to secure (1) that you'll be back, and (2) that the car won't be sold to someone else while you're gone. If the check is viewed as a deposit securing your right to buy the car, be certain that the words *contingent upon the approval of my mechanic* are clearly written into your letter of understanding. This agreement should be signed by both of you. Hopefully, you've already set up at least one inspection appointment with the repair shop. Time spent waiting has killed a lot of deals.

---

3 Remember that keeping good records does not, in and of itself, indicate that the car was well maintained, nor does spending large sums of money, even if done so routinely. It does indicate good intentions, but you're measuring a car here, not the prior owner's objectives.

18

❏ Be sure to share with your mechanic what you have learned, preferably in list form. Always ask him to inspect the car for signs of a major collision, and be willing to walk away from the deal if he finds them.[4] In addition, ask him to estimate the cost to repair the issues that you've raised, plus the ones he finds. This information should be on the shop's letterhead so that you can use it in your negotiations. Don't be surprised if this list encompasses $500-$1,000 worth of repairs; that might even be considered an average figure. Make sure that both you and your mechanic agree that the car is worth buying. If any issues are raised for which he cannot give a definite answer–Is that tire wear due simply to improper alignment, easily correctable once the tire has been replaced, or is that strut bent? Get both a best- and worst-case scenario with an estimated cost of repair.

❏ Finally, if you live in a jurisdiction that requires a state or local inspection prior to registration of the vehicle, definitely have it done before you settle down to negotiations, even if you know it's going to fail. Your mechanic might have missed something or the inspector might view an issue in a different light: The idea is to cover all the bases. Most state inspections allow a 30-day grace period for repairs.

If you're lucky, the first car that you run through this process will be lovely, well maintained, and reasonably priced. If so, negotiate the best deal possible, then nail it down. If you hesitate over a good car, God help you when you start in with the bad ones!

However, if the number of issues and their cost move past a certain point, drop it and move on. The amount of time, money, and emotional energy you have expended is nothing compared to what it could be if you get stuck with the wrong car. Even if you have to repeat this process four or five times with various cars, at least you will be learning something new with each visit, and hopefully getting more focused.

4. Where To Buy

If you're looking for a one-year-old car with low mileage, air conditioning, and automatic transmission, look no further than one of the national rental car companies. Because these cars are replaced yearly, these outfits do a big business in used car sales, complete with showrooms, 12-month warranties, the works. These cars are meticulously maintained and carry their service records to the showroom. Their only downside is that they are machine washed so often that their paint is usually quite scratched. (Nothing that a little dirt won't hide!) As always, have your choice checked over thoroughly before you sign on the dotted line.

Call it prejudice or knowledge of the odds, but I don't like used car lots. Many of the cars found their way there by way of auctions. Many have been wrecked, many have some fundamental problem that put them on the auction block. However, late-model, low-mileage cars improve the odds significantly; the larger lots: CarMax, Carvana, etc. have reputations to maintain and should offer better odds. In any case, it's essential to get your choice checked over by an independent shop.

---

4   Again, I'm not talking about fenders or bumpers here, I'm referring to frame or suspension damage. This kind of collision will almost invariably lead to costly repairs at a later date: struts, driveshaft seals, or worse.

18

Obviously, if you buy from an individual, you see only one car at a time. Furthermore, unless the car is still under its original manufacturer's warranty, you have no protection against future problems. On the other hand, the owner and his (her) repair orders can provide information that would be unavailable under most other circumstances. In addition, most individuals who sell their cars just want to get it over with, so accommodations can be made. Finally, because profit is usually not their main concern, the asking price is probably much closer to current Blue Book values.

Most new car dealerships limit their used car selection to relatively new, low-mileage cars. They offer you the convenience of looking at several cars in one location and can also arrange financing.[5] These cars were either taken in trade; came from their loaner, demonstrator, or rental car fleet; or were picked up from a wholesaler. Cars taken in trade are usually there because the original owner was either tired of the car (and has more money than the rest of us) or had just been given a dose of bad news from the service department. As for the fleet cars, they have usually not been maintained at all. But it's very hard to screw up a car in less than 10,000 miles, and most fleet cars fall into that category. The ones to watch out for are the cars bought at auction. They are almost always pretty, low mileage, repossessed, and full of worms. If you can determine which category a particular car fits into, you're ahead of the pack. In any case, it's essential to get your choice checked over by an independent shop.

As to price, the dealerships want top dollar, often asking a thousand dollars or more over NADA retail. They attempt to justify this by expounding on the benefits of their warranty, which I suppose is fair enough. But remember: Most of these warranties are limited to thirty days and apply only to the drive train (engine, transmission, and differential). Moreover, a warranty is a piece of paper: It doesn't imply that the car is in good shape, nor does it mean that the used car department will oblige your requests for repair. Most would prefer to stonewall on as many issues as they can. This doesn't mean that you can't get a square deal from a dealership, but used car departments are usually the weakest link in a dealer's reputation, simply because there is almost no used car in existence that doesn't have some problems. Used car departments are in business to make money; they can be notorious for not noticing problems that are not purely cosmetic.

One final note: If you are interested in an extended used car warranty (as distinguished from the dealer's 30- or 90-day warranty), have your mechanic look over the policy before deciding to buy. The extended warranty packages available on used cars cover significantly less than their new car cousins, and last for a shorter period of time.

18

---

5   If the car is still under warranty, be sure to register the change in ownership with the factory to ensure that the warranty remains valid. The original owner's manual usually includes a form or instructions on the changeover.

As for financing, your bank will usually have a better rate. This can be arranged after-the-fact: the next day, for example.